INTERNATIONAL CENTRE FOR MECHANICAL SCIENCES

COURSES AND LECTURES - No. 296

APPLICATION OF SYSTEM IDENTIFICATION IN ENGINEERING

EDITED BY

H.G. NATKE
UNIVERSITÄT HANNOVER

SPRINGER - VERLAG  WIEN - NEW YORK

Le spese di stampa di questo volume sono in parte coperte da contributi
del Consiglio Nazionale delle Ricerche.

This volume contains 196 illustrations.

ISBN 3-211-82052-3 Springer Verlag Wien-New York
ISBN 0-387-82052-3 Springer Verlag New York-Wien

PREFACE

System identification is a powerful tool in Engineering. Its various methods in the frequency and in the time domain have been extensively discussed in earlier CISM courses. The aim of this course is to describe the state of the art in specific application areas, e.g. estimation of eigenquantities (in the airplane and aerospace industry, in civil engineering, in naval engineering etc.), noise source detection, fault detection by investigation of dynamic properties, such as machine sound characteristics, and the identification of the dynamic behaviour of flow induced systems (e.g. aeroelastic problems). Geotechnical applications are also one of the fields of interest.

The lecture notes contain demonstrations of several methods and include a valuation of combining various kinds of experience. Such complex information includes not only theoretical aspects of identification but also advice on practical handling, e.g. concerning testing effort and data handling.

The course was announced as "The Boltzmann Session". Boltzmann, who was an excellent theoretical physicist as well as a very skilful experimenter, once said that "nothing is more practical than theory". I entirely agree with him. The reader will thus find an introductory review of the identification of vibrating structures, and, of course, some more theoretically orientated papers.

I wish to thank all the participants in the course for their contributions, especially the lecturers for the oustanding work they did in Udine and for preparing their final papers. Last but not least our thanks must go to the CISM cooperators for the excellent work they have done and for their kind hospitality.

H.G. Natke

Hannover.

CONTENTS

Page

INTRODUCTION TO SYSTEM IDENTIFICATION:
FUNDAMENTALS AND SURVEY

H.G. Natke, N. Cottin
Universität Hannover, Hannover, F.R.G.

1.   Fundamentals

The application of system identification to engineering problems requires certain knowledge of
- the inherent theoretical relations
- the test and measuring conditions (and their inevitably
  imperfect realization)
- the deterministic and statistical approaches in system
  identification (e.g. time series analysis).

The content of CISM course 272 in 1980 on "Identification of Vibrating Structures" /1.1/ and the existing
refs. dealing with time series /1.2,1.3/, time series and

stochastics /1.4/, and such books as contain time series
and experimental modal analysis /1.5,1.6/, mean that the
fundamentals of system identification only need to be summarized here. The term "system" is used as a synonym and
abbreviation for mechanical systems /1.7/ to which this
course is restricted.

1.1   General

The systems we are dealing with are assumed to be
time invariant and, in addition, linear and nonlinear. The
dynamic behaviour of the system, and the dynamic process
the system is subjected to, can often be described by
input/ output relations, which result in a system of equations: the mathematical model. In general the goal of
system (structure) analysis is the prediction of the dynamic behaviour of the system under investigation. This is
the well-known direct problem of system analysis that requires sufficiently accurate system modelling and the
knowledge of dynamic loads. On the other hand, the inverse
problem (Fig. 1.1) deals with the modelling and the design
of the system itself (system identification, design problem) and with input identification (Fig. 1.2). In the
design problem, the input and output quantities are given
and one is looking for a system (for its model) which

fulfills the input/output relations best. Input identification is defined by a given model description and a given
output, while system identification includes the determination of a system description by measured input and/or
output quantities.

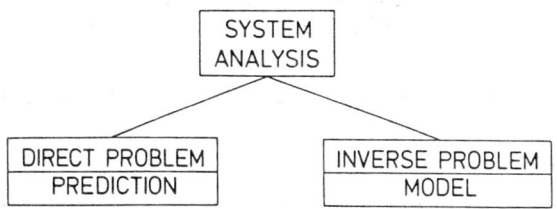

FIG. 1.1   CLASSIFICATION OF
SYSTEM ANALYSIS

As is known, we distinguish between the black-box
model and the parametric model (Fig. 1.3). The black-box
model is a non-structured mathematical model, e.g. the
frequency response function. Parametric system identification uses a structured model, so that only its parameters
are unknown, i.e. the identification problem is reduced to
parameter estimation.

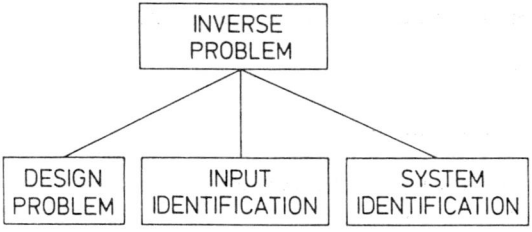

FIG. 1.2 CLASSIFICATION OF THE
INVERSE PROBLEM

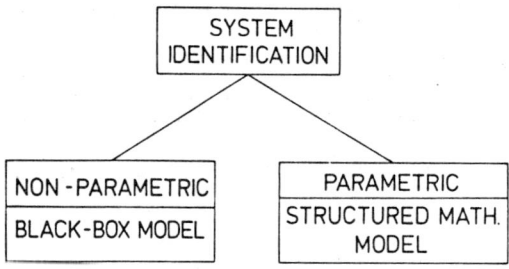

FIG.1.3 CLASSIFIGATION OF SI
CONCERNING THE MATH.
MODEL USED

"Estimation" is used in its mathematical sense: statistical methods have to be applied, because the measurements are often distorted by random errors which have to
be reduced in order to obtain results with a large information content and with high confidence. The identification problem is accompanied by the following practical and
theoretical problems: test conditions (cf. Fig. 1.4) concerning

- test environment and test equipment

- excitation

- measurement techniques

- data acquisition

- signal processing including data reduction

- noise reduction,

considerations concerning

- the choice of mathematical model

- the choice of appropriate estimation methods

- data processing (including algorithms and routines).

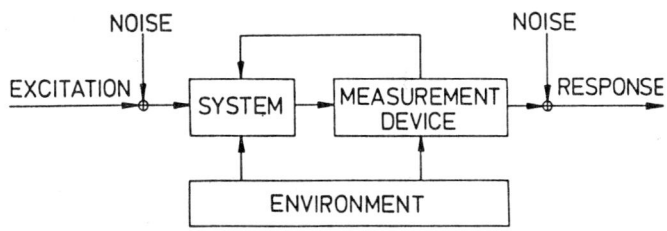

FIG.1.4 SYSTEM DISTURBANCES

The practical and theoretical problems mentioned above are closely connected as regards such terms as observability, controllability and identifiability /1.1/. In addition, one has to take into account that the model is only an approximate description of the system and its dynamic behaviour, i.e. the model is not unique. This means, for example in the linear case, that there may be a minimum order model instead of the model being worked with, and that in the nonlinear system behaviour, for example, the description in the form of a polynomial is done with the 5th power within the required accuracy without one knowing that a 3rd power may already be sufficient.

Dynamic investigations of complicated systems need system analysis: that means a computational model. If this model is not sufficient, i.e. if it cannot describe the system behaviour completely, tests have to be performed,

and this results in a test model (e.g. input/output measurements). Now two problems arise: 1) the use in general
of insufficient a priori knowledge (computational model
with its predictions) with regard to testing, and 2) how
to utilize the a priori knowledge within the estimation.
The first problem is a "philosophical" one: should verification be done while one is influenced by knowing the
results of the mathematical model (critical engagement)?

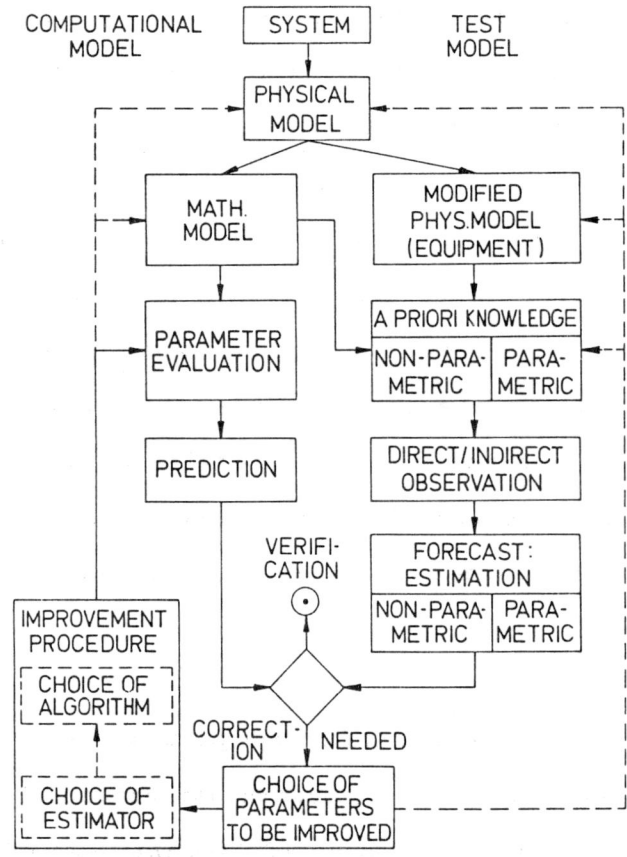

FIG.1.5   COMPARISON OF RESULTS OF SYSTEM
          ANALYSIS AND SYSTEM IDENTIFICATION WITH
          FOLLOWING PARAMETER IMPROVEMENT

All a priori knowledge should be included in the testing. The solution of the second problem is indicated in Fig. 1.5. The prediction of the computational model has to be compared with the corresponding quantities of the test model. These are, of course, estimated values (noise reduced) from the observations. If the predicted and forecast results do not lead to a verification within a given quality criterion, then the computational model generally has to be improved. In parameter estimation the Bayesian approach permits the inclusion of a priori knowledge.

In system identification the mathematical tools are statistical methods. With regard to data processing, for the reduction of measurement noise and for non-parametric identification use is made of the Fourier transform and Laplace transform with their numerical procedures for handling linear systems and of the Hilbert transform for detecting non-linearities.

1.2 Structural Model

The fundamentals of system analysis are given here from a practical point of view by a discretized (with respect to the local coordinates) model of n degrees-of-freedom:

$$\mathbf{M\ddot{u}(t) + B\dot{u}(t) + Ku(t) = p(t)} \tag{1.1}$$

with the quadratic physical parameter matrices of order n,

M   the inertia matrix

B   the viscous damping matrix and

K   the stiffness matrix,

u(t) is the vector of generalized coordinates with

t   the time parameter,

p(t) is the vector of external forces.

Dots indicate differentiation with respect to time. Eq. (1.1) is, for example, generated by finite element modelling. Here the assumption of viscous damping is arbitrary from the physical point of view. For hysteretic damping see /1.5/. Eq. (1.1) is a mathematical model with given structure and with given order n. The associated unstructured model in the frequency domain is defined by

$$\mathbf{S(j\omega)\ U(j\omega) = P(j\omega)\ resp.\ U(j\omega) = F(j\omega)\ P(j\omega)}$$

with

$\mathbf{U(j\omega) := F\{u(t)\}}$, F : symbol for Fourier transform

$\mathbf{P(j\omega) := F\{p(t)\}}$,

$\mathbf{S(j\omega)}$ dynamic stiffness matrix

$\mathbf{F(j\omega) := S^{-1}(j\omega)}$ frequency response matrix

and with initial conditions equal to zero.

$$\tag{1.2}$$

These equations do not imply only the definitions and interpretations of $S(j\omega)$ and $F(j\omega)$ but also their calculation in the case of non-erroneous data. If one denotes the Laplacian transform by   and the Laplacian variable by L s, i.e.

$U(s) := L \{u(t)\}$,

$P(s) := L \{p(t)\}$,

$S(s)$ dynamic stiffness matrix in the s domain,

$H(s) := S^{-1}(s)$ transfer function matrix,

(1.3)

the equations corresponding to Eq. (1.2) are

$S(s) \, U(s) = P(s)$ and $U(s) = H(s) \, P(s)$ respectively.

If the knowledge of the modal quantities of the associated undamped model and of the viscous damped model are required, a structured mathematical model has, of course, to be used:

$$(-\omega_{0i}^2 \, M + K) \, \hat{u}_{0i} = 0, \qquad i = 1(1)n, \tag{1.4}$$

$$(\lambda_{Bl}^2 \, M + \lambda_{Bl} B + K) \, \hat{u}_{Bl} = 0, \quad l = 1(1) \, 2n. \tag{1.5}$$

The mathematical properties of the matrix eigenvalue problems and their solutions are well-known /1.1,1.5,1.8/ and therefore not repeated here. Only the orthogonality relations are summarized in the undamped case (for symmetric K):

$$M_g := \text{diag} (m_{gi}), \qquad \hat{u}_{0i}^T M \hat{u}_{0k} = \delta_{ik} m_{gi} ,$$

$$K_g := \text{diag} (\omega_{0i}^2 m_{gi}), \qquad \hat{u}_{0i}^T K \hat{u}_{0k} = \delta_{ik} \omega_{0k}^2 m_{gi} , \qquad (1.6)$$

$$\delta_{ik} - \text{Kronecker symbol},$$

and in general for viscous damping:

$$\Lambda_B \hat{V}_B^T M \hat{U}_B + \hat{V}_B^T M \hat{U}_B \Lambda_B + \hat{V}_B^T B \hat{U}_B = I ,$$

$$\Lambda_B \hat{V}_B^T M \hat{U}_B \Lambda_B - \hat{V}_B^T K \hat{U}_B = \Lambda_B , \qquad (1.7)$$

where $\Lambda_B := \text{diag} (\lambda_{Bl})$, $(2n, 2n)$ diagonal matrix of eigenvalues,

$\hat{U}_B := (n, 2n)$ modal matrix of right eigenvectors,

$\hat{V}_B := (n, 2n)$ modal matrix of left eigenvectors

$(...)^T$ describes the transposed matrix $(...)$ and I stands for the identity matrix.

It can be shown that the physical parameter matrices M, B and K can be expressed by the modal quantities, therefore both representations are theoretically equivalent. For the undamped model with symmetric matrices it follows from the orthonormalization with the (n,n)-modal matrix $\hat{U}_0$ in the normalization $m_{gi} = 1$ that

$$M^{-1} = \hat{U}_0 \hat{U}_0^T = \sum_{i=1}^{n} \hat{u}_{0i} \hat{u}_{0i}^T ,$$

$$(1.8)$$

$$G := K^{-1} = \hat{U}_0 \Lambda_0^{-1} \hat{U}_0^T = \sum_{i=1}^{n} \frac{1}{\lambda_{0i}} \hat{u}_{0i} \hat{u}_{0i}^T , \quad \Lambda_0 := \text{diag} (\lambda_{0i}) ,$$

$$\lambda_{0i} := \omega_{0i}^2 .$$

The frequency response matrix can then be expressed by

$$F(j\omega) = (-\omega^2 M + K)^{-1} = \hat{U}_0 (-\omega^2 I + \Lambda_0)^{-1} \hat{U}_0^T = \sum_{i=1}^{n} \frac{\hat{u}_{0i} \hat{u}_{0i}^T}{\omega_{0i}^2 - \omega^2} . \quad (1.9)$$

The importance of the dyadic products of the eigenvectors can be seen. The flexibility influence coefficients are proportional to the square of reciprocal eigenfrequencies. The frequency response matrix shows the resonance phenomenon. In addition, because of its symmetry and
structure only one row is necessary in order to estimate
the eigenfrequencies and damping ratios.

Corresponding relationships exist for the damped system /1.1,1.5/ where a distinction should be made between
proportional and non-proportional damping. It may be
noted that these relations do not allow one to compute the
parameter matrices by means of identified eigenquantities
only, because the identified modal quantities are erroneous and incomplete.

The equations of motion with their physical parameter matrices can be taken for identification. In addition, the physical parameter matrices may be expressed by modal quantities, and these relations can also serve for identification. In addition, the solutions of the equations of motion can serve as mathematical models.

In the time domain the dynamic response of the viscous damped system, predicted by a model with symmetric matrices is

$$
\begin{aligned}
u(t) = \hat{U}_{B1} &\int_0^t e^{\Lambda_{B1}(t-\tau)} \hat{U}_{B1}^T \, p(\tau) \, d\tau \; + \\
+ \, \hat{U}_{B2} &\int_0^t e^{\Lambda_{B2}(t-\tau)} \hat{U}_{B2}^T p(\tau) \, d\tau \; + \\
+ \, \hat{U}_{B2}^* &\int_0^t e^{\Lambda_{B2}^*(t-\tau)} \hat{U}_{B2}^{*T} p(\tau) \, d\tau
\end{aligned}
\tag{1.10}
$$

with $e^{\Lambda t} := \text{diag}(e^{\lambda_i t})$, $\lambda_i^{re} < 0$ and index 1 denoting real eigensolutions and index 2 denoting complex eigensolutions.

The last two summands of Eq. (1.10) can be assembled:

$$u(t) = \hat{U}_{B1} \int_0^t e^{\Lambda_{B1}(t-\tau)} \hat{U}_{B1}^T p(\tau) \, d\tau \; +$$

$$\hspace{6cm} (1.11)$$

$$+ \; 2 \int_0^t Re\left[\hat{U}_{B2} e^{\Lambda_{B2}(t-\tau)} \hat{U}_{B2}^T\right] p(\tau) \, d\tau \quad .$$

The free vibrations after the excitation (say $t \geq T$) can be described by

$$u(t) = \hat{U}_B e^{\Lambda_B t} \int_{-\infty}^T e^{-\Lambda_B \tau} \hat{U}_B^T p(\tau) \, d\tau$$

$$\hspace{6cm} (1.12)$$

$$=: \hat{U}_B e^{\Lambda_B t} c \quad ,$$

It is a superposition of decaying exponential vector functions for a system with persistent damping.

For active systems (with nonsymmetric matrices) the dynamic response in the s domain is given by

$$U(s) = \hat{U}_B (I s - \Lambda_B)^{-1} \hat{V}_B^T P(s)$$

$$\hspace{6cm} (1.13)$$

$$= \sum_{l=1}^{2n} \frac{\hat{v}_{Bl}^T P(s)}{s - \lambda_{Bl}} \hat{u}_{Bl} \quad .$$

The dynamic response for weakly damped systems can be written as

$$U(s) = \sum_{i=1}^{2n} \frac{a_i}{s - \lambda_{Bi}} = \sum_{i=1}^{n} (\frac{a_i}{s - \lambda_{Bi}} + \frac{a_i^*}{s - \lambda_{Bi}^*}) \qquad (1.14)$$

with $(\ldots)^*$ equal to the conjugate complex term $(\ldots)$ and with in general

$$a_i := a_i(s) = \hat{u}_{Bi} (\hat{v}_{Bi}^T P(s)) , \qquad (1.15a)$$

for harmonic excitation with the amplitude vector p_0 it is

$$a_i := \hat{u}_{Bi} (\hat{v}_{Bi}^T p_0) \qquad (1.15b)$$

and for free vibrations it is

$$a_i := \hat{u}_{Bi} \left[\hat{v}_{Bi}^T \int_0^{T_0} e^{-\lambda_{Bi} t} p(t) \, dt \right] \qquad (1.16)$$

for impulse excitation $p(t) \neq 0$ within $0 \leq t \leq T_0$ and the response holds true for $t \geq T_0$.

The state space formulation of the model equation is also very common:

$$\dot{x} = Ax + f \qquad\qquad (1.17)$$

with the observation equation

$$y = Cx. \qquad\qquad (1.18)$$

In contrast to the state space formulation used in control engineering, here we have additional knowledge about the structure of the state vector and the state space matrix:

$$x(t) := \begin{pmatrix} u(t) \\ \dot{u}(t) \end{pmatrix}, \quad A := \begin{pmatrix} 0 & , & I \\ -M^{-1}K & , & -M^{-1}B \end{pmatrix}, \quad f(t) := \begin{pmatrix} 0 \\ M^{-1}p(t) \end{pmatrix}.$$
$$(1.19)$$

The advantages of the state space formulation are its reduction to the first derivative order and, in consequence, the simple matrix eigenvalue problem equation including its simple energetic orthogonality relations, and the fact that in addition to the state vector, only its first derivative is needed. The disadvantage is the order-doubling of the matrices.

The dynamic response both in the time and in the frequency domain of the systems considered can be expanded by eigenvectors of the corresponding systems. The participation factors of the eigenvectors depend on the work done by the excitation or on the initial conditions due to the eigenvectors, on the eigenfrequencies compared with the given frequency and on the modal damping ratios. As can be seen, by using orthogonality relations one can eliminate all the degrees-of-freedom except for the one to be investigated. This can be performed by multiplying the dynamic response in a suitable manner or by an appropriate excitation. Another way of eliminating one (generalized) degree-of-freedom may be by analytical separation using different responses due to different known excitations.

It is important to know the above relationships and properties of the model quantities for identification approaches, because the more one knows a priori, in trivial consequence the less has to be identified.

Resonance testing, which was very common in the past, serves for identifying the modal quantities of a single degree of freedom. It can be theoretically treated in a general way by the phase lag theory /1.9/. Instead of using it, let us explain the phase resonance by the usual procedure applied in practice. The input/output relation of the system harmonically excited by $p(t) = p_0 e^{j\Omega t}$ is described by Eq. (1.1)

$$(-\Omega^2 M + j\Omega B + K)(\hat{u}^{re} + j\hat{u}^{im}) = p_0 \ . \qquad (1.20a)$$

Separating it into the real and imaginary parts yields

$$(-\Omega^2 M + K)\, \hat{u}^{re} - \Omega B\, \hat{u}^{im} = p_0$$

$$\Omega B\, \hat{u}^{re} + (-\Omega^2 M + K)\, \hat{u}^{im} = 0 \; . \qquad (1.20b)$$

With the necessary and sufficient conditions

$$\Omega = \omega_{oi} \; ,$$

$$\hat{u}^{re}(\omega_{oi}) = 0 \; , \qquad (1.21a)$$

called phase resonance, the second equation of (1.20b) results in the matrix eigenvalue problem (1.4) with the eigenvector $\hat{u}^{im}(\omega_{oi}) = \hat{u}_{oi}$: the system vibrates in the i-th eigenmode due to an excitation given by the first equation of (1.20b):

$$p_{oi} = -\omega_{oi} B\, \hat{u}_{oi} = j\omega_{oi} B\, \hat{u}(\omega_{oi}) \; . \qquad (1.21b)$$

This equation means that a force proportional to the unknown viscous damping force in the i-th mode is necessary in order to excite the i-th, and only the i-th, natural mode of the system under test. This appropriate excitation compensates the damping force. Because the eigenvectors are linearly independent, n linear independent excitation vectors are necessary in order to excite all the modes. Transformation in generalized coordinates shows that when the system vibrates in the i-th natural mode it can be described in the i-th generalized coordinate like a one degree-of-freedom system. The above statement is true for non-proportional damping only for $\Omega = \omega_{oi}$. For proportional damping, in which the dynamic response can be expanded into the eigenvectors $\hat{u}_{oi}$, the same statement holds true for $0 \leq \Omega < \infty$. This enables the user to apply the well-known formulae of one degree-of-freedom systems to evaluate the modal damping (generalized damping) and the normalization factor of the i-th eigenvector (generalized mass). The determination of the complex generalized damping matrix would require the knowledge of all linear independent appropriate excitation vectors (see Eq. (1.21b)). - The calculation of the eigenvectors of the damped model is possible only with the knowledge of model parameter matrices or with the equivalent information.

In order to simplify the formalism (and/or the test) or to get a better insight into the system behaviour, it is very often advantageous to partition the system into subsystems and then to synthesize the models of the subsystems to the entire model. This can be done with the help of modal quantities (admissible vectors in the spirit of Rayleigh-Ritz) of the primary, generalized coordinates and by neglecting the remaining, secondary coordinates. This modal synthesis is done with a transformation which can be chosen so that approximate dynamic condensation is obtained /1.1,1.5,1.10/. Error considerations which can be used as corrections are discussed in /1.1,1.11 /.

If a system contains nonlinear elements the substructure synthesis presents itself: the investigation and identification of linear subsystems and of the nonlinear elements and then the coupling of the results by analysis.

The detection of nonlinearities is one problem, their parametrical description is another problem. There is a parametrical description for nonlinear elements using, for example, polynomial series for a given number of terms (power of polynomial), and thus the structure of the mathematical model is defined. Other models are described in /1.12/ and summarized in /1.13/, but these models are not able to reproduce hysteretic system behaviour. For hysteretic behaviour the reader's attention is drawn to /1.14/.

Finally, linear differential error analysis should be mentioned. It serves to detect model sensitivities, including those concerning the parameters. This information can give an indication of modelling errors and of the choice of subsystems to be corrected. The gain factors or functions coming from linearized Taylor expansions show

the influence of the parameters considered, and, for
example, dynamic variables etc. Sensitivity considerations
can also be made with respect to initial conditions etc.
Modifications based on linearized investigations /1.5/ can
use Taylor expansions, too. If the modifications concern
some modal degrees-of-freedom they can be taken into
account using the spectral decompositions of the frequency
response matrix or of the parameter matrices.

The basis for input identification in the time domain
is the deconvolution of system output signals. The mathematical procedure, however, can cause numerically large
difficulties. In the frequency domain the input can be
calculated easily if one knows the frequency response
function matrix of the system with sufficient accuracy and
also the corresponding output. If the system is partly
known one can try to determine the remaining model parameters (e.g. damping ratios) and the input quantities
/1.15/.

1.3   Comments on Acoustic Quantities

Acoustic emissions can be generated directly by an acoustic source or indirectly by structure-borne vibrations and radiation. The inverse is also true: acoustic loading can cause structural response. This fact can be used as a non-contact excitation of modes (acoustic modal analysis). The problem which arises is that only particular eigenvectors are excited measurably.

The determination of sound pressures, intensities, source locations, the elimination of noise etc. is based on well-known physical relationships, source characteristics and on signal properties /1.5/. The total sound power passing through a defined surface with area S is defined by /1.16/

$$W = \int_S \vec{I} \, d\vec{S} \, , \qquad\qquad (1.22)$$

where $\vec{I}$ is the sound intensity vector and $\vec{S}$ the normal vector at the surface. In an isotropic nonviscous homogeneous medium with zero mean flow and without field force effects, the time-averaged product of the instantaneous pressure and the corresponding particle velocity $\vec{v}$ at the same position equals the intensity

$$\vec{I} = \frac{1}{T} \int_0^T p(\vec{r}, t) \, \vec{v}(\vec{r}, t) \, dt \qquad\qquad (1.23)$$

with position vector $\vec{r}$ and t the time variable. T is the
period of waves.

$\vec{v}$ cannot be measured directly, therefore Euler's relationship

$$\mathbf{grad} \ p(\vec{r}, t) = -\rho_0 \frac{d\vec{v}(\vec{r}, t)}{dt} \qquad\qquad (1.24)$$

with ρ_0 the air density is used in order to obtain the
velocity

$$\vec{v}(\vec{r}, t) = -\frac{1}{\rho_0} \int_0^t \mathbf{grad} \ p(\vec{r}, \tau) \, d\tau \ . \qquad\qquad (1.25)$$

The integrand in (1.25) can be taken approximately as
the first order difference (measured with closely spaced,
distance Δr, microphones with identical (!) characteristics). The accuracy of this approximation is also determined by the choice of Δr compared with wavelength λ.
In addition to this direct method in the time domain, the
use of correlation functions and their Fourier transforms
(spectral density functions) leads to Eq. 1.16 (ergodicity
assumed)

$$I(r) = R_{pv}(r, \tau=0) = \int_{-\infty}^{\infty} S_{pv}(r, f) \, df \qquad (1.26)$$

(which can be handled by a two-channel FFT analyzer). The approximations known from time series analysis may be applied for solving Eq. (1.26).

Intensity measurement serves also for source detection, and its determination can now be carried out close to the radiating surfaces.

In order to overcome the general acoustic radiation problem one has to formulate the boundary value problem; e.g. as the Helmholtz integral equation /1.17/ (pressure field equal to the surface integral of surface pressure superimposed by the velocity over the radiating boundary). If the surface velocity is known, this equation is a Fredholm integral equation of the second kind for the unknown surface pressure. Apart from special cases with known solutions there are a lot of publications on approximations (e.g. /1.18/), which can be used as a basis for identification. The numerical formulation of the Helmholtz integral equation is used in /1.19/, taking the derivative of the free space Green's function. Similarly to the approaches in experimental modal analysis and finite element methods, an approximation is made by partitioning the source surface into a finite number of planar surface elements (chosen, for example, so that the velocity and pressure is distributed uniformly over the elements). The result is a system of linear inhomogeneous equations.

1.4   Test model

The definition of the test model is contained in Fig. 1.5. The test planning in general takes into consideration the a priori knowledge of system analysis and experience with comparable structures. It should contain an optimization of test signals and locations of pickups with respect to the test object and the (predicted) dynamic behaviour. The test is carried out with the real structure (or a dummy) which is adequately supported. The structure is generally equipped with attached exciters and pickups. First the mechanical effects (inertia, stiffness, damping) of this equipment have to be assessed (see the 2nd part of this Introduction). The electrical effect should be ascertained (e.g. calibration, direct controlling; feedback depending on vibration characteristics has to be avoided). All in all, deterministic errors including environmental influences have to be detected so that they can be prevented or corrected analytically. Fig. 1.4 indicates possible disturbances.

The measurement has to be planned carefully, and, of course, the excitation. A check of the observation is recommended (e.g. by a quick look and coherence measurement), and also a check of the assumptions made considering the physical model. The latter, for example, means examining the hypothesis of linearity (homogeneity and superposition) and of time invariance. Independent of the kind of identification used, statistical methods must be applied to the measured continuous signals in every case in order to reduce random errors: time series analysis.

Tasks arising in time series analysis are
- discretization (sampling and quantization)
- demand on frequency resolution
- windowing
- aliasing
- estimating of correlation and spectral density functions.

The digital processing of continuous signals demands a digitization which includes sampling and quantization. The quantization can be done without difficulty (overload does not have to be a problem), the sampling can be guided

by Shannon's theorem. The finite length T of each measurement record (rectangular time window $[0,T]$), sets limits
to the frequency resolution $\Delta f = 1/T$. The finite observation time and the chosen window cause leakage in the
Fourier transformed data. Analog filtering is necessary
before digitization in order to avoid aliasing errors. The
application of the finite DFT (often used) in the form of
fast working computer routines (FFT) or hardware treats
the signal as if it were periodical, i.e. the coefficients
of a Fourier series expansion of the signal are calculated. The routines are cyclic procedures (transformations) with which inverse DFT and convolution integrals
can be calculated. It can be shown that the DFT of the
discrete correlation function equals the discrete periodogram. The periodogram is defined by

$$\mathbf{S_{xy}}(\omega, T) := \frac{1}{T}\, \mathbf{X_T^*}(j\omega)\, \mathbf{Y_T}(j\omega)\,, \qquad\qquad (1.27)$$

where $X_T^*(j\omega)$ is the conjugate complex finite Fourier
transform of x(t) and $Y_T(j\omega)$ respectively, and T is the
measuring time (record length). This periodogram is a
rough approximation of the spectral density function because in addition it is necessary to calculate the expected value of Eq. (1.27) according to the definition of
spectral densities.

Another property within the Fourier and Laplace transformation should be mentioned. The measured signal transformed into the frequency domain gives the signal and the noise in the frequency domain. The covariance matrix of correlated noise signals transformed by DFT yields a diagonal matrix /1.20/. This is a great advantage when dealing with covariances of estimates in the frequency domain. The same statement does not hold true for the discrete Laplace transform, even if the noise signals are uncorrelated. The advantage of the Laplace transform is the improvement of the signal to noise ratio of transient signals due to the exponential window.

1.5   Parameter and State Estimation

In the classical non-Bayesian approach of parameter estimation the parameters to be estimated are considered as fixed but unknown constants (a). The "true" parameter values may be denoted by $\overset{\circ}{a}$. The prior knowledge here consists of an initial guess e of the "true" parameter values. The posterior knowledge is given by parameter estimates $\overset{\Delta}{a}$ which are realizations of the applied estimators that are statistical variables (Fig. 1.6). These variables satisfy the sampling distribution $p(\overset{\Delta}{a}, N)$, which depends on the estimator and the sample size N and is in

general unknown. Its most significant characteristics are
the expected values $E\{\overset{\triangle}{a}\}$, the bias $E\{\overset{\triangle}{a}\} - \overset{\circ}{a}$ and the
covariance matrix $\mathrm{cov}(\overset{\triangle}{a}) = E\{(\overset{\triangle}{a} - E\{\overset{\triangle}{a}\})(\overset{\triangle}{a} - E\{\overset{\triangle}{a}\})^T\}$. The
estimators considered here are the least squares (LS),
weighted LS (WLS) and the method of instrumental variables
/1.5,1.12/. The disadvantage of the LS methods are that
they can lead to biased estimates due to measurement
noise. Using the instrumental variables method is one
possible way of removing this disadvantage.

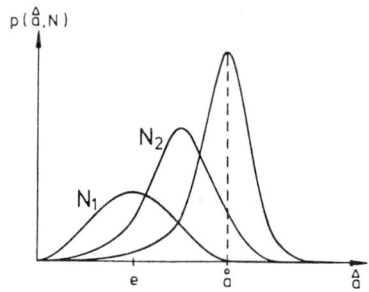

FIG.1.6   NON - BAYESIAN APPROACH
          TO PARAMETER IDENTIFICATION

In the Bayesian approach to parameter estimation
/1.21, 1.22, 1.23, 1.24/ the parameters are assumed to be
random variables. The prior knowledge consists of the corresponding prior probability density function (pdf) p(a)
which can be "informative" (cf. (1) in Fig. 1.7) or "non-
informative" (cf. (2) in Fig. 1.7) and reflects the stati-
stician's prior confidence in the prior model. The posterior knowledge consists of the conditional posterior pdf
of a due to the input/output measurements (U^M/P^M) obtained
from the system to be identified. It contains the total
information a posteriori available about the parameters.
In practice this information is often reduced to the Bayesian estimators denoted as mean

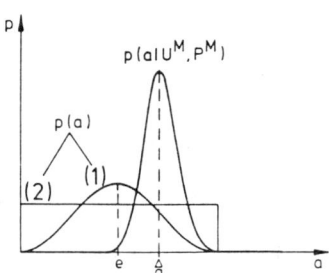

FIG.1.7 BAYESIAN APPROACH TO
PARAMETER IDENTIFICATION

$$E\left\{a \middle| U^M, P^M\right\} = \int a\, p\left(a \middle| U^M, P^M\right) da \qquad (1.28a)$$

and mode

$$\hat{a} := \max_a\, p\left(a \middle| U^M, P^M\right). \qquad (1.28b)$$

If one assumes normal distributions for both the measurement noise and a priori for the parameters, the pdf is normal too, and the most probable value (mode) can be obtained by minimizing the quadratic loss function of the extended WLS formulation:

$$J(a) = v^+(a)\, G_v v(a) + (a - e)^T G_e (a - e). \qquad (1.29)$$

The first term in (1.29) with the residuum vector $v(a)$ (and its conjugate complex and transposed v^+) describes the usual WLS. The second term contains the a priori knowledge. If the weighting matrix G_e predominates over the weighting matrix G_v (in the sense of its norm), then the estimates $\hat{a}$ are close to e and vice versa. In the first case the engineer's confidence in the prior values is large, in the second case he relies on the measurements. To obtain estimates of minimum variance the Hermitian and positive definite weighting matrices have to be chosen as inverse covariance matrices with respect to the measurement noise and the prior estimates e respectively. The advantage of this formulation is that it includes the prior knowledge, that missing test data can be replaced by prior knowledge, and that it can be used as a parameter perturbation method for solving the problem iteratively. In the case where the method is nonlinear in the parameters to be estimated, the perturbation is done by choosing the weighting matrices parametrically and by using the previous solution as starting values for the next iteration.

The maximum likelihood estimation and its classification remains to be discussed. Applying the maximum likelihood estimation is just the same as using the Bayesian estimator without any a priori pdf for the parameters when estimating the mode (Eq. (1.28b)). The remaining (conditional) joint pdf is called the likelihood function (L). For the sake of convenience, ln L is considered (ln L is monotonic, therefore the maximum of L and ln L occurs at the same value):

$$\nabla_a \ln L(a, U^M, P^M)\big|_{a=\hat{a}} = \frac{\partial}{\partial a} \ln L(a, U^M, P^M)\big|_{a=\hat{a}} = 0 \ . \quad (1.30)$$

It is obvious that the joint pdf of the samples requires knowledge of the covariance matrix of the measurement noise. If there is no a priori knowledge of the noise covariance matrix it can be estimated together with the parameters /1.20, 1.21, 1.23/. The classification of the estimation methods mentioned is shown in Fig. 1.8.

A state estimation may sometimes be necessary in addition to parameter estimation. The starting point is the linear(ized) process (1.17), (1.18) contaminated with the measurement and process noise:

$$\dot{x}(t) = Ax(t) + f(t) + r(t) \, ,$$
$$y(t) = Cx(t) + n(t) \, .$$

$$(1.31)$$

The observation of $y(t)$ is done during the time interval $[0,T]$ and we want to determine an estimate of $x(t)$ denoted by $\overset{\Delta}{x}(t|T)$. (The combined estimation of parameters and state variables is described in /1.25/). We have to distinguish three cases:

$t > T$: prediction, extrapolation

$t = T$: filtering

$t < T$: smoothing, interpolation, estimation.

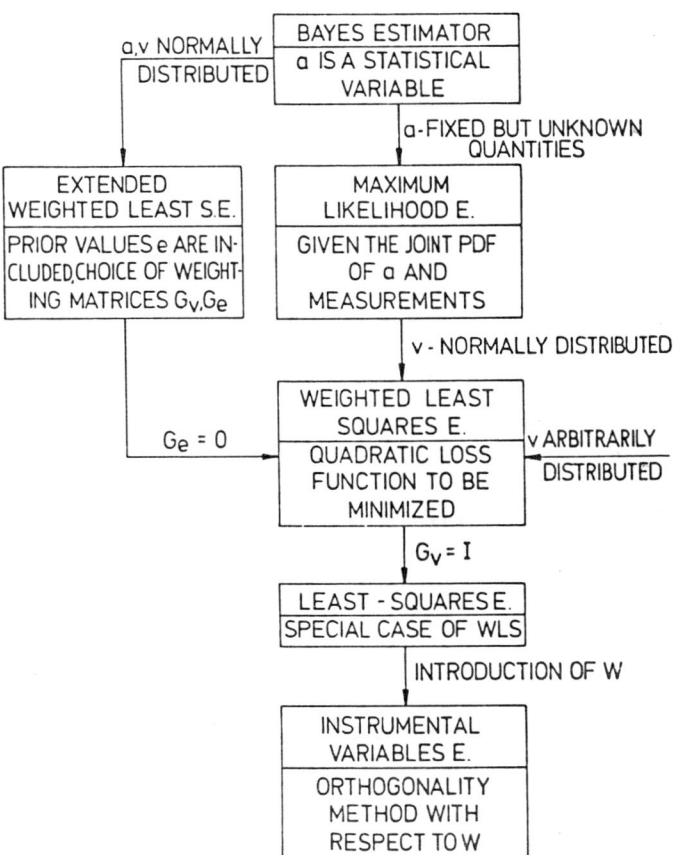

FIG.1.8 CLASSIFICATION OF USUAL ESTIMATORS

The Kalman (or Kalman-Bucy) filter /1.12/ starts, for example, with the Bayesian estimation. Assuming a conditional pdf $p(x|y)$ and using Bayes' rule

$$p(x|y) = \frac{p(x|y)}{p(y)} = \frac{p(y|x)\,p(x)}{p(y)} \; , \qquad\qquad (1.32)$$

after the observation of y, a new posteriori pdf for x can be determined. This leads to an iterative procedure in which the result of the previous calculation is the a priori knowledge for the next calculation when using the new observations. Taking into consideration normally distributed noise with zero mean, for the sampled case one obtains

$$x(k+1) = A\,x(k) + f(k) + r(k)$$
$$\qquad\qquad\qquad\qquad\qquad\qquad (1.33)$$
$$y(k) \quad = C\,x(k) + n(k) :$$

$$\hat{x}(k+1) = A\,\hat{x}(k) + f(k) + r(k+1)\,[y(k+1) - C\,\{A\hat{x}(k) + f(k)\}]\;,$$

$$r(k+1) = Q(k+1)\,C^{T}\,[C\,Q(k+1)\,C^{T} + C_{nn}]^{-1} = P(k+1)\,C^{T}\,C_{nn}^{-1}\;.$$

$$Q(k+1) = A\,P(k)\,A^{T} + C_{rr}\;,$$

$$P(k+1) = [Q^{-1}(k+1) + C^{T}\,C_{nn}^{-1}\,C]^{-1} =$$
$$\qquad\qquad\qquad\qquad\qquad\qquad\qquad (1.34)$$
$$\qquad = Q(k+1) - Q(k+1)\,C^{T}\,[C\,Q(k+1)\,C^{T} + C_{nn}]^{-1}\,C\,Q(k+1)\;,$$

where $\overset{\Delta}{x}(o)$ and P_o (= G_e if parameters are to be estimated) are given. Summarized, the following symbols are used:

$$\overset{\Delta}{x}(k) = A \overset{\Delta}{x}(k-1) + f(k-1) \qquad \text{filtered, estimated state}$$

$$C[A\overset{\Delta}{x}(k-1) + f(k-1)] \qquad \text{observation of filtered state}$$

$$y(k) - C[A\overset{\Delta}{x}(k-1) + f(k-1)] \qquad \text{error between observed and}$$
filtered value

$Q(k+1)$ covariance matrix of $x(k+1)$ based on k+1 observations

$P(k)$ a posteriori covariance matrix of $x(k)$ based on k observations

C_{nn}, C_{rr} covariance matrices of n and r respectively

If the process behaviour is nonlinear one can use the filter equations given above through linearizing around a known approximate solution of the state equations. If the linearization is done with the best estimate $\overset{\Delta}{x}$ of the state vector the approach is known as the extended Kalman filter /1.25/.

2.   On options and limitations for dynamic tests

As already mentioned, system identification needs both the mathematical model a priori obtained by system analysis and the experimental model from dynamic testing of the real structure. These tests are designed to validate the mathematical model by checking the basic assumptions made, such as linearity, stiffness distribution, mass distribution, damping distribution, and boundary conditions, and to reduce the prior uncertainties of model parameters. On the other hand, it often occurs that conditions correctly assumed in the model cannot be realized by the test procedure (e.g. boundary conditions during dynamic tests of spacecraft), so that the discrepancy between model and test results thus resulting has to be removed mathematically. System analysis and test procedure therefore have to be complementary.

Since dynamic tests for system identification are applied in civil engineering as well as in mechanical engineering or the aircraft industry, a variety of test methods have already been developed to standard procedures adapted to the different test items and test facilities /2.1/1.7/.

2.1   Environmental and boundary conditions

It is obvious that civil engineering structures
cannot be tested separately from their natural environment. Two important types of dynamic interaction should be
mentioned: ground-structure interaction and fluid-struc-
ture interaction (e.g. offshore structures, tall buildings, cooling towers, chimneys) /2.2/2.3/2.4/. These
interactions may severely affect the results of dynamic
tests and require increased effort in data processing to
reduce the interference /2.5/ and/or have to be taken into
account when designing the mathematical model /2.3/. However, such natural loads can also be regarded as random
excitation, and measured vibrations ("ambient response
data"), for instance due to wind, in some cases allow one
to estimate fundamental frequencies and damping ratios of
buildings /2.1/2.3/.

On the other hand, it can be difficult to simulate
the correct boundary conditions during the dynamic identification test, for instance in the aircraft and spacecraft
industry when ground-based tests ("ground vibration testing") have to simulate flight environments. A frequent
method of conducting such vibration tests on aircraft
under "free" boundary conditions is to suspend them in a
low-frequency support system comprised of long cables and/
or soft mechanical or air springs. The general requirement

on the support system is that it has no natural frequencies within a factor of four or five (or more for highly damped structures) of the test structure modes of interest. This often results in a pendulum suspension of the structure by attaching the cables to structural points of less deflection i.e. of relatively high stiffness. When complete aircraft are tested dynamically on the ground the boundary conditions are often not simulated by a soft suspension but by supporting the test structure directly with air (pneumatic) springs /1.5/. For Figs. with various suspensions see /2.21/.

Soft and pendulum suspensions are also used for testing spacecraft structures dynamically on the ground, i.e. in the earth-gravity environment. However, this ground test technique becomes questionable when lightweight constructions of large size have to be tested at frequencies below about 1 Hz /2.6/. Suspension frequencies about a factor of five below the first natural frequency of the structure can require impractically long cables or impractically soft springs. On the other hand, long cables and soft springs have their own natural frequencies which give rise to an upper frequency limit of the test setup which is below most of the modes of the space structure under test. Two alternatives are available: mathematically removing the effects of the test setup and/or the use of

special test setups and narrow-band testing for each mode
desired /2.6/.

In spite of these difficulties the low frequency suspension of such test structures seems to be more advantageous for simulating boundary conditions than "rigid"
fixtures, which are sometimes used when structures or
structural components (substructures) are tested on a
shaker table. Really rigid fixtures can never be obtained
in practice, but cause interactions between test setup and
test structure, which then have to be taken into account
in the mathematical model.

2.2   Excitation of test structures

If natural (ambient) loads or operating loads (e.g.
unbalance forces of rotating machines (turborotors etc.))
are not available or insufficient for dynamic tests of the
structure under investigation, the test structure has to
be excited by a suitably chosen test device. Which kind of
excitation (exciters) is optimum depends on the test object as well as on the time available for carrying out the
test. For dynamic tests of civil engineering structures it
is often easiest to apply step relaxation ("snapback testing"), i.e. the structure is preloaded with a measured

force through a cable which is then suddenly released
/2.5/. With this excitation technique in general only the
lowest natural modes of the structure are excited sufficiently /1.5/. In order to excite higher natural modes
eccentric mass exciters are used. They are driven by an
electric or hydraulic motor and mounted directly on the
structure generating almost sinusoidal forces
/2.4/2.7/2.8/. Excitation frequencies up to 20 Hz and
forces up to 200 kN are reported /2.7/ and also an accuracy of excitation frequencies of about 0.003 Hz /2.8/.

In /2.8/ hydraulic exciters are proposed for exciting
highway bridges in their natural modes. The hydraulic vibrator is a device which transforms the power of a high-
pressure flow of fluid from a pump to the reciprocating
motion of a piston in a hydraulic actuator. An electromagnetic (electrodynamic) driver unit and a power valve or
hydraulic amplifier are used. Depending on the power capacity of the system, large forces and large velocities of
motion can be generated, made possible with a large
stroke. For harmonic excitation without distortion the
operating frequency range has an upper limit of about 50
Hz /2.9/, while the low-frequency limit is usually zero
(for further details see also /1.7/2.10/). Since the
weight of a hydraulic vibrator is small relative to the
forces attainable, a heavy reaction mass or support is
necessary if such a vibratory force is to be introduced in

a large structure. This may be a complicating factor when large, fixed structures have to be excited.

If large forces and/or large vibration amplitudes are not necessary but higher excitation frequencies are required (e.g. when testing spacecraft structures), electromagnetic (electrodynamic) exciters (shakers) are used which are essentially constructed like the electrodynamic driver mentioned above. The driving force is generated by the interaction between an alternating current flow in an armature coil and an intensive magnetic dc field passing through the coil which is concentrically located (with radial clearances) in the annular air gap of the dc electromagnet. The armature coil is connected to the electromagnetic unit by leaf springs, which are relatively soft in the direction of force but stiff in the transverse direction, also limiting the peak-to-peak displacement. The generated force is limited by the available power amplifier current and the capacity of the cooling device for carrying off the dissipated heat caused by copper and iron losses in the electromagnetic unit. The power amplifier can also be the cause of the operating frequency range having a lower limit above zero cps, because not all power amplifiers are able to deliver extremely low frequencies (cf. Fig. 2.1).

The sinewave of the force output by electromagnetic vibrators has less distortion than that of other exciter types. This is an advantage for vibration tests when sinusoidal forces are required. This is also one of the reasons why electrodynamic shakers are so often applied in vibration tests and are available in various sizes. Because the generated force is usually less than the weight of the shaker, elaborate installations are not needed if the shakers can be suspended near and coupled to the structure being investigated.

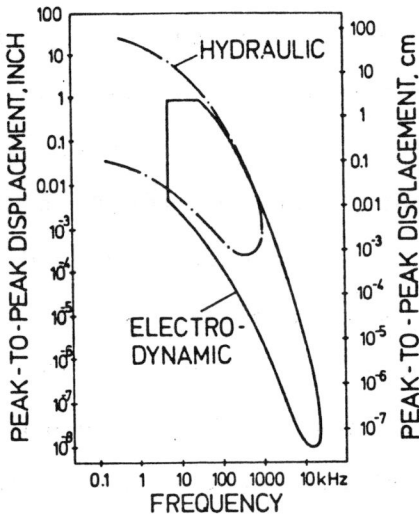

Fig. 2.1 TYPICAL OPERATING RANGES OF
       HYDRAULIC AND ELECTRODYNAMIC
       EXCITERS

Very often the applied forces have to be controlled in dynamic tests of structures. Therefore it is necessary to measure the imposed forces directly at the structural point of application, so that the interactions between structure - exciter attachment - electrodynamic/electro-hydraulic exciter system can be compensated by a suitable control system.

By using such electric control devices it is possible to apply not only forces with sinusoidal waveform but also various other kinds of deterministic and/or random excitations. Sine-wave generators and random noise generators have been well-known for a long time as voltage suppliers in control systems of vibration machines for qualification tests of structures /1.7/. However, with the implementation of digital signal processing, test signals are not limited to sine waves or pure random. With the discrete Fourier transform, any physically realizable signal can be used for excitation. So it is feasible to choose from a variety of excitation types and select the one that is optimum for the test in question.

Apart from the step relaxation (already mentioned above) and impact (impulse) techniques (see below), the most common deterministic excitation signals are derived from sinusoidal signals:

Stepped-sine excitation (also known as "sinusoidal dwell") is often applied in vibration tests (e.g. ground vibration tests of aircraft) based on sinusoidal signals of constant frequency. By "stepping" the sine frequency through the frequency range of interest, frequency response functions can be evaluated for a number of discrete frequency points. Or the structure under test can be excited in one of its normal modes, provided that several exciters are used simultaneously in an appropriate exciter configuration and "dwell" on the corresponding natural frequency of the structure (phase resonance technique /2.9/). As all excitation energy is located in one single frequency, a high signal-to-noise ratio is attained. The favourable peak-to-rms ratio allows one to study excitation level dependencies of system parameters. This is essential for the detection and the analysis of non-linear system behaviour. Another advantage of stepped-sine excitation is the variable frequency resolution, which can easily be adapted to the measured system response. The main disadvantage of this excitation method is that after a step in the frequency range, the experimentalist has ·to wait for the transient system response to die out before

he can measure the steady-state response. So vibration
tests with stepped-sine excitation are extremely time-
consuming. (For application in modal analysis cf.
/1.5/2.11/).

In order to diminish this disadvantage swept-sine
excitation /1.5/1.7/2.12/ is often used. When slow sine
sweeps are applied, the sweep rate is kept so slow that
the excited system can reach a quasi-steady state. If this
condition is violated and the frequency response function
is evaluated as usual for stepped sine excitation it is
distored, and the resonance frequencies seem to be shifted
in the sweep direction. (For a suitable sweep rate cf.
/1.7/). The advantages and disadvantages of slow sine excitation are therefore almost the same as for stepped sine
excitation, except for the shorter duration of the vibration test.

By applying a fast sine sweep (chirp, Fig. 2.2), the
sweep rate exceeds the quasi-steady state condition. In
order to avoid leakage errors when analyzing the system
response with a usual digital Fourier-analyzer, the fast
sine sweep is made periodic in the analyzer window (periodic chirp). The number of periods required to achieve
periodicity in the response is determined by the impulse
response time of the structure under test /2.13/. In
addition to its speed the advantages of a periodic fast
sine sweep are: good peak-to-rms ratio, usefulness in

characterizing structural non-linearities, good signal-to-noise ratio, controlled bandwidth. It is disadvantageous that periodic chirp generates periodic noise due to structural non-linearities, and that it is difficult to set transducer gains.

Pure (true) random excitation with ergodic signals (normally with Gaussian distribution) is well-known for its ability to obtain the best linear approximation of a non-linear system for a given level of random signal input by averaging successive records of frequency domain data, which results in a good linear estimate of the frequency response function /2.12/2.13/. This is important if parameter estimation techniques are to be utilized to extract the modal properties. The main disadvantage with pure random excitation is that neither input (force) nor response are periodic within the measurement time, giving rise to leakage error when an FFT-analyzer is used. This requires the use of weighting or window functions (e.g. Hanning window) to reduce the leakage errors. However, the computed frequency response function is distorted by windows of this kind, and the frequency resolution is reduced /2.13/2.14/.

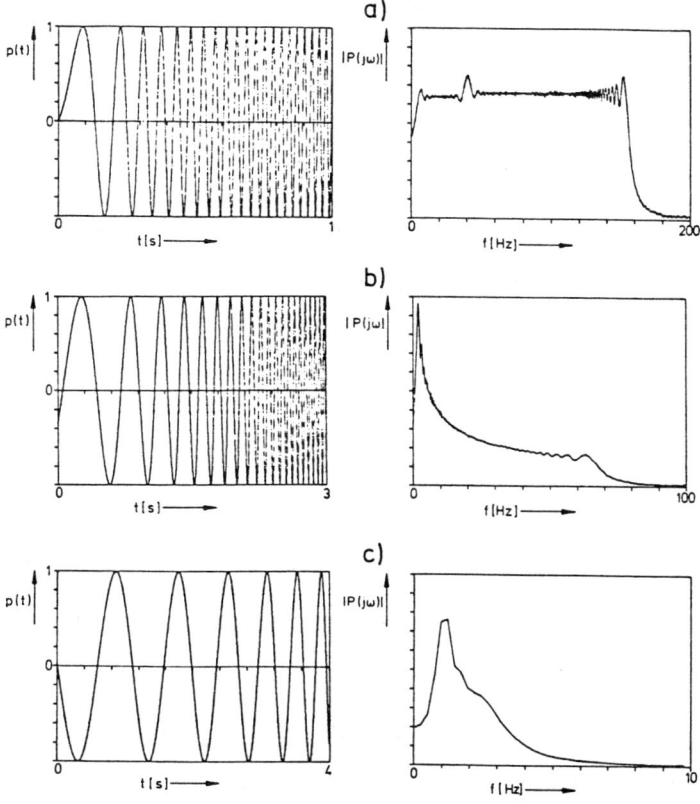

FIG.2.2   FAST SINE SWEEP EXCITATIONS :
          a) LINEAR
          b) EXPONENTIAL
          c) LOGARITHMIC

Periodic random excitation can be applied to avoid leakage errors, where a random number generator is used to create an array of values in the time domain, which when Fourier transformed yield a spectrum with random amplitude and phase /2.12/2.13/. Periodic random excitation has the same advantage concerning non-linear systems as pure random excitation, but the process of establishing periodicity (i.e. the test structure is excited with this input in a repetitive cycle until the transient response dies out and the stationary response becomes periodic with the input) is time-consuming, in particular when lightly damped structures are tested /2.12/2.13/. A further disadvantage is that more sophisticated hardware is required in order to generate periodic signals synchronously with the measurement process /2.15/.

Pseudo-random signals are created in the frequency domain, usually with uniform amplitude and random phase throughout the desired frequency range. Fourier transformed into the time domain, the signal is transmitted to the exciter system through a digital-to-analog converter. This process of signal generation leads to an advantage of pseudo-random excitation: the excitation signal is always periodic within the sample window and therefore does not suffer from leakage errors. One principal disadvantage of pseudo-random excitation, however, is that non-linearities of the test structure will generate periodic noise which cannot be averaged out /1.5/2.12/2.13/. To avoid this disadvantage, periodic pseudo-random excitation can be used /1.5/: during the first input of pseudo-random excitation one has to wait for the transient system response to die out so that the steady-state response (and the corresponding input) can be measured. After that another pseudo-random excitation is generated which is uncorrelated with the previous one, and the steady-state response of the system is measured again. This procedure is repeated several times, and averaging the resulting frequency response functions reduces not only random measurement noise but also distortions due to system non-linearities. Periodic pseudo-random excitation needs more measuring time than a pseudo-random one, but structural tests with periodic pseudo-random excitation can still be carried out about ten times faster than the same tests with harmonic excitation. (For the generation and application of pseudo-random binary test signals cf. /1.12/2.17/).

Burst random signals contain the properties of random and transient signals ("random transient"). The random input history function is truncated at a point so that the response history decays to zero within the sample period. So this excitation signal has all the advantages of periodic random without the disadvantage of increased test time to establish periodicity. Even for lightly damped structures the response history will decay to zero very quickly, due to the excitation system producing a damping force aiding the dissipation of the stored energy of the structure under test. This additional damping force is measured by the force transducer located between the structure and the shaker armature, and can thus be taken into account /2.13/2.15/.

Besides step relaxation (already mentioned), impact or hammer excitation is a second (very popular) type of transient excitation. Its main advantages are obvious: fastest test method for linear systems in low noise environment, minimum equipment requirements, excellent field test method and useful on operating systems. Its disadvantages are: very high peak-to-rms ratio and therefore not suitable for non-linear systems, limited control of frequency content, poor signal-to-noise ratio. (For the impact testing technique including local damping effects cf. /1.5/).

It is very common to use the test signals summarized above (cf. also Table 2.1) for single point excitation. But during the single input excitation of a system large differences may exist in the vibration amplitudes at various locations because the excitation power dissipates within the structure, particularly in the case of heavy damping. Multiple point excitation is expected to result in a better energy distribution, especially in large and/ or complicated structures /1.5/. The principal advantage of multiple point excitation is the increase in accuracy and consistency of structural parameter estimates /1.5/. In order to estimate the (matrix of) frequency response functions it is required that the inputs are not perfectly correlated at any frequency. (Theoretically, completely uncorrelated inputs would be best, but some degree of correlation cannot be avoided in practice due to the impedance mismatch between the excitation system and the structure).

	SINE STEADY STATE	TRUE RANDOM	PERIODIC IN ANALYZER WINDOW			TRANSIENT		
			PSEUDO RANDOM	RANDOM	FAST SINE	IM-PACT	BURST SINE	BURST RANDOM
MINIMIZE LEAKAGE	NO	NO	YES	YES	YES	YES	YES	YES
SIGNAL TO NOISE RATIO	VERY HIGH	FAIR	FAIR	FAIR	HIGH	LOW	HIGH	FAIR
RMS TO PEAK RATIO	HIGH	FAIR	FAIR	FAIR	HIGH	LOW	HIGH	FAIR
TEST MEASUREMENT TIME	VERY LONG	GOOD	VERY GOOD	FAIR	FAIR	VERY GOOD	VERY GOOD	VERY GOOD
CONTROLLED FREQUENCY CONTENT	YES	YES **	YES **	YES **	YES **	NO	YES **	YES **
CONTROLLED AMPLITUDE CONTENT	YES	NO	YES **	NO	YES **	NO	YES **	NO
REMOVES DISTORTION	NO	YES	NO	YES	NO	NO	NO	YES
CHARACTERIZE NONLINEARITY	YES	NO	NO	NO	YES	NO	YES	NO

** REQUIRES ADDITIONAL EQUIPMENT OR SPECIAL HARDWARE

TABLE 2.1 EXCITATION FUNCTIONS

For further excitations (e.g. by blast and cartridges) cf. /2.21/ and /2.22/. Table 2.1 summarizes some
excitation functions.

2.3   Some final remarks on measurement techniques

Some a priori knowledge of the expected system vibration is required for the suitable selection of transducers, i.e. if displacement, velocity, acceleration, or
strain-measuring transducers should be used (e.g. displacement transducers for low-frequency vibration, where
corresponding velocity or acceleration measurements yield
an impractically small output, or acceleration measurements may be useful, where suitable displacement or velocity pickups would be too large because of clearance requirements). The effect of added mass and the change in
stiffness of the test structure must also be considered.
It may be mentioned that a mass distribution of the picups
mounted proportional to the mass distribution of the
system under test will not change the natural modes; only
the eigenfrequencies will be influenced in a correctable
manner /1.5/.

The main requirement for a tranducer mounting is to couple the transducer to the system under test so that the transducer accurately follows the motion of the surface to which it is attached. So the effective stiffness of the transducer mounting must be large in the frequency range of interest. Typical methods of coupling piezo-electric acceleration transducers to a test item are shown and described in /2.18/. For some general rules to be observed in the mechanical design of transducer mountings (e.g. brackets) cf. /1.7/ (and also for special mounting methods, when, for example, a transducer has to be mounted on thin structures).

The use of electromechanical transducers requires the installation of wiring. Wiring installations should be designed to minimize electrical noise being self-generated or induced when electrical energy is coupled into the measurement circuits, e.g. by ground loops, varying magnetic or electric fields. So, for example, separate signal and power current return leads to reduce ground-loop coupling between signal and power circuits should be used, as well as twisted pair or coaxial cables to reduce inductive pickup in signal leads. Electrical noise may also be generated by the motion of some part of the wiring (e.g. by the variation of contact resistance in connectors because of the change in geometry of the wire). Such electrical noise may be reduced by fixing the cable to a structure at frequent intervals /1.7/.

Concerning calibration techniques for transducers and the field calibration of measurement systems the reader is referred to /1.7/2.16/2.19/2.20/.

3.   Survey of Identification Methods

The fundamental relationships and procedures are already provided in the first chapter. However, the equations must be suitably prepared in order to be applied for estimating the quantities desired.

3.1   Some Time Domain Methods

In structural engineering time domain methods have (so far) played a minor role, although the authors note a positive trend. If we restrict ourselves to stationary processes, there is one class of models that has to be mentioned: ARMA models (Auto Regressive Moving Average). The modelling describes the approach for developing the ARMA equations from the equation of motion: Fig. 3.1.

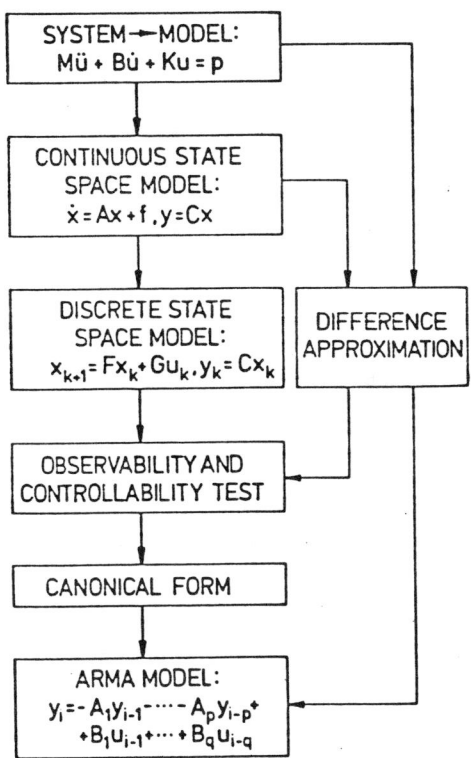

FIG.3.1 FORWARD MODELLING

From the state space model (1.17) and with the observation equation (1.18) one obtains the discrete-time state
space model /3.1/

$$x_{k+1} = F x_k + G u_k \qquad\qquad (3.01)$$

$$y_k = C x_k , \quad k = 0(1) N - 1 ,$$

with

$$F := e^{Ah} , \quad h := t_{k+1} - t_k$$

$$G := \int_0^h e^{A\tau} B \, d\tau \qquad\qquad (3.02)$$

$$x_k := x(kh)$$

$$f(t) =: B u(t) .$$

Other representations can be obtained by substituting
the differential quotient in (1.17) for finite differences. If the number of approximation points at a finite
interval tends towards infinity, then the matrix obtained
converges with the matrix e^{Ah}. It holds true that the
discrete model with the state matrix F is covariance equivalent, i.e. the values of covariance of the output at t_k
= kh for every k are the same for the discrete model as
for the continuous model. The models obtained by the

difference scheme can therefore be called asymptotically covariance equivalent.

Only fully controllable (C) and observable (O) systems can be completely identified. Therefore the controllability and observability (C/O) of the system have to be investigated before the identification procedures are applied. C/O depend on the state matrix F (the structural properties) as well as on the control matrix G and the observation matrix C. This means that the non-controllable (non-observable) system can be turned into a C/O system by the proper choice of the method of observation and excitation. This can be done, for example, by shifting the excitation points and measuring points, or by introducing additional ones. C/O tests can be done in the standard form or in the modal form /3.1-3.3/. The C/O result is theoretically two-valued: yes - no; but in practice, ill-conditioned matrices can be obtained as a consequence of weak C (excitation close to a modal line of a mode). - A check of the degree of the C/O can be made by the transformation of the state space representation to the uniformly balanced model /3.4-3.7/. The uniformly balanced representation allows one to display, and therefore to omit, the weakly controllable and observable state variables by order reduction.

The derivation of the ARMA representation from the discrete-time space model consists of two steps: first, the state equations are transformed to the observability canonical form /3.8, 3.9/, then the ARMA model is obtained by rearranging. The different models are described in /3.10/ from a uniform point of view.

The estimation of the parameters a is presented in Fig. 3.2. It is a model linear in the parameters to be estimated:

$$\mathbf{y} = \boldsymbol{\Phi}\,\mathbf{a} + \mathbf{v}\ ,\qquad\qquad(3.03)$$

y is the measurement vector $(N,1)$, $\boldsymbol{\Phi}$ is the data matrix (N,p), v is the vector of unknown measurement noise. The data matrix can be given in different forms: auto-correlation, covariance, pre- and post-windowed /3.11 - 3.14/.

The estimation can be done in batch or in recursive form, both with reference to time and order.

With the instrumental variables matrix W one obtains the Yule-Walker equation from (3.03)

$$\mathbf{W^T}\,\boldsymbol{\Phi}\,\hat{\mathbf{a}} = \mathbf{W^T}\,\mathbf{y}\ .\qquad\qquad(3.04)$$

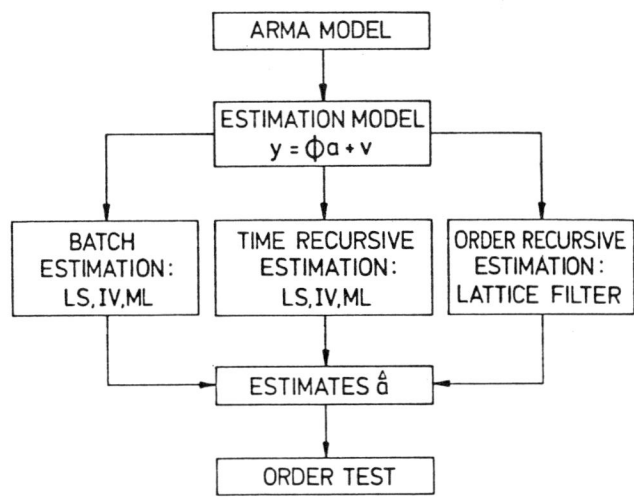

FIG. 3.2 ESTIMATION OF ARMA-PARAMETERS

When $W = \phi$ we obtain the LS estimator. The time recursive estimation can be done with the methods mentioned
previously (Chapter 1.5). Order recursive estimation is
only possible if the matrix $W^T\phi$ in the Yule-Walker equation has Toeplitz form, and then the lattice filter method
can be applied. If $W^T\phi$ is non-Toeplitz but near Toeplitz,
in the sense given in /3.15/, then the lattice filter can
be taken together with a recursive corrector /3.15/.

The order of the model is often not known in identification. Order determination can be performed by a

- determinant ratio test
- test of the linear independence of output signals
- loss function test
- signal error test
- transfer function error test
- F-test
- final prediction error test
- Akaike information criterion
- test of pole configuration.

The last step in ARMA identification is the backward modelling, as shown in Fig. 3.3.

In structural mechanics, time domain methods deal mainly with modal identification instead of estimating the physical parameter values directly with measured acceleration and, for instance, computed velocities and displacements. The usage of the free decay solution of (1.11) is very common in the form of the system eigensolution (1.12) (contaminated by noise).

The following problems are combined with using these equations

- solving a nonlinear system of equations
- determination of the effective number of degrees-of-
  freedom
- interference by noise.

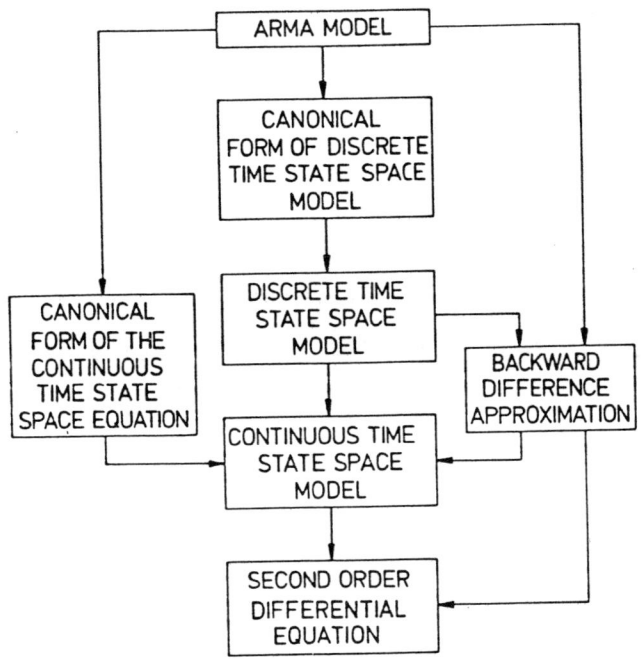

FIG.3.3 BACKWARD MODELLING

The first to deal with this identification technique was Prony in 1795 /3.16/. He transformed Eq. (1.12) into polynomial form. Modern researchers are Ibrahim et al. They convert Eq. (1.12) to an eigenvalue problem using an oversized model /3.17-3.20/. A Ph.D. thesis /3.21/ should be mentioned here. The polyreference method /3.22/ uses (multiple solutions) (1.12) for different initial conditions and combines these equations for various time increments.

Another approach is based on the singular value decomposition instead of least squares and is called the eigensystem approach /3.23/. This procedure is advantageous if it is possible to determine the order of the system from the decomposition, thus reducing the order of the parameter equations to be solved. A slightly different approach that is independent of /3.23/ is derived in /3.24/. A survey is contained in Ref. /3.25/.

3.2   Modal Identification in the Frequency Domain

Two principles have to be described. The first is phase resonance testing and the second is the phase separation technique. Both methods are discussed in detail in /1.5/. The phase resonance method as described in Chapter

1.2 involves the isolation of one natural mode by harmonical excitation with an appropriate force amplitude
vector (1.20b). Theoretically this vector is one of n
necessary linearly independent excitation vectors and is
often chosen approximately. Therefore the vibration in the
considered generalized one degree-of-freedom is biased a
priori. The eigenvector components are measured directly;
various observations can be repeated several times and
then averaged in order to reduce the random observation
errors. The remaining modal quantities follow from curve
fitting the real and imaginary parts of the frequency
response functions measured at various measurement points.
The variances of these estimates do not give an indication
of their errors, because the deterministic deviation due
to non-appropriation of the excitation is unknown. In
practice it is impossible to determine more than one generalized damping value for each natural mode. Determination of the generalized masses generally requires an additional test. This test is done with additional forces
shifted by $\pi/2$ compared with the excitation used, or with
additional stiffnesses or masses (for example, electrical-
ly simulated).

The detection of eigenfrequencies dependent on the
chosen excitation is accomplished by using the indicator
function defined by the weighted sum of phase shifts (instead of zeros). The weighting is done with the approxi-

mate local kinetic energies in the measuring points. This
sum is then normalized by the corresponding total kinetic
energy. An example is shown in Fig. 3.4. Besides the detection of eigenfrequencies one can also assess the appropriation of the excitation by the corresponding ordinates
deviating from zero.

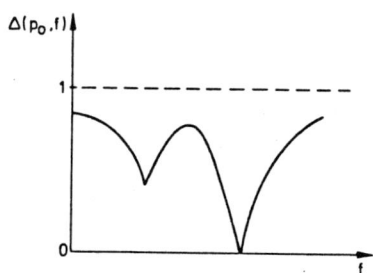

FIG.3.4 EXAMPLE OF AN INDICATOR FUNCTION
         FOR A GIVEN VECTOR p_0

One can try to eliminate non-resonant degrees-of-
freedom in the measured responses by error modelling (see,
for example, Nyquist plots and the shift of their cen-
tres).

The main advantage of this method is that it is
simple from the mechanical point of view (single degree-
of-freedom vibration). The disadvantages are the
- appropriation of excitation
- testing expenditure
- non-statistical approach a priori

- assumption of access to all the points necessary for attaching exciters.

The phase resonance method not infrequently fails when one of the previously summarized assumptions is not fulfilled, especially when there are closely neighboured eigenfrequencies.

In the second principle the excitation difficulties and the testing expenditure are shifted to the computational side because the modes are now separated by computation: the phase separation technique. It is based on the eigenvector decomposition of the measured and transformed (Fourier, Laplace) dynamic responses of the system due to measured and transformed excitation (including equal to zero), as given in Chapter 1.2. The choice of excitation (see Chapter 2.2) depends on several facts, starting with the availability, investigation of the frequency interval desired etc.

Assuming weakly damped systems and taking Eq. (1.14) into consideration, the estimation can be performed by choosing various parameters, e.g. the modal quantities themselves, elements of matrices. Various estimation procedures can also be applied. The diversity of the existing methods is not to be discussed here, and the reader's

attention must be drawn to the following fact. An (statistical) estimation procedure is a priori applied, which
means that it is also possible to estimate the variances
(standard deviations) of the estimates. The order determination of the response matrix used (number of effective
degrees-of-freedom in a partial frequency interval much
smaller than the interval one is interested in) can be a
difficult problem, but iteratively chosen numbers and
various estimations with them help to overcome this
problem (see /3.26, 1.5/).

As contained in /1.5/, the first fully developed
phase separation method was published in 1968 /3.27/ by
the senior author using frequency response functions and
an approximate instrumental variables method for proportional damping, and in 1971 /3.28/ for non-proportional
damping. These methods were based on and extended from the
publications of Stahle and Forlifer /3.29, 3.30/. In the
US the building block approach by Klosterman /3.31, 3.32/
has become very common. One additional procedure by Wittmeyer should be mentioned /3.33/. He uses the expansion of
the dynamic response of a structurally damped system with
respect to the eigenvectors in the frequency domain. Left
multiplication of the equation by one transposed left-
eigenvector with the use of the orthonormalization relationship provides the eigenquantities of one degree-of-

freedom. This equation is used to estimate these eigenquantities. This is a very simple procedure from a theoretical point of view.

With respect to Fig. 1.5, the methods belonging to
the test model alone can be classified. The measurements
are non-parametrical (input/output), but the result of the
test model as modal quantities requires a structured model
as a priori knowledge, namely the structured equation of
motion. No further prior knowledge is contained in this
identification method. Finally, the reader's attention is
drawn again to the survey /3.25/.

3.3   Identification of Physical Parameters without and
with Knowledge of the Mathematical Model within the
the Frequency Domain

With regard to Fig. 1.5, we now want to include as
much a priori knowledge as possible within identification
by improving (adjusting) the mathematical model through
the use of measured values /1.1, 1.5, 3.25/. For example,
if we look at Eq. (1.29), the prior knowledge is included
with $G_e \neq 0$. If we choose $G_e = 0$, the prior knowledge is
reduced to the structured model by the choice of resi-

duals, and the prior values can be taken as initial values for an iterative procedure.

Next, the two problems concerning the choice of parameters and the choice of residuals have to be solved. The parameter matrices of the prior mathematical model (1.11) (or an equivalent one) may be partitioned in the form

$$M = \sum_{\sigma=1}^{S} M_\sigma \ ,$$

$$B = \sum_{\rho=1}^{R} B_\rho \ ,$$

$$(3.05)$$

$$K = \sum_{\iota=1}^{I} K_\iota \ , \ \text{ or } K^{-1} = G = \sum_{\iota=1}^{I} G_\iota$$

due to subsystems. This subsystem modelling can be found
- by looking for real subsystems
- due to partitioning dynamic coordinates (e.g. bending, torsion)
- through matrices of the connection elements of subsystems
- through matrices related to design variables.

By introducing global, factorial parameters a_j with respect to the parameter matrices of the subsystems, which ought to be estimated, one obtains the parameter matrices

to be improved:

$$M^C(a_M) := \sum_{\sigma=1}^{S} a_{M\sigma} M_\sigma \ , \quad a_M := \{a_{M\sigma}\} \ ,$$

$$B^C(a_B) := \sum_{\rho=1}^{R} a_{B\rho} B_\rho \ , \quad a_B := \{a_{B\rho}\} \ ,$$

$$K^C(a_k) := \sum_{\iota=1}^{I} a_{k\iota} K_\iota \ , \quad a_k := \{a_{k\iota}\} \quad \text{or} \qquad (3.06)$$

$$G^C(a_G) := \sum_{\iota=1}^{I} a_{G\iota} G_\iota \ , \quad a_G := \{a_{G\iota}\} \ .$$

With the estimated values $\overset{\Delta}{a}_j$, where the a_j are
$a = \{a_j\} = (a_M, \ a_B, \ a_K)^T$ or $(a_M, \ a_B, \ a_G)^T$, $j = 1(1)J$,
$J = S+R+I$, the improved identified parameter matrices are

$$\overset{\Delta}{M} := M^C(\overset{\Delta}{a}_M) \ ,$$

$$\overset{\Delta}{B} := B^C(\overset{\Delta}{a}_B) \ , \qquad (3.07)$$

$$\overset{\Delta}{K} := K^C(\overset{\Delta}{a}_K) \quad \text{or} \quad \overset{\Delta}{G} = G^C(\overset{\Delta}{a}_G) \ .$$

As can be easily seen, the matrices (3.05) of the
prior computational model are obtained with $a = e = \{1\}$.
It can also be seen that a global as well as an element-
wise improvement can be carried out. The a_j can be interpreted as linearized dimensionless design variables.

The selection of the submatrices can be assisted by

- a priori knowledge

- sensitivity analysis, and

- if necessary by re-calculations.

The objective of the investigations concerning the dynamic behaviour of the system may be to predict the dynamic variables α_i, i = 1(1)n_α , within a given accuracy. The corresponding measured (identified) quantities are denoted by $\overset{\Delta}{\alpha}_r$, r = 1(1) N, and those of the computational model to be corrected by $\alpha_r^C(a)$. The residuals are then defined by

$$v_r(a) := \overset{\Delta}{\alpha}_r - \alpha_r^C(a) \tag{3.08}$$

which are assembled in the residuum vector

$$v(a) := \left\{ v_r(a) \right\} . \tag{3.09}$$

of Eq. (1.29).

Various possibilities exist for the choice of residuals: input residuals, output residuals, partial residuals with respect to eigenquantities, equation error of the eigenvalue problem and the decomposition of physical parameter matrices with respect to eigenvectors.

Input Residuals: Fig. 3.5

In the dynamic response problem the input (forcing) and output (response) signals are measured and transformed in the s-domain assuming zero initial conditions. For applications when using data corrupted by noise see /3.34, 3.35/. The computational model is then written in the stiffness formulation by

$$(s^2 M + s B + K) U(s) = P(s) , \qquad (3.10)$$

in which the matrix B is a first (rough) estimation of the "true" damping matrix /3.37/. The measured and transformed signals are denoted by $\overset{\Delta}{U}_r := \overset{\Delta}{U}(s_r)$, $\overset{\Delta}{P}_r := \overset{\Delta}{P}(s_r)$ with distinct values of s. The input residuals are then defined by

$$v_{Ir} := \overset{\Delta}{P}_r - P_r^C(a) \qquad (3.11)$$

with

$$P_r^C(a) := \left[s_r^2 M^C(a_M) + s_r B^C(a_B) + K^C(a_k) \right] \overset{\Delta}{U}_r \qquad (3.12)$$

and the parametrization (3.06). As can be inferred from Eq. (3.12) in association with Eq. (1.29) and Eq. (3.06), the residuum vector is linear in the parameters, and therefore the estimation procedure leads to a linear system of equations. This fact is advantageous, but experience with this form of identification procedure shows its

sensitivity to measuring errors /3.34/3.35/.

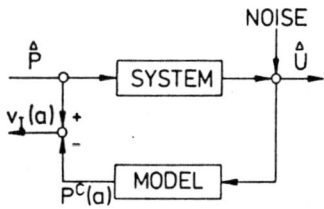

FIG.3.5 INPUT RESIDUALS

Output Residuals: Fig. 3.6

Output residuals are defined by

$$v_{0r} := \overset{\Delta}{U}_r - U_r^C(a) \qquad (3.13)$$

with

$$U_r^C(a) := \left[s_r^2 M^C(a_M) + s_r B^C(a_B) + K^C(a_M) \right]^{-1} \overset{\Delta}{P}_r \qquad (3.14)$$

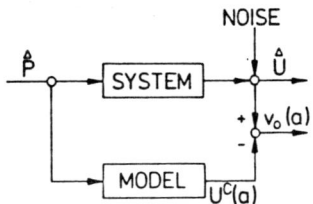

FIG.3.6 OUTPUT RESIDUALS

Because the inverse of the dynamic stiffness matrix appears in (3.14) and therefore in the residuum vector (3.09), the identification procedure is non-linear in the parameters. This procedure is discussed in /1.5, 3.36/ for undamped systems and in /1.5/ for damped systems. It is dealt with extensively in /3.34/.

If one considers the transfer functions of a multi-degree-of-freedom system, it can be seen that the choice of frequencies ω_r in the neighbourhood of the system resonances with $N \geq n$ provides the information which one obtains through the use of complete eigenquantities.

Residuals from Eigenquantities:

Full information which could be gathered from identification for each degree-of-freedom under consideration (phase resonance) is obtained from

$\overset{\Delta}{\omega}_{0r}$ the measured eigenfrequency

$\overset{\Delta}{u}_{0r}$ the measured eigenmode

$\overset{\Delta}{m}_{gr}$ the measured generalized mass (norm),

$$m_{gr} := \hat{u}_{0r}^T M \hat{u}_{0r}$$

$2\overset{\Delta}{\alpha}_r$ the measured damping ratio,

$$2\alpha_r := b_{Er} / (\omega_{0r}\, m_{gr})\,,$$

$$b_{Er} := \hat{u}_{0r}^T\, B\, \hat{u}_{0r}\,,$$

$$B_E := \hat{U}_0^T\, B\, \hat{U}_0 \overset{!}{=} \mathrm{diag}\,(b_{Ei})$$

$\overset{\Delta}{p}_{0r}$ the measured (in general approximately) appropriate force vector: in order to excite the dynamic response equal to Eq. (1.20b)

$\Delta\overset{\Delta}{u}_r^{re}$ the measured phase shifts, which ought to be as small as possible.

We will restrict ourselves to the eigenquantities of the associated undamped system; for the damped systems see /1.5/. The index r will run from 1 to N, with, in general, N < n. Because parts of the measured eigenquantities are the dynamic responses of special known forcings we can define input residuals

$$v_{Ir} := \overset{\Delta}{p}_{0r} - p_{0r}^C \tag{3.15}$$

and partial residuals

$$v_{mr} := \overset{\Delta}{m}_{gr} - m_{gr}^C\,, \tag{3.16}$$

$$v_{\alpha r} := \overset{\Delta}{\alpha}_r - \alpha_r^C \tag{3.17}$$

with

$$p_{0r}^C := (-\overset{\Delta}{\omega}_{0r}^2 M^C + j\overset{\Delta}{\omega}_{0r} B^C + K^C)\overset{\Delta}{u}_r , \tag{3.18}$$

$$m_{gr}^C := \overset{\Delta}{u}_{0r}^T M^C \overset{\Delta}{u}_{0r} , \tag{3.19}$$

$$\alpha_r^C := \overset{\Delta}{u}_{0r}^T B^T \overset{\Delta}{u}_{0r} / (2\overset{\Delta}{\omega}_{0r} \overset{\Delta}{m}_{gr}) . \tag{3.20}$$

The residuum vector may then be defined by

$$v_A^T := (v_{I1}^T , \ldots , v_{IN}^T , v_{m1} , \ldots , v_{mN} , \ldots , v_{\alpha 1} , \ldots , v_{\alpha N}) \tag{3.21}$$

The following formalism conforms to the explanations of Section 1.5. The formalism may, of course, be executed for a number of measured generalized masses different from those of eigenmodes etc.

The calculation of examples with simulated test results clearly show the large influence of erroneous measured generalized masses and the effect of chosen physically non-compatible submatrices (subsystems) on the estimation /3.37/. If the errors of the measured generalized masses are large, these magnitudes should be ignored in the estimation of the parameters a_j.

The identification of eigenmagnitudes using the phase resonance method seldom implies the measurement of phase shifts. The input residuals as employed above cannot be used without these data. In this case, the error of the matrix eigenvalue equation is taken instead:

$$v_{er} := (-\overset{\Delta}{\omega}{}^2_{0r} M^C + K^C) \overset{\Delta}{u}_{0r} .$$
(3.22)

The residual

$$v'_{pr} := \overset{\Delta}{p}_{0r} + \overset{\Delta}{\omega}_{0r} B^C \overset{\Delta}{u}_{0r}$$
(3.23)

is introduced in order to use the measured excitation. The residuals (3.19) and (3.20) complete the use of measured quantities. By reducing the number of residuals (3.23) by left-multiplication with $\overset{\Delta}{u}{}^T_{0r}$ and obtaining

$$v'_{pr} := \overset{\Delta}{u}{}^T_{0r} v'_{pr}$$
(3.24)

one can define a residuum vector v_B similar to v_A. The residuals again only depend linearly on the parameters a_j. Moreover, v_{er} and v_{mr} depend only on the parameters $a_{M\sigma}$ and $a_{K\iota}$. The residuals v_{pr} and $v_{\alpha r}$ depend only on $a_{B\rho}$. Therefore the estimation problem can be split and handled in two parts.

Calculations with this method provided similar experience as with the method described above. It was also possible to see that the results are less accurate than with the previous method.

It is possible to reduce the information basis of the estimation procedure even further. If only measured eigenfrequencies are available, the estimation procedure is based on partial residuals

$$v_{\omega r} := \overset{\Delta}{\omega}{}^2_{0r} - \omega^{C\,2}_{0r} \qquad (3.25)$$

or on the equation error

$$v_{0\omega r} := (-\overset{\Delta}{\omega}{}^2_{0r} M^C + K^C)\,\hat{u}^C_{0r} \qquad (3.26)$$

in combination with the normalization condition. In both cases the method is non-linear in the parameters because of the dependency of the eigenvectors $\hat{u}^C_{0r}(a)$ on a. The method based on the residuals (3.25) is published in /3.38/ and applied in /3.39/. A detailed discussion of the method with (3.25) and (3.26) can be found in /1.5/.

If eigenfrequencies and eigenvectors are measured, the partial residuals

$$\left\{ \begin{matrix} \overset{\mathbf{v}_{\omega r}}{\underset{\mathbf{u}_{0r}}{A}} - \hat{\mathbf{u}}_{0r}^{C} \end{matrix} \right\} \tag{3.27}$$

can be taken /1.5/.

Good experience has been gained using the equation error (3.22) combined with the normalization equation, which yield a linear WLS-estimator as in the case of input residuals. No one-to-one relationship between measured and calculated modes is necessary with this linear method. The experience mentioned above was gained from simulations with a viscous damped system /3.35/ for which these considerations hold true.

Another approach without using estimation methods directly, but indirectly through the application of pseudo-inverses, employs the modal decomposition of the physical parameter matrices for improvement /3.40/. In addition, one can introduce a norm of the matrices and "minimize" it together with some restrictions /3.41, 3.42/. For the application of these methods in structural optimization see /3.43/.

For a brief review see /3.25/. A discussion of when modal quantities and when physical parameter values should be estimated is discussed in /3.44/. The conversion of modal quantities of the damped system into those of the associated undamped system by using improvement methods is described in /3.45/.

It should be mentioned that the enumerated updating methods can be used for the order reduction of a mathematical model. Instead of measured quantities one may take computed ones and use these data for updating or estimating the parameters of a reduced order model. However, problems of physical compatibility will then arise.

3.4   Input Identification

The knowledge of operational loads is important for the dynamic assessment of systems to be designed as well as for existing systems. The direct measurement of input forces is often difficult or impossible, and then input identification has to be carried out (under working conditions).

The equation of motion in the time domain is the convolution integral. For causal systems with dynamic response $u(t)$ with initial conditions equal to zero and an excitation $p(t)$, the convolution sum holds true for the one degree-of-freedom system

$$u_i \doteq \tilde{u}_i = h \sum_{k=0}^{i} g_{i-k} \, p_k = h \sum_{k=0}^{i} g_k \, p_{i-k} \qquad\qquad (3.28)$$

where index i denotes the instant t_i, h is the time incre-
ment and g_k the response function due to the unit impulse
$\delta(t)$. As can be seen, the deconvolution is possible if u_i
and g_k are known, but carrying out the deconvolution numerically is often an unstable process /1.5/.

Input identification is thus mainly performed in the
frequency domain using Eq. (1.2) or (1.3) as basic equations. One idea for determining the excitation is to
measure it mainly statically with a dynamic correction.
Assuming the structured model (1.11) in the frequency
domain and neglecting the damping forces one obtains

$$P(j\omega) = K \, U(j\omega) - \omega^2 M \, U(j\omega) \; . \qquad\qquad (3.29)$$

This equation can be modified by expanding the dynamic term by modal quantities. Without using statistical
methods this approach is used in /3.46/.

By applying the non-structured equation (1.2) or
(1.3) the transformed dynamic response signal corrupted by
noise is used directly with a known frequency response
function matrix or a transfer function matrix respectively

(which may be estimated by using artificial test signals):

$$P(j\omega) = S(j\omega)\, U(j\omega) = F^{-1}(j\omega)\, U(j\omega)\; .$$

Noise reduction can be achieved using estimation procedures and with regard to the output signals when estimates of the outputs (spectral densities) are made and estimates of the inputs are determined:

$$S_{pp}(\omega) = S_{pu}(\omega) \left[F^T(j\omega) \right]^{-1} \tag{3.30}$$

or

$$S_{pp}(\omega) = \left[F^*(j\omega) \right]^{-1} S_{uu}(\omega) \left[F^T(j\omega) \right]^{-1} \tag{3.31}$$

when F^T is the transposed and F^* the conjugate complex frequency response function matrix, $S_{pp}(\omega)$ is the matrix of excitation spectral densities, $S_{uu}(\omega)$ is the matrix of output spectral densities and $S_{pu}(\omega)$ the matrix of cross spectral densities between input and output. Approximations of the spectral densities obtained from time series analysis can be introduced. The inversion of the previously identified matrix $F(j\omega)$ may cause difficulties in the case of lightly damped modes, but here the use of the Laplace transform can improve the results by dealing with additional damping (which means taking the transfer functions). In addition, the modal decomposition of $F(j\omega)$ can be used.

Instead of assigning the equation of motion (e.g. the state space formulation) directly, a modal transformation can be used /1.14, 3.47/. In general the modal transformation has to be applied with a smaller number of modes than degrees-of-freedom, and in consequence the pseudo-inverse of the modal matrix appears. The assumption that the number of non-zero components of the force vector must be smaller than the number of used modes has to be fulfilled /3.47/. In general, the truncation error of the modal decomposition is unknown.

3.5   Identification of Non-Linearities

The structure (of the model!) identification when dealing with linear systems is a relatively easy task, but the choice of damping and the assumption of the order of the system can involve difficulties. The modelling of non-linear behaviour is difficult in various ways. First of all, nonlinearities have to be detected. This is done mainly by testing the system harmonically (as the most effective when compared with other excitations) with different levels of response (!) amplitudes (and by looking for violated assumptions due to linear behaviour: Ny-

quist curves response spectrum, beat frequency locus
/3.48/), see Fig. 3.7. Another approach is the use of
indicator functions such as the SIG-function /3.49/. In
addition, the Hilbert transform serves for showing the
presence of nonlinearities. The real and imaginary parts
of the frequency response function of a linear system are
directly related by the (discrete) Hilbert transform (H):

$$F(j\omega_k) = F^{re}(j\omega_k) + jF^{im}(j\omega_k) = H\left\{F^{im}(j\omega_k)\right\} + jH\left\{F^{re}(j\omega_k)\right\}.$$

If nonlinearities appear in the measurements, the
real and imaginary parts of the above equation are not
identical any more.

When nonlinearities have been detected, the last step
is structure identification. This step is based on
- physics
- the pattern
- experience
- the high order correlation function
- the dispersion function /3.52/
- using special test signals.

With the known structure of the model the identification is
reduced to parameter estimation (Fig. 3.7).

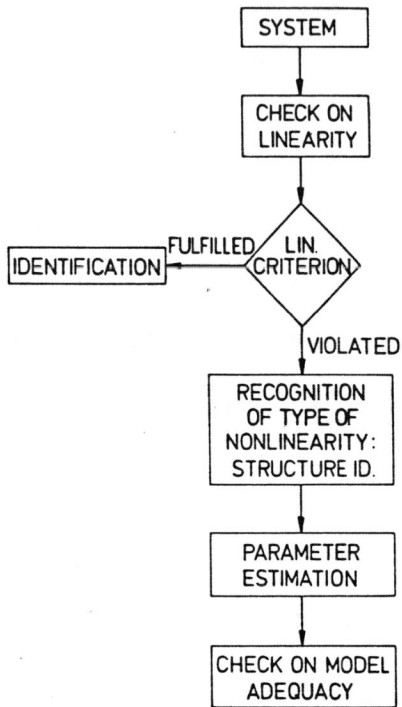

FIG. 3.7 GENERAL IDENTIFICATION
PROCEDURE

The usual (classical) approaches to the identification of nonlinear systems are
- Volterra series with high computation and storage expenditure, and the disadvantage that no hysteretic terms
  can be represented,

- the Wiener kernel approach, in which no hysteretic terms
  can be handled and which is restricted to white Gaussian
  input.

Instead of general function series one can, for example, take Tshebyshev polynomials. However, in every case the order of the structure is given by the a priori choice of the number of terms. Therefore identification is reduced to parameter identification. Modifications of the procedures previously mentioned have been published, as well as special integral operators (concerning subsystems), see the refs. in /3.51/. Another well-known approach uses

- dispersion functions /3.52/.

The following are common in practice

- statistical linearization and the

- piecewise linearized and displacement-dependent (parametrical) investigation of systems mainly using harmonic excitation and the phase resonance technique.

- Subsystem modelling and testing are also used, in which nonlinear elements between subsystems are replaced by linear elements with known dynamic properties (synthesis). The nonlinear elements are tested (identified) separately and combined with the linear subsystems by computation.

The reader can find a review of the identification of nonlinear systems in /3.51, 3.53, 3.54/, and the cited refs. there. Investigations into the identification of a system with modelled cubic stiffness and damping term can be found in /3.55, 3.56, 3.57/.

All refs. have in common the fact that the model structure has been selected. The validation of the model will then be proven by its significance (physical and statistical).

Order determination by iterative and recursive procedures are known, and for representations of nonlinear forces by polynomials this is described in /3.51, 3.58/. However, the estimation of the power of the polynomial together with the polynomial coefficients is not known in the literature as far the authors are aware. The reader can find first considerations in this direction in /1.12, 3.58/. The basic idea of this new development should be mentioned here. The basic equation is the recursive equation of economization in approximation theory:

$$P_{n-1}(x) = P_n(x) - \frac{a_n}{2^{n-1}} T_n(x) , \qquad (3.32)$$

$P_n(x)$ - polynomial of power n

$T_n(x)$ - Tshebyshev polynomial of power n

a_n - n-th coefficient of the Tshebyshev polynomial.

It is assumed that the nonlinear behaviour can be described by a polynomial. In Eq. (3.32) $P_n(x)$ is substituted by measurements $P^M(x)$ and the equation solved for n. This can be performed in different ways /3.58/, e.g. by taking into consideration an integral formulation with respect to x or a discretized version and then estimating n by averaging $n_i = n(x_i)$.

An example with a simulation of a polynomial

$$P_f(u) = u(1 + 0.5u^2 - 0.01u^4 + 0.00001u^6), \quad u := x/a, \quad a = 3,$$

is presented in /1.12/ for various uniformly distributed measuring errors. In order to avoid the residuals $P^M(x) - P_{n-1}(x)$, an iterative procedure is formulated depending on x_i and averaged: $\overset{\Delta}{n}$. Table 3.1 shows the result for assumed uniformly distributed measuring errors, parametrically assumed n and the mean value $\overset{\Delta}{n}$ with its standard deviation $\sigma_{\overset{\Delta}{n}}$. It can be concluded that the smallest value of $\overset{\Delta}{n}$ with the smallest standard deviation can be chosen (including the result with only one true decimal digit) as a result. Other calculations with different abscissae give similar results.

The last approach to the power estimation of polynomial-like nonlinearity mentioned here is associated with the extended WLS term in model improvement (Section 3.3):

$$J(a,n) = v^T(a,n)\, G_v\, v(a,n) + (a - a^{(0)})^T G_e (a - a^{(0)}) + g_n(n - n^{(0)})^2 \quad (3.33)$$

TABLE 3.1 POWER ESTIMATION

UNIFORMLY DISTR.ERR.	n	$\hat{n}$	$\pm \sigma_{\hat{n}}$
± 0	5	5.36	0.22
	7	7.03	0.01
	9	9.00	0.00
± 0.0005	5	5.36	0.22
	7	7.02	0.01
	9	8.97	0.01
± 0.005	5	5.36	0.22
	7	6.98	0.05
	9	8.82	0.05
± 0.05	5	5.20	0.15
	7	5.85	0.30
	9	7.47	0.34

$a^{(o)}$, $n^{(o)}$ are the prior knowledge of the polynomial representing the nonlinearity. The 3rd term is like a penal-
ty function, which makes the problem (3.33) dependent on
g_n into a convex one. The power estimation is controlled
here by the choice of the weighting matrices. One difficulty is how to choose g_n, and the second difficulty lies
in the mixed problem of continuous and integer variables.

References

1.1   Natke, H.G. (Editor); Identification of Vibrating
        Structures; CISM courses and lectures No. 272,
        Springer-Verlag, Wien, New York, 1982

1.2   Bendat, J.S. and A.G. Piersol; Random-Data: Analy-
        sis and Measurement Procedures; Wiley-Interscience,
        New York, London, Sydney, Toronto, 1971

1.3   Otnes, R.K. and L. Enochson; Applied Time Series
        Analysis; John Wiley and Sons, New York, 1978

1.4   Box, G.E.P. and G.M. Jenkins; Time Series Analysis
        Forecasting and Control; Holden-Day, San Franscis-
        co, 1970

1.5   Natke, H.G.; Einführung in Theorie und Praxis der
        Zeitreihen- und Modalanalyse; Vieweg & Sohn, Braun-
        schweig, Wiesbaden, 1983

1.6   Ewins, D.J.; Modal Testing: Theory and Practice;
        John Wiley & Sons, New York, 1984

1.7   Harris, C.M. and C.E. Crede (Editors): Shock and
        Vibration Handbook, Sec. Edition; McGraw-Hill, 1976

1.8   Lancaster, P.; Lambda-matrices and Vibrating
          Systems; Pergamon Press Oxford, London, Edinburgh,
          New York, Toronto, Paris, Braunschweig, 1966

1.9   Veubeke Fraejis de, B.M.; A Variational Approach to
          Pure Mode Excitation Based on Characteristic Phase
          Lag Theory; AGARD Rep. 39, April 1956

1.10   Meirovitch, L.; Computational Methods in Structural
          Dynamics; Sijthoff & Nordhoff, Alphen aan den Rijn,
          the Netherlands, Rockville, Maryland, USA, 1980

1.11   Natke, H.G.; Angenäherte Fehlerermittlung für Mo-
          dalsynthese-Ergebnisse innerhalb der Systemanalyse
          und Systemidentifikation; ZAMM 61, 1981, 41-53

1.12   Eykhoff, P.;System Identification - Parameter and
          State Estimation; John Wiley and Sons, London, New
          York, Sydney, Toronto, 1974

1.13   Natke, H.G. (Hrsg.); Dynamische Probleme - Model-
          lierung und Wirklichkeit; CRI-Rep. CRI-K 1/84, 1984

1.14   Cifuentes, A.O.; System Identification of Hystere-
          tic Structures; California Institute of Technology,
          EERL 84-04, Pasadena, Cal., 1984

1.15   Natke, H.G. and N. Cottin; On the Input Identifica-
        tion of Tall Buildings without the Usual Limiting
        Assumptions; 3rd Internat. Conf. on Computational
        Methods and Experimental Measurements, Sept. 1986
        in Porto Caras, Greece (Eds: G.A. Keramidas, C.A.
        Brebbia), Springer-Verlag Berlin, Heidelberg, New
        York, Tokyo, 1986

1.16   Faky, F.; Sound and Structural Vibration - Radiation,
        Transmission and Response; Academic Press, 1985

1.17   Pierce, A.D.; Acoustics: Introduction to its Physical
        Principles and Applications; McGraw-Hill, 1981

1.18   Kress, R.; Integralgleichungsmethoden bei direkten
        und inversen Randwertproblemen aus der Theorie aku-
        stischer und elektromagnetischer Schwingungen; Lehr-
        stuhl für Numerische und Angewandte Matehamtik, Uni.
        Göttingen

1.19   Sas, P. and P. Vandeponseele; Combined Use of Dynamic
        Analysis and Acoustic Radiation Models, a Potential
        Tool for the Design of Silent Mechanical Structures:
        Proc. of the 10th Internat. Seminar on Modal Analy-
        sis, Part II, Leuven 1985

1.20   Cottin, N.; Parameterschätzungen mit Hilfe des
        Bayesschen Ansatzes bei linearen elastomechanischen
        Systemen; Diss. U Hannover, 1983, CRI-F-2/1983

1.21   Isenberg, J.; Processing from Least Squares to Baye-
        sian Estimation; J.H. Wiggins Co., Redondo Beach, CA,
        ASME paper No. 79-WA/DSC-16, 1979

1.22   Box, G.E.P. and G.C. Tiao; Bayesian Interference in
        Statistical Analysis; Addison-Wesley Publishing Com-
        pany Reading, Massachussets; Menlo Park, California;
        London; Don Mills, Ontario, 1973

1.23   Bard, Y.; Nonlinear Parameter Estimation; Academic
        Press New York and London, 1974

1.24   Peterka, V.; Bayesian Approach to System Identifica-
        tion; in: P. Eykhoff (Editor): Trends and Progress in
        System Identification; Pergamon Press Oxford / New
        York / Toronto / Sydney / Paris / Frankfurt, 1981

1.25   Balakrhishnan, A.V.; Kalman Filtering Theory; Sprin-
        ger-Verlag, New York, Berlin, Heidelberg, Tokyo, 1984

2.1   Ibanez, P.; Review of Analytical and Experimental
        Techniques for Improving Structural Dynamic Models;
        Applied Nucleonics Company Inc. Santa Monica, Cali-
        fornia, Paper No. 1149-1, January 1977

2.2   Warburton, G.B.; The Dynamical Behaviour of Struc-
        tures; Pergamon Press, Oxford /New York/Toronto/
        Sydney/Paris/Frankfurt, 1976

2.3   Hart, G.C. (Ed.); Dynamic Response of Structures:
        Experimentation, Observation, Prediction and Con-
        trol, Proceedings of the 2nd Specialty Conference
        in Sheraton, Atlanta, Georgia, January 15-16, 1981

2.4   Hart, G.C. and R.B. Nelson (Eds.); Dynamic Response
        of Structures, Proceedings of the 3rd ASCE Confer-
        ence, University of California, Los Angeles, Cali-
        fornia, March 31 - April 2, 1986

2.5   Natke, H.G. and H. Schulze; Parameter Adjustment of
        a Model of an Offshore Platform from estimated Ei-
        genfrequencies Data; Journal of Sound and Vibra-
        tion, Vol. 77, No. 2, pp 271-285, 1981

2.6   Hanks, B.R.; Dynamic Verification of very large
        Space Structures; Second Int. Symp. on Aeroelasti-
        city and Structural Dynamics, DGLR-Report 85-02, pp
        648-655; collected papers of the symposium, held in
        Aachen, April 1-3, 1985

2.7   Kolb, M., W.-D. Longree, A. Westram; Untersuchungen
        von Eigenschwingungen der Forschungsplattform Nord-
        see bei natürlicher und künstlicher Anregung, VDI-
        Bericht Nr. 269, S. 97-106, 1976

2.8   Kernbichler, K., R. Flesch, G. Rauscher; Dynamische
        Untersuchungen von Großbrücken (Massivbrücken), In-
        situ-Versuche und Rechenmodelle; in: Natke, H.G.
        (Herausgeber): Dynamische Probleme - Modellierung
        und Wirklichkeit -, Vorträge der Tagung am 4. und
        5. Okt. 1984 in Hanover, Mitteilung des Curt-Risch-
        Instituts der Universität Hannover, CRI-K 1/84, Bd.
        II, S. 379-398, 1984

2.9   Natke, H.G.; Beschreibung einer Schwingungsanlage
        und deren Einsatz für die dritte Stufe der EUROPA
        I; Bericht EV-B 11, Vereinigte Flugtechnische Werke
        GmbH, 1969

2.10   Tustin, W.; A Comparison of Techniques and Equip-
        ment for Generating Vibration; The Shock and Vibra-
        tion Digest Vol. 9, No. 10, pp 3-10, Shock and
        Vibration Information Center Washington, D.C., 1977

2.11   Van der Auweraer, H., P. Vanherck, P. Sas, R.
        Snoeys; Experimental Modal Analysis with Stepped-
        Sine Excitation; Proceedings of the 10th Interna-
        tional Seminar on Modal Analysis, K.U. Leuven, Bel-
        gium, 30 Sept. - 4 Oct. 1985

2.12   Brown, D., G. Carbon, K. Ramsey; Survey of Excita-
        tion Techniques Applicable to the Testing of Auto-
        motive Structures; International Automotive Engi-
        neering Congress and Exposition, Detroit, Febr. 28
        - March 4, 1977, SAE-Paper No. 770029, 15 pp, 1977

2.13   Olsen, N.; Excitation functions for structural fre-
        quency response measurements; Proceedings of the
        2nd International Modal Analysis Conference, pp
        894-902, (IMAC-II), Orlando, Febr. 1984

2.14   Van Brussel, H.; Comparative assessment of har-
        monic, random, swept sine and shock excitation
        methods for the identification of machine tool

structures with rotating spindles; Annuals of the CIRP, Vol. 24/1/ 1975, pp 291-296

2.15   Allemang, R.J., R.W. Rost, D.L. Brown; Multiple Input Estimation of Frequency Response Functions: Excitation Considerations; ASME-Paper No. 83-DET-73, 11 pp

2.16   Halvorsen, W.G., D.L. Brown; Impulse Technique for Structural Frequency Response Testing; Sound and Vibration, November 1977, pp 8-21

2.17   Strobel, H.; Experimentelle Systemanalyse, Akademie-Verlag, Berlin 1975

2.18   Broch, J.T.; Messungen von mechanischen Schwingungen und Strößen; Brüel & Kjaer, Naevum, Dänemark, 1970

2.19   Vanherck, P.; On the calibration of accelerometers; Proceedings of the 10th International Seminar on Modal Analysis, K.U. Leuven, Belgium, 30 Sept.-4 Oct., 1985

2.20   Lally, R.W.; Trends in transducer technology; Proceedings of the 10th International Seminar on Modal

Analysis, K.U. Leuven, Belgium, 30 Sept. - 4 Oct., 1985

2.21   Natke, H.G.; Application in Aerospace and Airplane Engineering; within these lecture notes

2.22   Natke, H.G.; Transiente Anregungen von mechanischen Schwingungen in der Versuchstechnik (Transient excitations of mechanical vibrations in testing techniques); Technisches Messen tm, 52. Jahrgang, Heft 11/1985, S. 393-398, Aufsatz V 170-11

3.1   Director, S.W. and R.A. Rohrer; Introduction to System Theory; McGraw Hill, New York, 1972

3.2   Brogan, W.L.; Modern Control Theory; Quantum, New York, 1974

3.3   Luenberger, D.G.; Introduction to Dynamic Systems, Wiley, New York, 1979

3.4   Moore, B.C.; Principal Component Analysis in Linear Systems; Controllability, Observability and Model

Reduction; IEEE Trans. Autom. Control, Vol. AC-26, No. 1, 1981

3.5   Shokoohi, S., L.M. Silverman and P.M. Van Dooren; Linear Time-variable Systems: Balancing and Model Reduction; IEEE Trans. Autom. Control, Vol. AC-28, No. 8, 1983

3.6   Fernando, K.V. and H. Nicholson; On the Structure of Balanced and Other Principal Represenations of SISO Systems; IEEE Trans. Autom. Control, Vol. AC-28, No. 2, 1983

3.7   Gawronski, W. and H.G. Natke; Balancing Linear Systems, will be published in Int. J. Systems Sci. 1987

3.8   Jaquot, R.G.; Modern Digital Control Systems; Dekker, New York, 1981

3.9   Luenberger, D.G.; Introduction to Dynamic Systems; Wiley, New York, 1979

3.10   Gawronski, W. and H.G. Natke; On ARMA Models for Vibrating Systems; Probabilistic Engineering Mechanics, 1986, Vol. 1, No. 3, 150-156

3.11   Makhoul, J.; Linear Prediction: A Tutorial Review;
        Proc. IEEE, Vol. 63, No. 4, 1975

3.12   Friedlander, B.; Lattice Filters for Adaptive Pro-
        cessing; Proc. IEEE, Vol. 70, No. 8, 1982

3.13   Friedlander, B.; Lattice Methods for Spectral Esti-
        mation; Proc. IEEE, Vol. 70, No. 9, 1982

3.14   Gawronski, W. and H.G. Natke; Lattice Filters in
        the Identification of Linear Systems; CRI-Report,
        No. CRI-B-1/85

3.15   Gawronski, W. and H.G. Natke; Lattice Filters and
        Non-Toeplitz Systems of Equations; Proc. 13th Int.
        Conf. Modelling and Simulation, Lugano, 1985

3.16   Prony, R.; Experimental and Analytical Work on the
        Laws of Dilatibility of Elastic Fluids and on the
        Expansive Force of Water and Alcohol Vapours of
        Different Temperatures; J. l'Ecole Polytechnique,
        Vol. 1, No. 2, 1795

3.17   Ibrahim, S.R. and Mikulcik, E.C.; A Time Domain
        Modal Vibration Test Technique; The Shock and
        Vibration Bulletin, Bulletin 43, June 1973, 21-37

3.18   Ibrahim, S.R. and E.C. Mikulcik; The Experimental
        Determination of Vibration Parameters from Time Re-
        sponses; The Shock and Vibration Bulletin, Bulletin
        46, August 1976, 187-196

3.19   Ibrahim, S.R. and E.C. Mikulcik; A Method for the
        Direct Identification of Vibration Parameters from
        the Free Response; The Shock and Vibration Bulle-
        tin, Bulletin 47, September 1977, 183-198

3.20   Ibrahim, S.R. and R.S. Pappa; Large Modal Survey
        Testing Using the Ibrahim Time Domain Identifica-
        tion Technique; Journal of Spacecraft and Rockets
        (AIAA), Vol. 19, No. 5, Sept.-Oct. 1982, 459-465

3.21   Jahn, K.-D.; Rechnergestützte Auswertung von
        Schwingungsversuchen; Diss. TU Hannover, 1978

3.22   Vold, H., J. Kundrat, G.T. Rocklin and R. Russel; A
        Multi-Input Modal Estimation Algorithm for Minicom-
        puters SAE paper No. 820194, 1982

3.23   Juang, J.N. and R.S. Pappa; An Eigensystem Realiza-
        tion Algorithm (ERA) for Modal Parameter Identifi-
        cation; Workshop on Identification and Control of
        Flexible Space Structures; Pasadena, CA, June 1984

3.24   Gawronski, W. and H.G. Natke; On Realizations of
        the Transfer Function Matrix; will be published in
        Internat. J. of Systems Science 1987

3.25   Kozin, F. and H.G. Natke; System Identification
        Techniques; in Structural Safety, 3 (1986) 269-316

3.26   Natke, H.G.; Anwendung eines versuchsmäßig - rech-
        nerischen Verfahrens zur Ermittlung der Eigen-
        schwingungsgrößen eines elastomechanischen Systems
        bei einer Erregerkonfiguration; Z. Flugwiss. 18
        (1970), Heft 8, 290-303

3.27   Natke, H.G.; Ein Verfahren zur rechnerischen Er-
        mittlung der Eigenschwingungsgrößen aus den Ergeb-
        nissen eines Schwingungsversuches in einer Erreger-
        konfiguration; Diss. TH München, 1968; English
        Translation: NASA--TT-F-12446, 1969

3.28   Natke, H.G.; Die Berechnung der Eigenschwingungs-
        größen eines gedämpften Systems aus den Ergebnissen
        eines Schwingungsversuches in einer Erregerkonfigu-
        ration; Jahrbuch 1971 der DGLR, 98-120

3.29   Stahle, C.V. and W.R. Forlifer; Ground Vibration
        Testing of Complex Structures; AIA - AFOSR Flight

Flutter Testing Symposium, Washington, D.C., May 1958

3.30   Stahle, C.V.; Phase Separation Technique for Ground Vibration Testing; Aerospace Engineering, July 1962

3.31   Klosterman, A.L.; On the Experimental Determination and Use of Modal Representations of Dynamic Characteristics; Ph.D. Diss., U. of Cincinnati, 1971

3.32   Klosterman, A.L.; Dynamic Design Analysis via Building Block Approach; Shock and Vibration Bulletin, No. 42, Part 1, June 1972

3.33   Wittmeyer, H.; Parameteridentifikation bei Strukturen mit benachbarten Eigenfrequenzen speziell bei Flugschwingungsversuchen; Z. Flugwiss. Weltraumforsch. 6, 1982, Heft 2, 80-90

3.34   Cottin, N., H.-P. Felgenhauer and H.G. Natke; On the Parameter Identification of Elastomechanical Systems Using Input and Output Residuals; Ingenieur-Archiv 54, 1984, 378-387

3.35   Cottin, N. and H.G. Natke; On the Parameter Identification of Elastomechanical Systems Using Weighted

Input and Modal Residuals; Ingenieur-Archiv 56,
1986, 106-113

3.36   Natke, H.G.; Die Korrektur des Rechenmodells eines
        elastomechanischen Systems mittels gemessener er-
        zwungener Schwingungen; Ingenieur-Archiv 46 (1977)
        168-184

3.37   Natke, H.G.; Deliberations on the Improvement of
        the Computational Model with Measured Eigenmagni-
        tudes; Rev. Roum. Sci. Techn.-Me. Appl. Tome 28,
        No. 2, Bucarest 1983, 159-173

3.38   Natke, H.G., D. Collmann and H. Zimmermann; Beitrag
        zur Korrektur des Rechenmodells eines elastomecha-
        nischen Systems anhand von Versuchsergebnissen;
        VDI-Berichte No. 221, 1974, 23-32

3.39   Zimmermann, H., D. Collmann and H.G. Natke; Erfah-
        rungen zur Korrektur des Rechenmodells mit gemesse-
        nen Eigenschwingungen am Beispiel des Verkehrsflug-
        zeuges VFW 614; Z. Flugwiss. Weltraumforsch. 1,
        1977, Heft 4, 278-285

3.40   Link, M.; Theory of a Method for Identifying In-
        complete System Mechanic Matrices from Vibration

Test Data; Z. Flugwiss. Weltraumforsch. 9 (1985), Heft 2, 76-82

3.41   Baruch, M.; Methods of Reference Basis for Identification of Linear Dynamic Structures; AIAA Journal 22, No. 4, 1984

3.42   Caesar, B.; Update and Identification of Dynamic Mathematical Models; Proc. of the 4th Internat. Modal Analysis Conf., Los Angeles, CA, 1986, 394-401

3.43   Natke, H.G.; Minimaländerungen an Teilsystemen aufgrund von dynamischen Anforderungen; will be published in Ingenieur-Archiv 57 (1987)

3.44   Natke, H.G.; Improvement of Analytical Models with Input/Output Measurements contra Experimental Modal Analysis; Proc. 4th Internat. Modal Analysis Conf., Los Angeles, CA, 1986, 409-413

3.45   Natke, H.G. and D. Rotert; Determination of Normal Modes from Identified Complex Modes; Z. Flugwiss. Weltraumforsch. 9. 1985, Heft 2, 82-88

3.46   Öry, H., H. Glaser and D. Holzdeppe; Quality of
        Modal Data Analysis and Reconstruction of Forcing
        Functions Based on Measured Output Data; Proceed-
        ings of the 4th Intern. Modal Analysis Conf., Los
        Angeles, CA, 1986, 850-857

3.47   Desanghere, D. and R. Snoeys; Indirect Identifica-
        tion of Excitation Forces; Proceedings of the 10th
        International Seminar on Modal Analysis, Part IV,
        Leuven, 1985

3.48   Tomlinson, G.R.; Detection, Identification and
        Quantification of Nonlinearity in Modal Analysis -
        A Review; Proceedings of the 4th Intern. Modal
        Analysis Conf., Los Angeles, CA, 1986, 837-843

3.49   Mertens, M. et al.; Basic Rules of a Reliable De-
        tection Method for Non-linear Dynamic Behaviour;
        Proceedings of the 10th Intern. Seminar on Modal
        Analysis, Part IV, Leuven, 1985

3.50   Simon, M. and G.R. Tomlinson; Use of the Hilbert
        Transform in Modal Analysis of Linear and Non-li-
        near Structures; Journal of Sound and Vibration
        (1984) 96(4), 421-436

3.51   Natke, H.G. and W. Gawronski; On Structure Identi-
       fication within Nonlinear Structural Systems; Rep.
       CRI-B-2/1984 Curt-Risch-Institut, U Hannover, 1984

3.52   Rajbman, N.S. and V.M. Cadeev; Identifikation - Mo-
       dellierung industrieller Prozesse; VEB-Vortrag
       Technik, Berlin 1980

3.53   Billings, S.A.; Identification of Nonlinear
       Systems, a Survey; IEE Proc. Vol 127, Pt. D, Nr. 6,
       Nov. 1980, 272-285

3.54   Natke, H.G. and J.T.P. Yao; Research Topics in
       Structural Identification; Proc. 3rd Conf. on Dyna-
       mic Response of Structures, EM Div./ASCE, Univ. of
       California, L.A., April 1986, 542-550

3.55   Dat, R.; L'essai de Vibration d'une Structure Im-
       perfaitement lineaire; La Rech. Aerospatiale 1975,
       No. 4, 223-227

3.56   Merritt, P.H.; A Method of System Identification
       with an Experimental Investigation; Shock and
       Vibration Bulletin 47, 1977, No. 4, 175-181

3.57   Natke, H.G.; Fehlerbetrachtungen zur parametrischen
        Identifikation eines Systems mit kubischem Steifig-
        keits- und Dämpfungsterm; Czerwenka-Festschrift
        1979, TU München

3.58   Natke, H.G.; Addendum 1 and 2 to CRI-Report CRI-B-
        2/1984, Curt-Risch-Institut, U. Hannover, CRI 2.1,
        2.2/1984

BALANCED STATE SPACE REPRESENTATION
IN THE IDENTIFICATION OF DYNAMICAL SYSTEMS

W. Gawronski
NASA Langley Research Centre, Hampton, Virginia, U.S.A.
H.G. Natke
Universität Hannover, Hannover, F.R.G.

1.   Introduction

In a static analysis of a structure, dividing the
structure into a large number of finite elements in most
cases gives a satisfactory solution. In a dynamic analysis,
however, division into more units does not necessarily give
better results. Such factors as damping, excitation forces,
support stiffnesses etc., often cannot be determined accurately, and even sophisticated measurements do not give satisfactory results, because of the random nature of these
quantities.

A useful method for solving this problem consists of parameter estimation using measurements of vibrations and excitations. For example, one can assume a structured discrete model with lumped masses for a vibrating off-shore structure, and from the recorded vibratory data of the structure one can determine the damping, mass, and stiffness matrices. One may also assume a mathematical model, such as the ARMA model, of the structure, and from the recorded vibrational data estimate its parameters.

In such an approach one of the important questions that arises is the order of the model. What is the best number of degrees of freedom of the model? What is the best order of the ARMA model? Experience shows that increasing the order of a model increases the accuracy of the results, but starting from the order number, which we can call critical, the accuracy increases slowly, or not at all.

Several methods for the evaluation of the system order have been developed /1,9,14-21/. Here we shall discuss only one, briefly called balancing. The method was originally developed by Moore /14/, and further developed by Parnebo, Silverman /15/, Shokoohi, Silverman, Van Dooren /16/, Gawronski and Natke /1,2/.

2.   System Balancing - Moore's Approach

In this paragraph we shall consider the continuous-time system only (the discrete-time case is similar and can be found in /14/). The system is given by the equation

$$\dot{x} = Ax + Bu, \quad y = Cx \tag{2.1}$$

where $x \in R^n$, $y \in R^q$, $u \in R^p$. In the following the triple (A,B,C) is called the representation of the system.

It is well known that the system is fully controllable and observable if the controllability $\mathcal{C}$ and observability $\mathcal{O}$ matrices

$$\mathcal{C} = [B \quad AB \quad \ldots \quad A^{n-1}B] \tag{2.2}$$

$$\mathcal{O} = [C^T \quad (CA)^T \quad \ldots \quad (CA^{n-1})^T]^T \tag{2.3}$$

are of full rank. In like manner, the system is fully controllable and observable if the controllability and observability grammians

$$W_c = \int_0^\infty e^{At} B B^T e^{A^T t} dt \tag{2.4}$$

$$W_o = \int_0^\infty e^{A^T t} \, C^T \, C \, e^{At} \, dt \tag{2.5}$$

are positive definite.

Let R be the linear nonsingular transformation of the state coordinates

$$x = R\bar{x} \, , \tag{2.6}$$

then the new state representation $(\bar{A}, \bar{B}, \bar{C})$ is obtained from (A, B, C) as

$$\bar{A} = R^{-1} A R, \quad \bar{B} = R^{-1} B, \quad \bar{C} = C R \tag{2.7}$$

The input-output relationship is invariant under the transformation R. However, the controllability and observability properties of the system depend on the chosen state variables, and the new controllability and observability matrices are given by

$$\bar{\mathcal{C}} = R^{-1} \mathcal{C} \, , \quad \bar{\mathcal{O}} = \mathcal{O} R \tag{2.8}$$

and the controllability and observability grammians by

$$\bar{W}_c = R^{-1} W_c R^{-T} \qquad\qquad (2.9)$$

$$\bar{W}_o = R^T W_o R. \qquad\qquad (2.10)$$

Moore's approach was as follows: for the controllable and observable system the grammians W_c, W_o should be nonsingular. He chose the singular values of W_c and W_o as a measure of the controllability and the observability. Let W_c have the following singular values d_1, d_2, ... , d_n ordered such that $d_i \geq d_{i+1}$, $i = 1$, ...,$n-1$. If for some m<n one has $d_m \gg d_{m+1}$ one can say that m state variables are strongly controlled, and n-m state variables are weakly controlled. If the same is valid for the observability grammian, one can say that m state variables are strongly observed and n-m state variables are weakly observed. But which one is strongly, and which one is weakly observed and controlled? And the other question follows: does a transformation R of state variables exist, such that the controllability and observability grammians are equal and diagonal? If so, they have the same singular values, and, at the same time, it is known which state variable "belongs" to which singular value. Of course, the answer is positive as regards the latter question. There is a nonsingular matrix R such that in the new state coordinates $\bar{x}$ given by (2.6) the controllability and observability grammians are

$$W_c = W_o = \Gamma^2 \tag{2.11}$$

where Γ is a diagonal matrix

$$\Gamma = \text{diag}(\gamma_1, \ldots, \gamma_n). \tag{2.12}$$

These coordinates are called balanced coordinates, since the system is now equally controllable and observable.

The balancing of symmetric systems was specifically investigated. A system is symmetric if its transfer function matrix is symmetric, or its Hankel matrix H

$$H = \mathcal{O}\,\mathcal{C} \tag{2.13}$$

is symmetric. The necessary (but not sufficient) condition for the system symmetry is the equality of its number of inputs and outputs, i.e. $p = q$. The single input and single output (SISO) system is always symmetric. The mechanical system with pick-ups and excitations located at the same place simultaneously, and for which the Betti principle is satisfied, is symmetric.

The transformation matrix R to the balanced representation is determined as follows /2,14/:

$$R = V_c \Gamma_c U_{o1} \Gamma_{o1}^{-\frac{1}{2}} \tag{2.14}$$

where V_c, Γ_c are obtained from the singular value decomposition of the matrix $\mathcal{C}$, and this decomposition is as follows:

$$\mathcal{C} = V_c \Gamma_c^2 U_c^T \tag{2.15}$$

where U_c and V_c are orthonormal matrices, Γ_c is a diagonal, positive definite matrix, and Γ_{o1}, U_{o1} are determined from the singular value decomposition of the matrix $\mathcal{O}_1 = \mathcal{O} V_c \Gamma_c$, which is in the form

$$\mathcal{O}_1 = V_{o1} \Gamma_{o1}^2 U_{o1}^T. \tag{2.16}$$

Balancing is a tool for the system reduction. If a system is balanced, i.e. (2.11) is fulfilled, and additionally, for some m<n one has $\gamma_m >> \gamma_{m+1}$ then the last n-m state variables are weakly controllable and, at the same time, weakly observable. For this reason they can be omitted in the state representation, since they have little influence on the input-output relationship.

Example 2.1. /2/ A system with the following state space representation

$$
A = \begin{bmatrix} 0 & 1 & 0 \\ 0 & 0 & 1 \\ -0.3250 & 0.0202 & 0.1500 \end{bmatrix}, \quad B = \begin{bmatrix} 1 \\ 0.65 \\ 0 \end{bmatrix}, \quad C = \begin{bmatrix} 1 & 0 & 0 \end{bmatrix}
$$

is considered. Its balancing coefficients are

$$
\Gamma = \text{diag}(1.2937, \quad 0.8614, \quad 0.0070)
$$

and its balanced representation (A,B,C) is obtained

$$
\bar{A} = \begin{bmatrix} 0.4515 & -0.5853 & 0.0006 \\ 0.5853 & 0.3486 & 0.0053 \\ 0.0006 & -0.0053 & -0.6501 \end{bmatrix}, \quad \bar{B} = \begin{bmatrix} -1.0385 \\ 0.2801 \\ 0.0027 \end{bmatrix},
$$

$$
\bar{C} = \begin{bmatrix} -1.0385 & -0.2801 & 0.0027 \end{bmatrix}.
$$

For this system $\gamma_2 \gg \gamma_3$, it can therefore be reduced, leaving only the first two state variables in the balanced representation, and the reduced representation is as follows

$$
\bar{A}_r = \begin{bmatrix} 0.4515 & -0.5853 \\ \\ 0.5853 & 0.3486 \end{bmatrix}, \qquad \bar{B}_r = \begin{bmatrix} -1.0385 \\ \\ 0.2801 \end{bmatrix},
$$

$$
\bar{C}_r = \begin{bmatrix} -1.0385 & -0.2801 \end{bmatrix}.
$$

3. Alternative Approach to System Balancing

The alternative approach to balancing was presented in /1/.

3.1 Continous time balanced systems.

In the Appendix and in /3/ it is shown that the grammians can be expressed by the observability and controllability matrices as follows:

$$
W_c = \mathcal{C}\,(Q(t) \otimes \tfrac{I}{p})\,\mathcal{C}^\mathsf{T}, \quad W_o = \mathcal{O}^\mathsf{T}(Q(t) \otimes \tfrac{I}{q})\,\mathcal{O} \qquad (3.1)
$$

where $\otimes$ denotes a Kronecker product, and the grammians are defined now as

$$
W_c(t) = \int_o^t e^{A\tau} B\, B^\mathsf{T}\, e^{A^\mathsf{T}\tau}\, d\tau \qquad (3.2)
$$

$$W_0(t) = \int_0^t e^{A^T \tau} c^T c \, e^{A\tau} \, d\tau .$$ (3.3)

The matrix $Q(t)$ is obtained from

$$Q(t) = V^{-1} D(t) V^{-T}$$ (3.4)

with the matrices V and $D(t)$ as follows:

$$V = \begin{bmatrix} 1 & \lambda_1 & \lambda_1^2 & \cdots & \lambda_1^{n-1} \\ 1 & \lambda_2 & \lambda_2^2 & \cdots & \lambda_2^{n-1} \\ \cdots\cdots\cdots\cdots\cdots\cdots \\ \cdots\cdots\cdots\cdots\cdots\cdots \\ 1 & \lambda_n & \lambda_n^2 & \cdots & \lambda_n^{n-1} \end{bmatrix}$$ (3.5)

$$D(t) = [d_{ij}(t)], \quad d_{ij}(t) = \frac{e^{(\lambda_i + \lambda_j)t} - 1}{\lambda_i + \lambda_j} \quad i,j=1,\ldots,n.$$ (3.6)

Let the matrix $Q(t)$ be decomposed in the following way:

$$Q(t) = L(t) L^T(t)$$ (3.7)

(for example, the Cholesky decomposition or the singular value decomposition can be used). Defining further

$$L_c(t) = L(t) \otimes I_p, \quad L_o(t) = L(t) \otimes I_q \tag{3.8}$$

and

$$\ell^*(t) = \mathcal{C} L_c(t), \quad \sigma^*(t) = L_o^T(t) \sigma \tag{3.9}$$

we obtain the grammians in the form

$$W_c(t) = \ell^*(t) \ell^*(t), \quad W_o(t) = \sigma^{*T}(t) \sigma^*(t). \tag{3.10}$$

Note that $L_c(t)$ and $L_o(t)$ are invariant under the linear transformations of state coordinates; therefore in the new coordinates we have

$$\bar{\ell}^*(t) = R^{-1} \mathcal{C} L_c(t) = R^{-1} \ell^*(t)$$
$$\bar{\sigma}^*(t) = L_o^T(t) \sigma R = \sigma^*(t) R. \tag{3.11}$$

The above notation allows us to introduce the following definition of the balanced system:

Definition 3.1. The realization $(\bar{A}, \bar{B}, \bar{C})$ of the linear system (2.1) is balanced over the interval $[0,T]$ if there is such a nonsingular matrix R that the singular value decomposition of the matrices $\bar{\ell}^*(T) = R^{-1} \ell^*(T)$ and $\bar{\sigma}(T) = \sigma^*(T) R$ are in the form

$$\bar{\ell}^*(T) = \Gamma(T) U^T(T), \quad \bar{\sigma}^*(T) = V(T) \Gamma(T) \tag{3.12}$$

where $V(T) = [v_1(T) \ldots v_n(T)]$, $U(T) = [u_1(T) \ldots u_n(T)]$,
$V^T V = U^T U = I$, and $\Gamma(T) = \text{diag}(\gamma_i(T))$, $i=1, \ldots, n$ is a
positive definite matrix. The vectors $v_i(t)$ and $u_i(t)$ are
called the balanced modes, and $\gamma_i(T)$ are called the balancing coefficients over the interval $[0,T]$.

For the balanced system the controllability and observability grammians are positive definite diagonal matrices

$$\bar{c}^*(T) \ \bar{c}^{*T}(T) = \bar{o}^{*T}(T) \ \bar{o}^*(T) = \Gamma^2(T). \tag{3.13}$$

If $T \to \infty$ the system is balanced over the interval $[0,\infty)$, or
simply balanced (in the sense of Moore). If the system is
stable, the balanced representation over $[0,\infty)$ always
exists. In this case the matrices ℓ^* and o^* are obtained
from (3.9), with L_c and L_o determined from (3.4) to (3.8),
and the matrix D is given by

$$D = [d_{ij}], \quad d_{ij} = -\frac{1}{\lambda_i + \lambda_j} \quad , \quad i,j = 1,\ldots,n. \tag{3.14}$$

The definition (3.12) of the balanced representation is
valid for any linear system, not necessarily a symmetric
one.

For the symmetric system the balanced representation has the property that the balancing modes are connected by the relationship /2/

$$U = V \Sigma \qquad\qquad (3.15)$$

where Σ is the sign matrix, $\Sigma = \text{diag}\ (\sigma_i)$, $i=1,\ldots,n$, and $\sigma_i = 1$ or $\sigma_i = -1$; the balanced representation $(\bar{A},\bar{B},\bar{C})$ also shows the sign symmetry, i.e.

$$\bar{A}^T = \Sigma \bar{A} \Sigma, \quad \bar{B} = \Sigma \bar{C}^T \quad . \qquad\qquad (3.16)$$

3.2 <u>Discrete-time balanced systems.</u> The system considered here has the following representation (A,B.C)

$$x_{i+1} = A x_i + B u_i, \quad y_i = C x_i \qquad\qquad (3.17)$$

with dim $x_i = n$, dim $u_i = p$, dim $y_i = q$, as before. In addition, we assume the system to be completely reachable and observable, with A having distinct eigenvalues $\lambda_1, \ldots, \lambda_n$. The linear transformation $\bar{x} = R x$ gives the new representation $(\bar{A},\bar{B},\bar{C})$ which is connected with the representation (A,B,C) by (2.7). The reachability $\mathcal{C}$ and observability σ matrices of the representation (A,B,C) are given by (2.2) and (2.3), and of $(\bar{A},\bar{B},\bar{C})$ by (2.8).

We further consider the system within the time interval $[0,N]$, and assume $N \geqslant n$. The matrices $\overset{*}{c}(N)$ and $\overset{*}{\sigma}(N)$ are defined as follows:

$$\overset{*}{c}(N) = c\, L_c(N), \qquad \overset{*}{\sigma}(N) = L_o(N)\, \sigma \qquad (3.18)$$

where

$$L_c(N) = L(N) \otimes \frac{I}{p}, \qquad L_o(N) = L(N) \otimes \frac{I}{q}. \qquad (3.19)$$

The matrix $L(N)$ is obtained from the decomposition of the matrix $Q(N)$

$$Q(N) = L(N)\, L^T(N) \qquad (3.20)$$

and $Q(N)$ is obtained from

$$Q(N) = V^{-1}\, D(N)\, V^{-T}. \qquad (3.21)$$

The matrix V is given by (3.5), and $D(N)$ is determined from

$$D(N) = [\, d_{ij}(N)\,], \qquad d_{ij}(N) = \frac{(\lambda_i \lambda_j)^N - 1}{\lambda_i \lambda_j - 1}, \quad i,j=1,\ldots,n. \qquad (3.22)$$

Similarly to the continuous time system, in the new coordinates we have

$$\bar{e}^{*}(N) = R^{-1} e^{*}(N), \quad \bar{\sigma}^{*}(N) = \overset{*}{\sigma}(N) R. \tag{3.23}$$

From /3/ and the Appendix it follows that the reachability grammian over the interval $[0,N]$ can be determined from

$$W_{c}(N) = \overset{*}{e}(N) \; \overset{*T}{e}(N) \tag{3.24}$$

where the grammian is defined as follows:

$$W_{c}(N) = \sum_{i=o}^{N} A^{i} B B^{T} (A^{i})^{T}. \tag{3.25}$$

The observability grammian over the interval $[0,N]$ is defined similarly as

$$W_{o}(N) = \sum_{i=o}^{N} (A^{i})^{T} C^{T} C A^{i} \tag{3.26}$$

and can be determined from the following formula:

$$W_{o}(N) = \sigma^{*T}(N) \sigma^{*}(N). \tag{3.27}$$

The system (2.1) has be representation $(\bar{A},\bar{B},\bar{C})$ balanced, if the following conditions are fulfilled:

Definition 3.2. The realization $(\bar{A},\bar{B},\bar{C})$ of the linear discretetime system is balanced over the interval $[0,N]$, if a nonsingular matrix R exists such that the singular value decomposition of the matrices $\bar{\mathcal{C}}^*(N) = R^{-1} \mathcal{C}^*(N)$ and $\overset{*}{\bar{\sigma}}(N) = \sigma^*(N) R$ are in the form

$$\bar{\mathcal{C}}^*(N) = \Gamma(N) \; U^T(N), \quad \bar{\sigma}^*(N) = V(N) \; \Gamma(N) \tag{3.28}$$

where $V(N) = [v_1(N) \; \ldots \; v_n(N)]$, $U(N) = [u_1(N) \; \ldots \; u_n(N)]$, $V^T V = U^T U = I$, and $\Gamma(N) = \text{diag}(\gamma_i(N))$, $i=1, \ldots, n$, is a positive definite matrix. The vectors $v_i(N)$ and $u_i(N)$ are called the balancing modes, and $\gamma_i(N)$ the balancing coefficients over the interval $[0,N]$, and $v_i \in R^{np}$, $u_i \in R^{nq}$.

For the balanced system the reachability and observability grammians are positive definite diagonal matrices

$$\bar{\mathcal{C}}^*(N) \; \bar{\mathcal{C}}^{*T}(N) = \bar{\sigma}^{*T}(N) \; \bar{\sigma}^*(N) = \Gamma^2(N) \quad . \tag{3.29}$$

If $N \rightarrow \infty$ the system is balanced over the interval $[0,\infty)$, or simply, balanced (in the sense of Moore). If the system is stable the balanced representation over $[0, \infty)$ always exists. In this case the matrices $\mathcal{C}^*$ and $\mathcal{O}^*$ are obtained from (3.18), with L_C and L_O determined from (3.19), (3.20). However, in this case the matrix D is given by

$$D = [d_{ij}], \quad d_{ij} = \frac{1}{1 - \lambda_i \lambda_j} \quad , \quad i,j,=1,\ldots,n \qquad (3.30)$$

As before, the above definition is valid for arbitrary systems, and not necessarily symmetric ones.

Let us define the extended reachability and observability matrices

$$\mathcal{C}(N) = [B \ AB \ \ldots \ A^{N-1} B], \quad \mathcal{O}(N) = [C^T \ (CA)^T \ \ldots \ (CA^{n-1})^T]^T \qquad (3.31)$$

where $N \geqslant n$ and N-n is the extension number. Using this definition the reachability and the observability grammians over $[0,N]$ can be obtained simply as

$$W_C(N) = \mathcal{C}(N) \ \mathcal{C}^T(N), \quad W_O(N) = \mathcal{O}^T(N) \ \mathcal{O}(N) \quad . \qquad (3.32)$$

This gives us the possibility of introducing the alternative

definition of the balanced representation over the interval
$[0,N]$.

Definition 3.3. The realization $(\bar{A},\bar{B},\bar{C})$ of the linear
discretetime system is balanced over the interval $[0,N]$ if a
nonsingular matrix R exists such that the singular value
decomposition of the matrices $\bar{\mathcal{C}}(N) = R^{-1}\mathcal{C}(N)$, and
$\bar{\sigma}(N) = \sigma(N) R$ are in the form

$$\bar{\mathcal{C}}(N) = \Gamma(N) U^T(N), \quad \bar{\sigma}(N) = V(N) \Gamma(N) \qquad (3.33)$$

where $V(N) = [v_1(N) \ \ldots \ v_n(N)]$, $U(N) = [u_1(N) \ \ldots \ u_n(N)]$,
$V^T V = U^T U = I$, and $\Gamma(N) = \text{diag}(\gamma_i(N))$, $i=1,\ldots,n$, is a positive definite matrix. The vectors $v_i(N)$ and $u_i(N)$ are called
the balancing modes, and $\gamma_i(N)$ are called the balancing
coefficients over the interval $[0,N]$, and $v_i \in R^{Np}$, $u_i \in R^{Nq}$

.

This definition, however, is not useful over the infinite interval $[0,\infty)$, since the matrices in (3.31) as well
as the balancing vectors in this case have infinite dimensions.

4. Transformation to the Balanced Representation

In the following discussion no distinction has been
made between continous and discrete time systems, nor
between finite and infinite interval balancing, since the
transformation given below is valid for any of the cases.

Let H^* be the transformed Hankel matrix of dimensions
nq x np, obtained from

$$H^* = \sigma^* \ell^* \tag{4.1}$$

This matrix is invariant under linear transformations of the
state coordinates, that is

$$H^* = \sigma^* \ell^* = \bar{\sigma}^* \bar{\ell}^* \tag{4.2}$$

Let H^* have the following singular value decomposition

$$H^* = V_n \Gamma^2_n U^T_n \tag{4.3}$$

Now we introduce ℓ^* and σ^* from the definition of the
balanced system (see (3.12) or (3.28) to (4.2), obtaining

$$H^* = V \, \Gamma^2 \, U^T \tag{4.4}$$

Since V_h, U_h, and Γ_h fulfil the conditions of the balanced system, and the singular value composition is unique, the comparison of (4.3) and (4.4) gives $V = V_h$, $U = U_h$, $\Gamma = \Gamma_h$, i.e. the balanced representation is obtained from the singular value decomposition of the transformed Hankel matrix.

In order to determine the balanced representation of the system, the transformation matrix R (and possibly R^{-1} without the inversion of R) is determined. It is done as follows: the introduction of (3.11) to (3.12), or (3.23) to (3.28), gives

$$\Gamma U^T = R^{-1} e^* \; , \; V \Gamma = \sigma^* \, R$$

and therefore

$$R = e^* \, U \, \Gamma^{-1} \; , \; R^{-1} = \Gamma^{-1} \, V^T \, \sigma^* \tag{4.5}$$

The method is summarized as follows:
1. Form the transformed Hankel matrix $H^* = \sigma^* \, e^*$.
2. Find the singular value decomposition of H^*.
3. Determine the transformation matrices R and R^{-1} from (4.5).
4. Determine the balanced representation from (2.7).

Consider the special case - the system in the observability canonical form. For this case $\sigma = I$, therefore $H^* = L_o^T \ell L_c$. The transformation matrices in this case are obtained from the following formulas:

$$R = L_o^{-T} V \Gamma, \quad R^{-1} = \Gamma^{-1} V^T L_o^T \tag{4.6}$$

where U, V, Γ are obtained from the singular value decomposition of $H^* = L_o^T \ell L_c$.

<u>Example 4.1</u>. The balanced representation of the system with the following triple (A,B,C)

$$A = \begin{bmatrix} 0 & 1 \\ -100 & -10 \end{bmatrix}, \quad B = \begin{bmatrix} 0 \\ 1 \end{bmatrix}, \quad C = \begin{bmatrix} 1 & 0 \\ 0 & 1 \end{bmatrix}$$

is determined.

We follow the procedure given above.

1. The matrix H^* is determined. Since the eigenvalues of A are

$$\lambda_1 = -5.0000 + j\ 8.6603, \quad \lambda_2 = -5.0000 - j\ 8.6603$$

the matrix Q, determined from (3.4), is

$$Q = \begin{bmatrix} 1.0000 & 0.0050 \\ 0.0050 & 0.0005 \end{bmatrix}$$

so that L, obtained from (3.7) by the Cholesky factorization of Q, is

$$L = \begin{bmatrix} 0.3162 & 0.0000 \\ 0.0158 & 0.0158 \end{bmatrix}.$$

The controllability and observability matrices of the system, are as follows:

$$\mathcal{C} = \begin{bmatrix} 0 & 1 \\ 1 & -10 \end{bmatrix}, \; \sigma^T = \begin{bmatrix} 1 & 0 & 0 & -100 \\ 0 & 1 & 1 & -10 \end{bmatrix},$$

and the transformed matrices $\mathcal{C}^*$ and σ^*, obtained from (3.9) are

$$c^* = c \, L = \begin{bmatrix} 0.0158 & 0.0158 \\ 1.5811 & -0.1581 \end{bmatrix}$$

$$\sigma^{*T} = \sigma(L \otimes I_2) = \begin{bmatrix} 0.3162 & -1.5811 & 0.0000 & -1.5811 \\ 0.0158 & 0.1581 & 0.0158 & -0.1581 \end{bmatrix}.$$

The matrix H^* is therefore determined

$$H^* = \sigma^* c^* = \begin{bmatrix} 0.0075 & 0.0025 \\ 0.0000 & -0.0500 \\ 0.0025 & -0.0025 \\ -0.0500 & 0.0000 \end{bmatrix}$$

2. The singular value decomposition of H^* gives

$$\Gamma = \mathrm{diag}(0.2251, \ 0.2238), \ U = \begin{bmatrix} -0.9733 & -0.2298 \\ -0.2298 & 0.9733 \end{bmatrix}$$

$$V^T = \begin{bmatrix} -0.1555 & 0.2268 & -0.0367 & 0.9608 \\ 0.0142 & -0.9714 & -0.0600 & 0.2293 \end{bmatrix}$$

3. The transformation matrices R and R^{-1} are determined from (4.5), and we obtain

$$
R = \begin{bmatrix} -0.0845 & 0.0525 \\ \\ 0.5224 & -0.849? \end{bmatrix}, \quad R^{-1} = \begin{bmatrix} -8.5616 & -0.5291 \\ \\ 5.2623 & -0.8515 \end{bmatrix}.
$$

4. Finally, the balanced representation is obtained from (2.7)

$$
\bar{A} = \begin{bmatrix} -2.7639 & 5.584 \\ \\ -14.3931 & -7.2361 \end{bmatrix}, \quad \bar{B} = \begin{bmatrix} -0.5291 \\ \\ -0.8515 \end{bmatrix},
$$

$$
\bar{C} = \begin{bmatrix} -0.0845 & 0.0525 \\ \\ -0.5224 & -0.8498 \end{bmatrix}.
$$

Example 4.2. The balanced representation is determined for the system with matrices A and B in Example 4.1, and the matric C is as follows

$$
C = \begin{bmatrix} 1 & 0 \end{bmatrix}.
$$

This system is symmetric and in the observability canonical form. We continue to follow the procedure given above.

1. The determination of the matrix H*. Since the considered system has the same matrices A and B as the system in Example 4.1, the matrix L, and the matrices $\mathscr{C}$, $\mathscr{C}^*$ are therefore as in the above example. The observability matrix is a unit matrix, since the system is in the observability canonical form. For this reason we have

$$H^* = L^T \, \mathscr{C} \, L = \begin{bmatrix} 0.0075 & 0.0025 \\ & \\ 0.0025 & -0.0025 \end{bmatrix}.$$

2. The singular value decomposition of H* gives

$$\Gamma = \text{diag}(0.0899, \quad 0.0556)$$

$$U = \begin{bmatrix} -0.9733 & -0.2298 \\ & \\ 0.2298 & 0.9733 \end{bmatrix}, \quad V = U\Sigma = \begin{bmatrix} -0.9733 & 0.2298 \\ & \\ -0.2298 & -0.9733 \end{bmatrix},$$

$$\Sigma = \text{diag}(1, \quad -1).$$

3. The transformtion matrices are obtained from (4.6)

$$R = \begin{bmatrix} -0.2115 & 0.2115 \\ & \\ -1.3070 & -3.4217 \end{bmatrix}, \quad R^{-1} = \begin{bmatrix} -3.4217 & -0.2115 \\ & \\ 1.3070 & -0.2115 \end{bmatrix}.$$

4. The balanced representation is obtained from (2.7)

$$\bar{A} = \begin{bmatrix} -2.7639 & 8.9443 \\ \\ -8.9443 & -7.2361 \end{bmatrix}, \quad \bar{B} = \begin{bmatrix} -0.2115 \\ \\ -0.2115 \end{bmatrix},$$

$$C = \begin{bmatrix} -0.2115 & 0.2115 \end{bmatrix},$$

and is sign-symmetric.

5.   Properties of the Balanced System

We determine the new properties, leaving aside the well-known properties of the balanced system, which are shown elsewhere, see /14 - 16/. In further considerations the input u is considered to be random uncorrelated stationary, with zero mean value, i.e. $E(u) = 0$, $R_u = E(u\,u^T) = \frac{I}{p}$, $\frac{I}{p}$ is the unit matrix.

Property 1. If the system is balanced over the interval $[0, \infty)$, then its steady state response $x(t)$ is uncorrelated, with the covariance matrix as follows:

$$R_x = E(x \ x^T) = \Gamma^2 \tag{5.1}$$

where $\Gamma = \text{diag}(\gamma_i)$, and γ_i is the i-th balancing coefficient. Proof can be found in /1/.

Property 2. For the stationary output, with input being uncorrelated noise, the output index J_y

$$J_y = \| y \|^2 = E(y^T \ y) = \text{tr } R_y \tag{5.2}$$

is majorized as follows:

$$J \leq 2|a| \ (\text{tr } \Gamma^2)^2 \tag{5.3}$$

where a = tr(A).

Proof can be found in /1/.

If the input is a correlated stationary noise with the symmetric positive definite covariance matrix $R_u \neq I$, then the stationary response of the system with the representation (A,B,C) is equal to the response of the system with the representation $(A,\tilde{B},C)$ and uncorrelated input, where

$$\tilde{B} = B \ M \tag{5.4}$$

and M is obtained from the decomposition of R_u

$$R = M \ M^T \tag{5.5}$$

(see /1/).

Similarly, if the output index is weighted with the weight matrix P being symmetric positive definite

$$J_y = \|y\|_p^2 = E(y^T \ P \ y) \tag{5.6}$$

then the weighted index of (A,B,C) is equal to the un-weighted index of $(A,B,\tilde{C})$, where

$$\tilde{C} = S \ C \tag{5.7}$$

and S is obtained from the decomposition of the weight matrix

$$P = S^T \ S \ . \tag{5.8}$$

6. Model Reduction

It was shown /1,14,15/ that the system order can be reduced if some of the state variables are slightly in-

fluenced by input and, at the same time, weakly observed by
the output.

The influence of the input on the state variables, and
the state variables on the output, is evaluated by the
balancing coefficients. Let the balancing coefficients be
ordered such that $\gamma_i \geqslant \gamma_{i+1}$, and $\gamma_m \gg \gamma_{m+1}$ then one concludes
that the state variables $\bar{x}_i$, for i=m+1,...,n are much less
affected by the input. Similarly, the state variables $\bar{x}_i$ for
i=m+1,...,n affect the output much less than the variables
$\bar{x}_i$ for i=1,...,m. For that reason the system can be reduced
by deleting the state variables $\bar{x}_i$ for i=m+1,...,n. Denoting

$$J = [I_m \quad 0] \tag{6.1}$$

we obtain the reduced balanced realization $(\bar{A}_r, \bar{B}_r, \bar{C}_r)$ as
follows

$$\bar{A}_r = J \bar{A} J, \quad \bar{B}_r = J \bar{B}, \quad \bar{C}_r = \bar{C} J , \tag{6.2}$$

which means it is obtained from the balanced realization by
deleting the last n-m rows and columns of $\bar{A}$, the last n-m
rows of $\bar{B}$, and the last n-m columns of $\bar{C}$.

Next we determine how the reduction influences the
output of the system. Let the reduction index i_r be defined
as

$$i_r = J_e/J_y \tag{6.3}$$

where J_y is the output index as defined by (5.2), J_e is the
approximation index given by

$$J_e = \| e \| = E(e^T e) = tr(e\ e^T) \tag{6.4}$$

and e is the approximation error

$$e = y - y_r \tag{6.5}$$

where y is the output of the original system and y_r is the
output of the reduced system. From /1/

$$i_r \leqslant \frac{tr\ \Gamma_d^2}{tr\ \Gamma^2} = \frac{\sum\limits_{i=m+1}^{n} \gamma_i^2}{\sum\limits_{i=1}^{n} \gamma_i^2} \quad , \quad 0 \leqslant i_r \leqslant 1 \tag{6.6}$$

where Γ_d is the balancing coefficient matrix of the deleted
state variables, $\Gamma_d = diag(\gamma_{m+1}, \ldots, \gamma_n)$.

Example 6.1. The system with the following triple

(A,B,C)

$$A = \begin{bmatrix} 0 & 0 & 1 & 0 \\ 0 & 0 & 0 & 1 \\ -1.0000 & 0.1000 & -0.1000 & 0.0100 \\ 0.1000 & -0.1000 & 0.0100 & -0.0100 \end{bmatrix}, \quad B = \begin{bmatrix} 0 \\ 0 \\ 1 \\ 0 \end{bmatrix}$$

$$C = \begin{bmatrix} 0 & 1 & 0 & 0 \\ 0 & 0 & 0 & 1 \end{bmatrix}$$

is considered in this example.

The following balancing coefficients are obtained

$$\Gamma = \text{diag}(14.0281, \quad 13.8372, \quad 0.2916, \quad 0.2844)$$

as well as the following balanced representation

$$\bar{A} = \begin{bmatrix} -0.0044 & -0.2980 & 0.0005 & -0.0003 \\ 0.2987 & -0.0045 & 0.0003 & -0.0006 \\ 0.0006 & -0.0006 & -0.0472 & 0.9791 \\ 0.0006 & -0.0006 & -1.0300 & -0.0539 \end{bmatrix}, \quad \bar{B} = \begin{bmatrix} -1.3137 \\ 1.3152 \\ 0.0896 \\ 0.0934 \end{bmatrix}$$

$$\bar{C} = \begin{bmatrix} -1.2604 & -1.2587 & 0.0632 & -0.0656 \\ -0.3705 & 0.3813 & 0.0635 & 0.0665 \end{bmatrix},$$

The reasonable gap is observed between the second and the third balancing coefficients, and so the system can be reduced, leaving the first two state variables in the balanced representation. The reduced model is then obtained by deleting the last two rows and columns in $\bar{A}$, the last two rows in $\bar{B}$, and the last two columns in $\bar{C}$, obtaining

$$\bar{A}_r = \begin{bmatrix} -0.0044 & -0.298 \\ 0.2987 & -0.0045 \end{bmatrix}, \bar{B}_r = \begin{bmatrix} -1.3137 \\ 1.3152 \end{bmatrix}, \bar{C}_r = \begin{bmatrix} -1.2604 & -1.2587 \\ -0.3705 & 0.3813 \end{bmatrix}$$

The reduction index is evaluated from (6.6), obtaining $i_r \leqslant 0.0207$.

7.   Balancing Vibrating Systems

The balancing of vibrating systems was considered in /5,8,10/. The results of Gregory /8/ and of Jonckheere /10/ require further study. The Gregory balanced representation of the symmetric vibrating system is not sign-symmetric (a balanced representation of the symmetric system is sign-symmetric, as we have seen above). The Jonckheere model is derived from the property that the matrices A and B of the balanced system are in the form

$$A = [\, a_{ij} \,], \quad B = [\, b_1, \ldots, b_n \,]^T, \quad a_{ij} = -b_i b_j / (\, \gamma_i + \sigma_i \sigma_j \gamma_i \,)$$

where $\sigma_i = +1$ or -1, and γ_i is the i-th balancing coeffi-cient. From this property the balanced representation can be determined only for a special choice of matrices A,B,C. Furthermore, the state space representation considered by Jonckheere does not represent the single-degrees-of-freedom (SDOF) vibrating system. Both developments consider the balancing of the separate modes. Generally, the system as a whole cannot be balanced in this way.

7.1 Balancing a single-degree-of-freedom vibrating system. The system considered is given by the following representation

$$A = \begin{bmatrix} 0 & 0 \\ -\omega^2 & -2\zeta\omega \end{bmatrix}, \quad B = \begin{bmatrix} 0 \\ b \end{bmatrix}, \quad C = [\, c_1 \quad c_2 \,] \qquad (7.1)$$

where ω is the natural frequency and ζ is the damping coefficient. In particular, the system can be considered as a single vibrating mode of a multi-degree-of-freedom (MDOF) system.

For simplicity of notation we denote two cases

Case 1: $\mu = c_2/c_1$, with the notation $c_1 = c$

Case 2: $V = c_1/c_2$, with the notation $c_2 = c$.

In Case 1 it is $C = c \begin{bmatrix} 1 & \mu \end{bmatrix}$, in Case 2 $C = c \begin{bmatrix} V & 1 \end{bmatrix}$ holds true.

Simple calculations shows /5/ that the transformed Hankel matrix is, in Case 1

$$H^* = bc/(4\omega^2 \zeta) \begin{bmatrix} \mu\omega & 1 \\ 1 & 2\zeta - \mu\omega \end{bmatrix} \tag{7.2}$$

and in Case 2

$$H^* = bc/(4\omega^2 \zeta) \begin{bmatrix} \omega & V \\ V & 2\zeta V - \omega \end{bmatrix} \tag{7.3}$$

The eigenvalues α_1 and α_2 (which are real) and the eigenvectors x_1, x_2 of H^* give the balancing coefficients (γ_i) and modes (u_i, v_i)

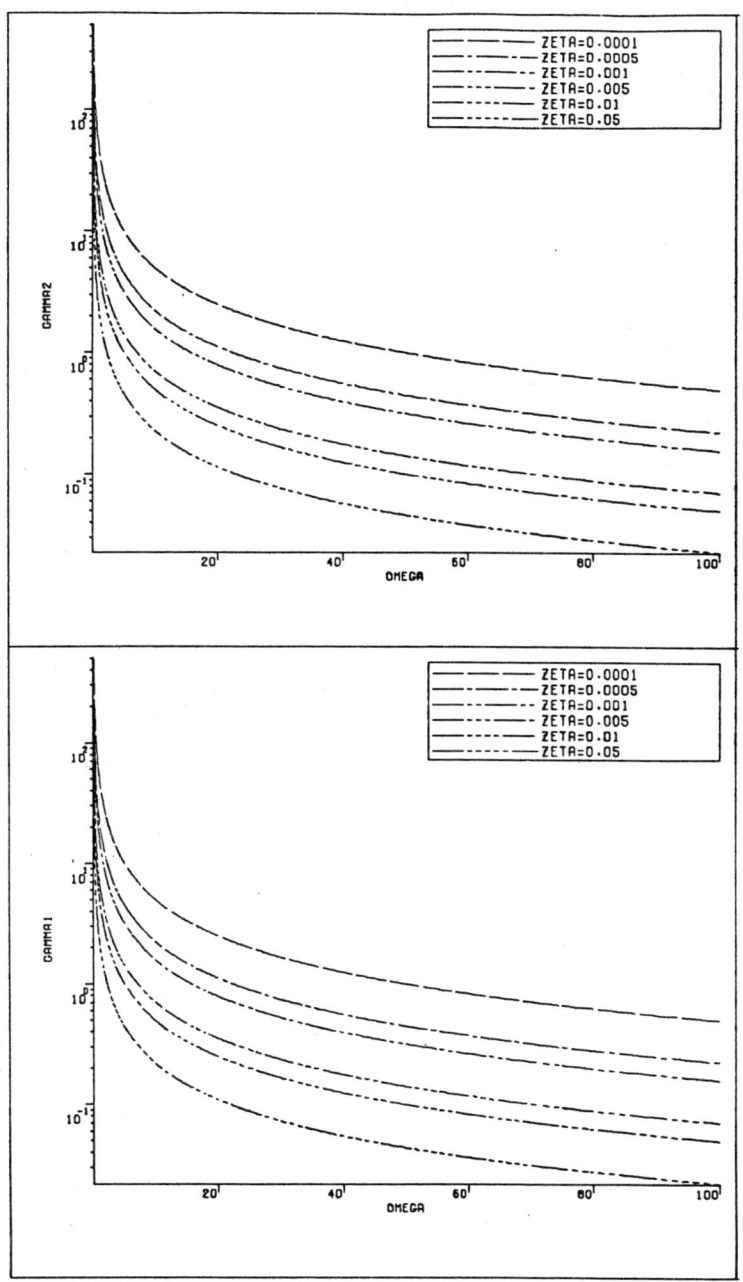

Fig.7.1.a.   The balancing coefficients γ_1 and γ_2 for
small damping ($0 < \zeta < 0.1$) and $\mu = 0$.

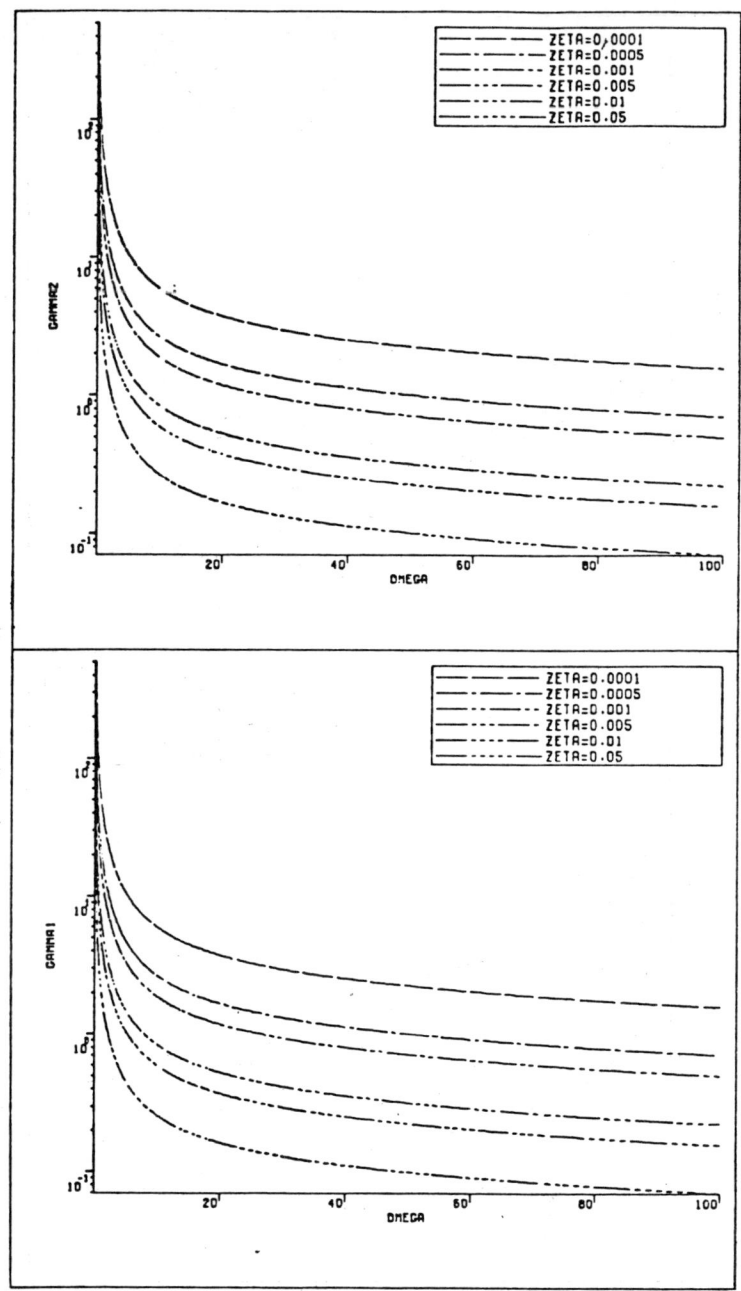

Fig.7.1.b.   The balancing coefficients γ_1 and γ_2 for small
            damping $(0<\zeta<0.1)$ and $\mu = 0.1$.

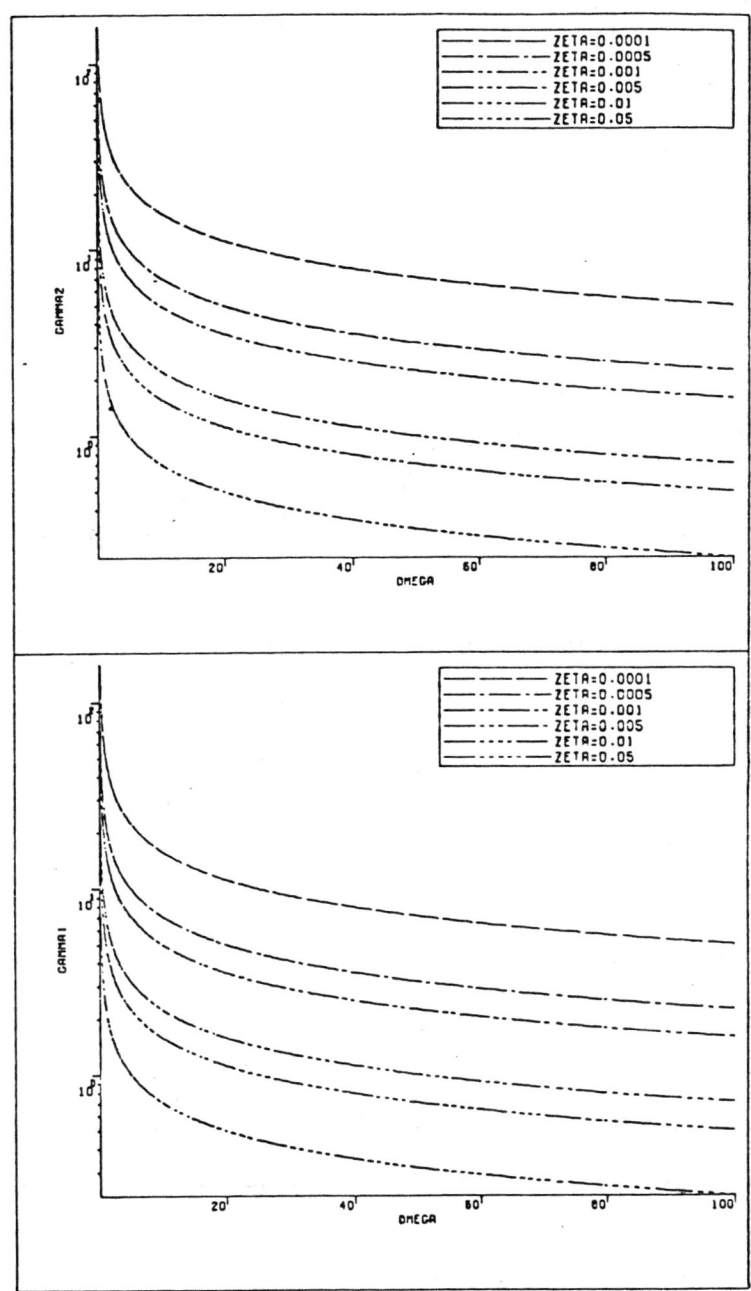

Fig.7.1.c.   The balancing coefficients γ_1 and γ_2 for small
damping $(0 < \zeta < 0.1)$ and $V = 0$.

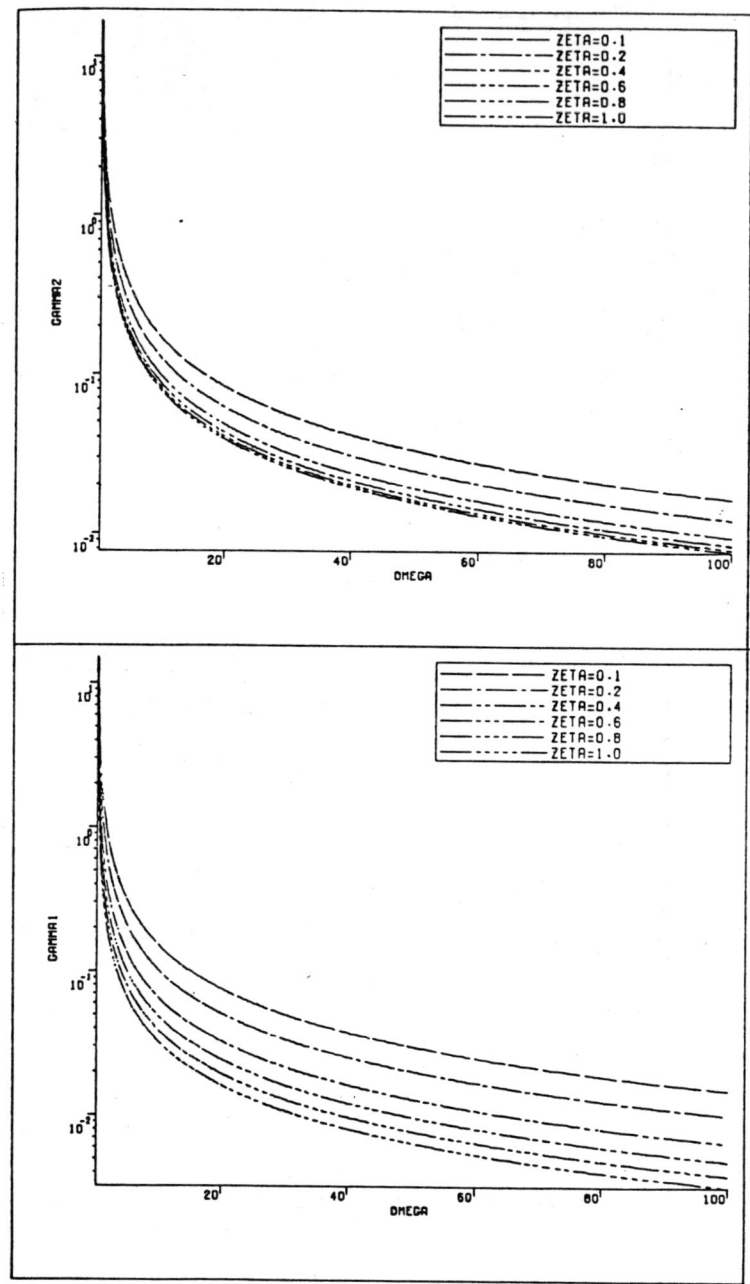

Fig.7.2.a. The balancing coefficients γ_1 and γ_2 for larger
damping $(0.1 \leq \zeta \leq 1)$ and $\mu = 0$.

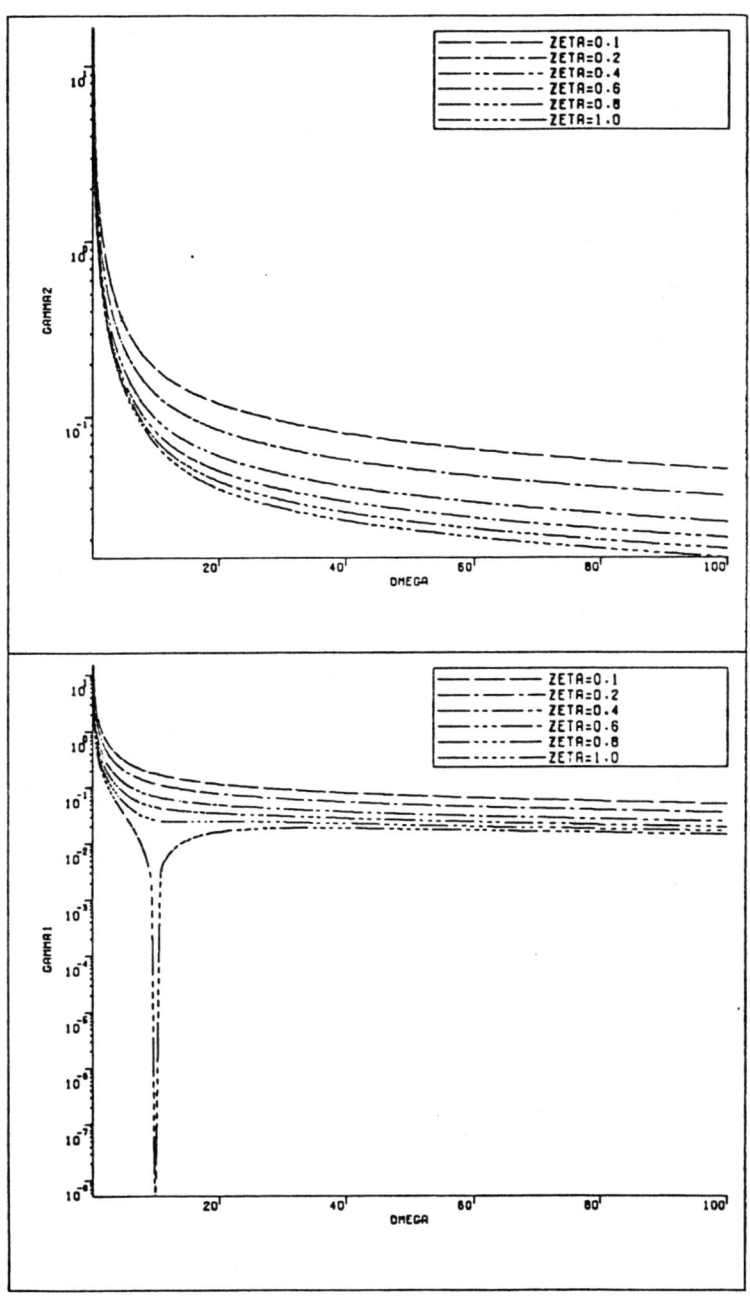

Fig.7.2.b.   The balancing coefficients γ_1 and γ_2 for larger
             damping $(0.1 \leqslant \zeta \leqslant 1)$ and $\mu = 0.1$.

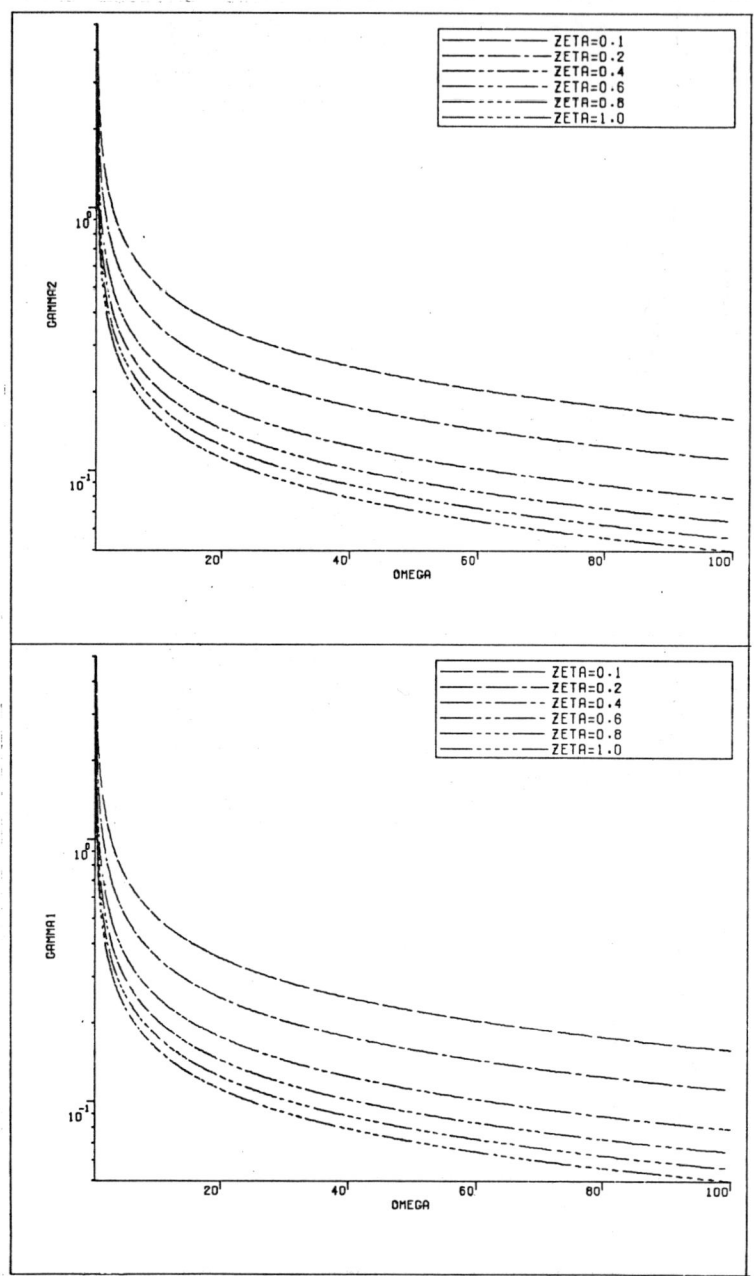

Fig.7.2.c. The balancing coefficients γ_1 and γ_2 for larger
            damping $(0.1 \leq \zeta \leq 1)$ and $\nu = 0$.

$$\gamma_1 = |\alpha_1|^{\frac{1}{2}}, \quad \gamma_2 = |\alpha_2|^{\frac{1}{2}}, \quad u_1 = v_1 = x_1, \quad u_2 = -v_2 = x_2 \qquad (7.4)$$

and

$$\sigma_1 = 1, \quad \sigma_2 = -1. \qquad (7.5)$$

The matrices Γ, Σ, U, V are as follows:

$$\Gamma = \begin{bmatrix} \gamma_1 & 0 \\ 0 & \gamma_2 \end{bmatrix}, \quad \Sigma = \begin{bmatrix} 1 & 0 \\ 0 & -1 \end{bmatrix}, \quad U = [\, u_1 \quad u_2 \,], \quad V = [\, v_1 \quad v_2 \,]$$

Next we consider the eigenvalues of H^* for $bc \neq 1$. If $bc \neq 1$, the obtained eigenvalues for $bc = 1$ should be multiplied by bc. The plots of γ_1 and γ_2 for different values of ω, ζ, μ, and ν are given in Figs. 7.1 and 7.2. Note that for $\mu = 0$ only displacement is measured, and for $\nu = 0$ only velocity is measured (as we see later, for $\nu = 0$ we have $\gamma_1 = \gamma_2$).

Having determined the matrices Γ, U and Σ we obtain the transformation matrices R, R^{-1}

$$R = \mathscr{C}^* U \Gamma^{-1}, \quad R^{-1} = \Sigma \Gamma^{-1} U^T \sigma^* \qquad (7.6)$$

as a symmetric system version of (4.5). The balanced representation is determined from (2.7).

The analytical formulas for the balanced representation
are too complex to be worth deriving, so we consider explicitly only the case $V = 0$ (velocity measured only). For this
case the matrix H^* is

$$H^* = 1/(4\omega\zeta) \begin{bmatrix} 1 & 0 \\ 0 & -1 \end{bmatrix}$$

The eigenvalues and eigenvectors of this matrix are $\alpha_1 = -\alpha_2$
$= (4\omega\zeta)^{-1}$, $u_1 = [1 \quad 0]$, $u_2 = [0 \quad 1]$ therefore the balancing coefficients and modes are $\gamma_1 = \gamma_2 = \gamma = 1/(2\sqrt{\omega\zeta})$,
$\Gamma = \gamma I_2$, $U = I_2$, $\Sigma = \text{diag}(1, -1)$.

The transformation matrices are found from (7.6)

$$R = \begin{bmatrix} 0 & \omega^{-1} \\ 1 & 0 \end{bmatrix}, \quad R^{-1} = \begin{bmatrix} 0 & 1 \\ \omega & 0 \end{bmatrix},$$

and then from (2.7) we obtain the balanced realization

$$\bar{A} = R^{-1} A R = \omega \begin{bmatrix} -2\zeta & -1 \\ & \\ 1 & 0 \end{bmatrix}, \qquad \bar{B} = R^{-1} B = b \begin{bmatrix} 1 \\ \\ 0 \end{bmatrix},$$

$$\bar{C} = C R = C \begin{bmatrix} 1 & 0 \end{bmatrix}.$$

Example 7.1. The balancing of the SDOF vibrating system is considered with the following parameters: stiffness k = 10, damping l = 0.1, mass m = 1, and μ = 0.

From these parameters we obtain the eigenfrequency ω = $\sqrt{k/m}$ = 3.1623, and damping ratio = $1/2\omega$ = 0.0158. From the singular value decomposition of H (which is given by (7.2)), we find

$$\gamma_1 = 1.2674, \quad \gamma_2 = 1.2475, \quad \Sigma = \text{diag}(1, \quad -1)$$

and

$$U = \begin{bmatrix} -0.7237 & -0.6902 \\ & \\ -0.6902 & 0.7237 \end{bmatrix}.$$

The balanced state space representation is given by the formulas (2.7) and (7.6), and the following results are obtained:

$$\bar{A} = \begin{bmatrix} -0.0492 & 3.1619 \\ \\ -3.1619 & -0.0508 \end{bmatrix}, \quad B = -0.3976 \begin{bmatrix} 1 \\ \\ 1 \end{bmatrix},$$

$$\bar{C} = \begin{bmatrix} -0.3976 & 1 & -1 \end{bmatrix}.$$

7.2 Balancing multi-degree-of-freedom vibrating systems

Here we consider a MDOF vibrating system represented by the state-space representation

$$A = \begin{bmatrix} 0 & I \\ \\ -M^{-1}K & -M^{-1}L \end{bmatrix}, \quad B = \begin{bmatrix} 0 \\ \\ B_o \end{bmatrix}, \quad C = \begin{bmatrix} C_q & C_v \end{bmatrix} \quad (7.7)$$

where K, L, M are the stiffness, damping and mass matrices respectively of dimensions mxm, and m is the number of degrees of freedom of the system. The matrix A is nxn, where n = 2m, and B, C are nxp and qxn, where p is the number of inputs and q is the number of outputs.

In order to obtain the balanced representation of the above system, the procedure given in Chapter 4 should be applied to the triple (A,B,C) given by (7.7). Balancing every mode separately does not guarantee that the whole system is being balanced. Also, the equations (7.7) in modal form do not make the balancing procedure simpler.

The following examples illustrate the balancing of non-symmetric and symmetric MDOF vibrating systems.

Example 7.2. The balancing of the two-degree-of-freedom system is considered. The following data are available: the stiffness matrix

$$
K = \begin{bmatrix} 1000 & -1000 \\ -1000 & 4000 \end{bmatrix} ,
$$

the mass matrix is unitary $M = I$, the damping matrix is proportional to K, $L = 0.01 \ K$, the matrix C_v is a zero-matrix, and the matrices B_o, C_q are

$$
B_o = \begin{bmatrix} 1 \\ 1 \end{bmatrix} , \quad C_q = \begin{bmatrix} 1 & 1 \\ 1 & -1 \end{bmatrix} .
$$

For this system the balancing coefficients

$$
\Gamma = 0.1 \ \mathrm{diag}(7.3916, \quad 6.4801, \quad 1.4200, \quad 1.0611)
$$

and the balancing modes

$$U = \begin{bmatrix} -0.9424 & 0.3282 & -0.0637 & 0.0124 \\ -0.3296 & -0.9438 & 0.0179 & 0.0197 \\ 0.0567 & -0.0345 & -0.9805 & 0.1851 \\ 0.0078 & 0.0213 & 0.1851 & 0.9825 \end{bmatrix} ,$$

$$V = \begin{bmatrix} -0.8321 & -0.2867 & 0.1904 & 0.1314 \\ -0.4415 & -0.1622 & -0.3699 & -0.2283 \\ -0.2895 & 0.8298 & -0.1080 & 0.0714 \\ -0.1593 & 0.4483 & 0.1537 & -0.1252 \\ 0.0556 & 0.0216 & -0.4197 & 0.0355 \\ 0.0151 & 0.0310 & 0.7844 & -0.0673 \\ 0.0027 & -0.0181 & 0.0078 & -0.4487 \\ 0.0114 & -0.0111 & -0.0135 & 0.8383 \end{bmatrix}$$

are determined, and the following balanced representation is obtained:

$$\bar{A} = \begin{bmatrix} -3.0355 & -26.1742 & 0.0495 & -0.2089 \\ 26.1882 & -3.9503 & 0.0512 & 0.2356 \\ 2.1862 & -2.8496 & -10.6834 & 62.8901 \\ 2.8485 & -3.7073 & -62.8901 & -32.3313 \end{bmatrix} , \quad \bar{B} = \begin{bmatrix} -0.1821 \\ 0.1821 \\ 0.0656 \\ 0.0853 \end{bmatrix}$$

$$\bar{C} = \begin{bmatrix} -0.1651 & -0.1575 & 0.0312 & -0.0401 \\ -0.0769 & -0.0914 & -0.0577 & 0.0753 \end{bmatrix} .$$

Example 7.3. The two-degree-of-freedom system as in Example 7.2 is considered with the same parameters as before, except that C = [1 0]. This is a symmetric system (every single input, single-output system is symmetric).

The following balancing parameters are obtained:

$$\Gamma \;=\; 0.01 \; \mathrm{diag}(6.0781, \quad 5.3358, \quad 0.5968, \quad 0.4661)$$

$$U \;=\; \begin{bmatrix} -0.9422 & 0.3292 & -0.0618 & -0.0076 \\ -0.3308 & -0.9432 & 0.0211 & -0.0221 \\ 0.0523 & -0.0388 & -0.9947 & -0.0795 \\ 0.0104 & 0.0215 & 0.0793 & -0.9966 \end{bmatrix},$$

$V = U\Sigma$, and $\Sigma = \mathrm{diag}(1, \quad -1, \quad 1, \quad -1)$. The following balanced representation obtained is sign-symmetric

$$\bar{A} \;=\; \begin{bmatrix} -2.9389 & -26.2542 & -0.8987 & -1.5699 \\ 26.2542 & -4.0013 & -1.1683 & -2.1150 \\ 0.8987 & -1.1683 & -6.9897 & -63.6133 \\ -1.5699 & 2.1150 & 63.6133 & -36.0702 \end{bmatrix}, \quad \bar{B} \;=\; \begin{bmatrix} -0.1474 \\ 0.1509 \\ 0.0223 \\ -0.0396 \end{bmatrix}$$

$$\bar{C} \;=\; [\,-0.1474 \quad -0.1509 \quad -0.0223 \quad -0.0396\,].$$

7.3 The reduction of vibrating systems via balancing.

Here we show the application of the system balancing to the

model reduction, according to the explanation given in

Chapter 6. This is shown in two examples.

Example 7.4. Let us consider the possibility of reducing

the SDOF vibrating system with b=c=1, $\zeta = 1$, $\mu = \omega^{-1}$.

Its matrix H^* is obtained from (7.2)

$$H^* = 0.25\, \omega^{-2} \begin{bmatrix} 1 & 1 \\ 1 & 1 \end{bmatrix}$$

and has the following eigenvalues and eigenvectors:
$\alpha_1 = 0.5\omega^{-2}$, $\alpha_2 = 0$, $x_1 = 0.7071 [\, 1 \quad 1\,]$,

$x_2 = 0.7071 [\, 1 \quad -1\,]$. The balancing coefficients and modes

are obtained from (7.4)

$$\Gamma = \text{diag}(0.7071\, \omega^{-1}, \quad 0),$$

$$V = U = 0.7071 \begin{bmatrix} 1 & 1 \\ 1 & -1 \end{bmatrix}, \quad \Sigma = \text{diag}(1, \quad 1).$$

Since $\gamma_2 = 0$, the second variable in the balanced representation can be reduced. Moreover, it follows from (6.6) that
the reduction index is 0, i.e. the reduction error is zero.

It is easy to see that the reduced system has the
representation $\bar{A}_r = -\omega$, $\bar{B}_r = \bar{C}_r \omega^{-\frac{1}{2}}$.

The plots in Fig. 7.3. show the dependence of the
reduction index i_r (in this case $i = \gamma_2/(\gamma_1+\gamma_2)$) on ω and ζ,
for $\mu = 0.05$. From the plots one can see the possibility of
reduction without considerable errors for ζ close to 1, and
natural frequency close to 20 (i.e. for ω close to μ^{-1}).

Example 7.5. Reduction of the MDOF system. The system
is shown in Fig. 7.4. In this figure $m_1 = m_2 = 1$, $k_1 = k_2$
$= 1$, $l_i = 0.0346$ k_i, $i = 1.2$. Its state representation is

$$A = \begin{bmatrix} 0 & 0 & 1 & 0 \\ 0 & 0 & 0 & 1 \\ -2 & 1 & -0.0693 & 0.0346 \\ 1 & -1 & 0.0346 & -0.0346 \end{bmatrix}, \quad B = \begin{bmatrix} 0 \\ 0 \\ 1 \\ 0 \end{bmatrix}, \quad C = [1\ 0\ 0\ 0].$$

For this system the balancing coefficients are obtained

$$\Gamma = \text{diag}(4.1352, \quad 4.0911, \quad 1.5922, \quad 1.5483).$$

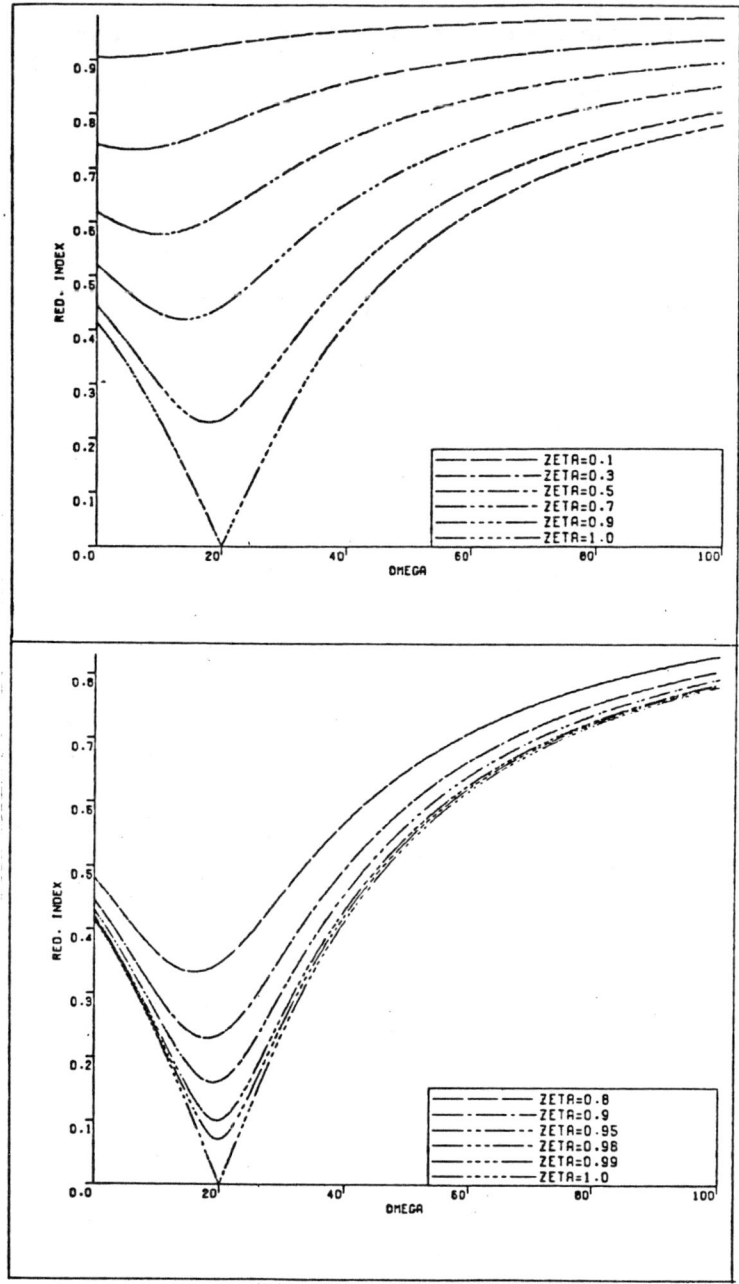

Fig.7.3.a. The reduction index for the SDOF vibrating
            system.

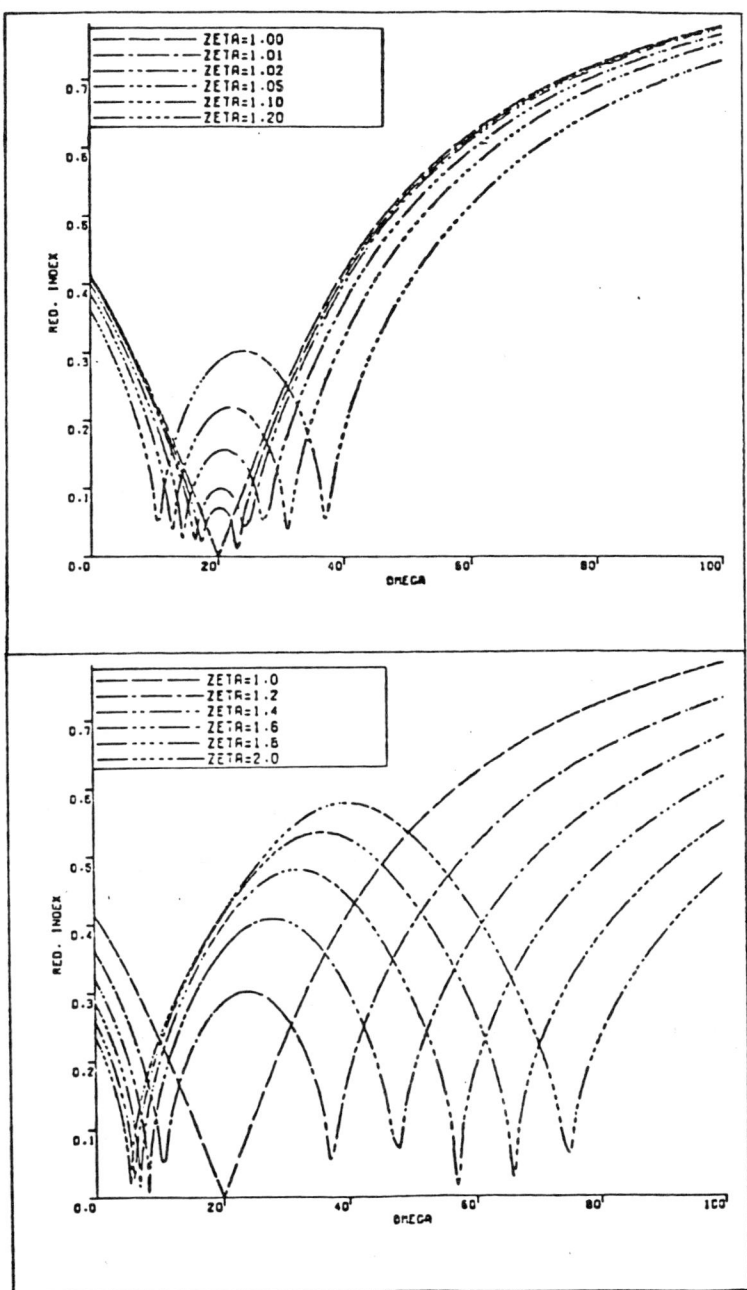

Fig.7.3.b.   The reduction index for the SDOF vibrating
            system.

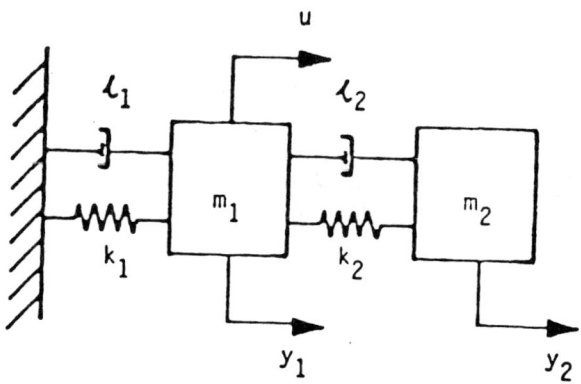

Fig.7.4.   MDOF system considered in Example 7.5.

Obviously, the matrix   shows no significant gap between the
balancing coefficients, therefore the reduction of the
system is not recommended.

Now consider the situation when the stiffness k_2 is
larger than before, namely k_2 = 100, and let the remaining
parameters be as previously. The system triple (A,B,C) is
now as follows:

$$A = \begin{bmatrix} 0 & 0 & 1 & 0 \\ 0 & 0 & 0 & 1 \\ -101 & 100 & -3.4946 & 3.4600 \\ 100 & -100 & 3.4600 & -3.4600 \end{bmatrix}$$

B, C as before. For this system we obtained the following balancing coefficients

$$\Gamma = \text{diag}(4.5455, \quad 4.4903, \quad 0.0570, \quad 0.0448)$$

and the balanced representation

$$\bar{A} = \begin{bmatrix} -0.0085 & -0.7062 & 0.0038 & 0.0038 \\ 0.7062 & -0.0087 & 0.0038 & 0.0039 \\ 0.0038 & -0.0038 & -2.6279 & -13.7542 \\ -0.0038 & 0.0039 & 13.7542 & -4.3095 \end{bmatrix}, \quad \bar{B} = \begin{bmatrix} -0.5935 \\ 0.5934 \\ 0.1308 \\ -0.1315 \end{bmatrix}$$

$$\bar{C} = \begin{bmatrix} -0.5935 & -0.5934 & 0.1308 & 0.1315 \end{bmatrix} .$$

The matrix Γ now shows a split between the balancing coefficients, namely $\gamma_2 >> \gamma_3$, therefore the last two state variables can be reduced. The new, reduced representation is as follows:

$$
\bar{A}_r \;=\; \begin{bmatrix} -0.0085 & -0.7062 \\ \\ 0.7062 & -0.0087 \end{bmatrix}, \qquad \bar{B}_r \;=\; \begin{bmatrix} -0.5935 \\ \\ 0.5934 \end{bmatrix},
$$

$$
\bar{C}_r \;=\; \begin{bmatrix} -0.5935 & -0.5934 \end{bmatrix}.
$$

The reduction index is $i_r \leqslant 0.0113$, i.e. the impulse response of the reduced system differs from the impulse response of the original system by about 1 %. It is easy to see that the reduced representation is close to the balanced representation of the SDOF vibrating system with the following parameters: m = 2, k = 1, l = 0.0346. The latter system is obtained from the system in Fig. 7.4 by setting $k_2 \longrightarrow \infty$, and it has the following balanced representation $(\bar{A}_1, \bar{B}_1, \bar{C}_1)$:

$$
\bar{A}_1 \;=\; \begin{bmatrix} -0.0085 & -0.7070 \\ \\ 0.7070 & -0.0088 \end{bmatrix}, \qquad B_1 \;=\; \begin{bmatrix} -0.5946 \\ \\ 0.5946 \end{bmatrix},
$$

$$
C_1 \;=\; \begin{bmatrix} -0.5946 & -0.5946 \end{bmatrix}.
$$

8.   Order determination of the AR and ARMA models /4/

Here we consider an auto-regressive (AR) model of
order p, as well as an auto-regressive-moving-average (ARMA)
model of order (p, q) where p $\geq$ q. The AR model is described
by the equation

$$y_i = -a_1 \, y_{i-1} - a_2 \, y_{i-2} - \ldots - a_p \, y_{i-p} + b_1 \, u_{i-1} + v_i$$
$$(8.1)$$

the ARMA model is given by

$$y_i = - a_1 \, y_{i-1} - a_2 \, y_{i-2} - \ldots - a_p \, y_{i-p} +$$

$$+ b_1 \, u_{i-1} + b_2 \, u_{i-2} + \ldots + b_q \, u_{i-q} + v_i \qquad (8.2)$$

and y_i is the measured output, u_i is the measured input, v_i
is the noise corrupting the output. We assume that the noise
is uncorrelated with the input.

It is easy to check that AR(p) = ARMA(p,1). Further we
consider the ARMA(p,p) model only, which we denote ARMA(p)
for simplicity. If q < p, we set b = 0 for q < i $\leq$ p. This
assumption gives the order of the ARMA model dependent on
one parameter only, namely p.

The method of evaluating the order of the identified system consists of the following stages:

1. The determination of the state space representation of the ARMA model.

2. The transformation of the state variables to the balanced coordinates.

3. The evaluation of the usefulness of each balanced state variable from the point of view of its reachability and observability, and possible reduction of the system order.

4. Return from the reduced balanced coordinates to the new reduced ARMA model.

The state space representation for the ARMA model is described in /7,13/, and is given by the equation (3.17), where

$$
A = \begin{bmatrix}
0 & 1 & 0 & \cdots & 0 \\
0 & 0 & 1 & \cdots & 0 \\
\cdots\cdots\cdots\cdots\cdots\cdots\cdots\cdots\cdots\cdots\cdots\cdots \\
\cdots\cdots\cdots\cdots\cdots\cdots\cdots\cdots\cdots\cdots\cdots\cdots \\
-a_p & -a_{p-1} & -a_{p-2} & \cdots & -a_1
\end{bmatrix} , \qquad (8.3)
$$

$$C = [\,1 \quad\quad 0 \quad\quad 0 \;\text{....}\; 0\,], \tag{8.4}$$

$$B = \Lambda^{-1} B_o, \tag{8.5}$$

and

$$B_o = [\,b_1 \quad\quad b_2 \;\text{....}\; b_p\,], \tag{8.6}$$

$$\Lambda = \begin{bmatrix} 1 & 0 & 0 & \text{....} & 0 \\ a_1 & 1 & 0 & \text{....} & 0 \\ a_2 & a_1 & 1 & \text{....} & 0 \\ \multicolumn{5}{c}{\text{..............................}} \\ \multicolumn{5}{c}{\text{..............................}} \\ a_{p-1} & a_{p-2} & a_{p-3} & & \end{bmatrix}. \tag{8.7}$$

The extended reachability and abservability matrices for this system are defined by (3.31). The system Hankel matrix is defined as

$$H = \mathcal{O}(N) \quad \mathcal{C}(N). \tag{8.8}$$

The balanced representation is obtained by the singular value decomposition of the Hankel matrix, which is as follows:

$$H = U \Gamma^2 \Sigma U^T \tag{8.9}$$

and the transformation matrices to the balanced representation are

$$R = \mathcal{C}(N) U \Gamma^{-1} \quad , \quad R^{-1} = \Sigma \Gamma^{-1} U^T \sigma(N). \tag{8.10}$$

If for some number $p_o < p$ we have a gap within the balancing coefficient row, i.e. $\gamma_{Po} >> \gamma_{Po+1}$, then the state
variables $\bar{x}_i$ in the balanced representation can be reduced
for $i = p_o+1, \ldots ,p.$

For the rest of this chapter, instead of the reduction
index i_r, see (6.3) and (6.6), we use the logarithmic index
r_1

$$r_1 = - \log i_r \quad , \quad r_1 \geq 0. \tag{8.11}$$

Example 8.1. In this example we test how the order of
extension N influences the balancing coefficients. Let the
ARMA(2) model be given

$$y_i = -0.5 \, y_{i-1} + 0.3 \, y_{i-2} + u_{i-1} + 2.7 \, u_{i-2} + v_i$$

It has the state space representation (A,B,C) as follows:

$$
A = \begin{bmatrix} 0 & 1 \\ 0.3 & -0.5 \end{bmatrix}, \quad B = \begin{bmatrix} 1 \\ 2.2 \end{bmatrix}, \quad C = [1 \quad 0] .
$$

The balanced representation is determined for the extension number N=6. The extended reachability and observability matrices are

$$
\mathcal{C}\ (6) = \begin{bmatrix} 1.0000 & 1.7000 & -0.5500 & 0.7850 & 0.5143 & 0.5143 \\ 1.7000 & -0.5500 & 0.7850 & -0.5575 & 0.5143 & -0.4244 \end{bmatrix}
$$

$$
\mathcal{O}^{T}(6) = \begin{bmatrix} 1.0000 & 0.0000 & 0.3000 & -0.1500 & 0.1650 & -0.1275 \\ 0.0000 & 1.0000 & -0.5000 & 0.5500 & -0.4250 & 0.3775 \end{bmatrix}
$$

therefore the symmetric Hankel matrix is obtained (only its lower part is shown)

$$
H = \begin{bmatrix} 1.0000 & & & & & \\ 1.7000 & -0.5500 & & & & \\ -0.5500 & 0.7850 & -0.5575 & & & \\ 0.7850 & -0.5575 & 0.5143 & -0.4244 & & \\ -0.5575 & 0.5143 & -0.4244 & 0.3665 & -0.3105 & \\ 0.5143 & -0.4244 & 0.3665 & -0.3105 & 0.2652 & -0.2258 \end{bmatrix}
$$

The singular value decomposition of H gives the following
balancing matrices:

$$\Gamma = \text{diag}(1.8000, \quad 1.4736), \qquad \Sigma = \text{diag}(-1, \ 1)$$

$$U = \begin{bmatrix} -0.4281 & 0.5731 & -0.4150 & 0.3794 & -0.3142 & 0.2709 \\ -0.8567 & -0.4866 & -0.0137 & -0.1391 & 0.0654 & -0.0745 \end{bmatrix}.$$

The transformation matrices are obtained

$$R = \begin{bmatrix} 0.7705 & -1.2642 \\ -1.0316 & -0.7170 \end{bmatrix}, \qquad R^{-1} = \begin{bmatrix} 0.3866 & -0.6806 \\ -0.5562 & -0.4154 \end{bmatrix},$$

and the balanced representation is as follows:

$$A = \begin{bmatrix} -0.9072 & -0.2635 \\ 0.2635 & -0.7170 \end{bmatrix}, \qquad B = \begin{bmatrix} -0.7705 \\ -1.2624 \end{bmatrix},$$

$$C = \begin{bmatrix} 0.7705 & -1.2625 \end{bmatrix}.$$

The plot in Fig. 8.1 shows the dependence of the balancing
coefficients on the extension number N. The balancing coefficients and the balanced representation for the infinite

interval balancing (N→∞) are

$$\Gamma = \text{diag}(1.952, \quad 1.4818)$$

$$A = \begin{bmatrix} -0.8965 & -0.2355 \\ 0.2355 & 0.3965 \end{bmatrix}, \quad B = \begin{bmatrix} -0.7977 \\ -1.2792 \end{bmatrix}$$

$$C = \begin{bmatrix} 0.7977 & -1.2792 \end{bmatrix}.$$

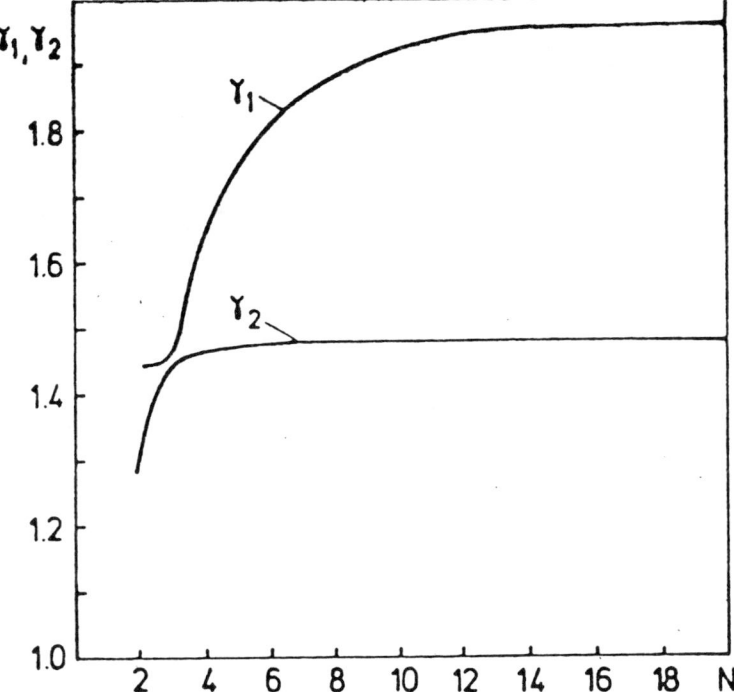

Fig.8.1.   Dependence of the balancing coefficients on the extension number N.

Example 8.2. Given the AR(4) model

$$y_i = 0.8\ y_{i-1} - 0.5\ y_{i-2} - 0.3\ y_{i-3} + 0.3\ y_{i-4} + u_{i-1} + v_i$$

where v_i is non-measurable noise. The system is identified from the measured data y_i, u_i, i=1,...,4000, for the different noise level σ_v. We assume that the system order is not known, therefore the systems of order from p=2 to 7 are identified. The logarithmic reduction index r_1 is determined for the order of extension N=50. Fig. 8.2 shows the logarithmic reduction index for the different noise levels. The jump in the values of r_1 for p=5 indicates that the system order is 4.

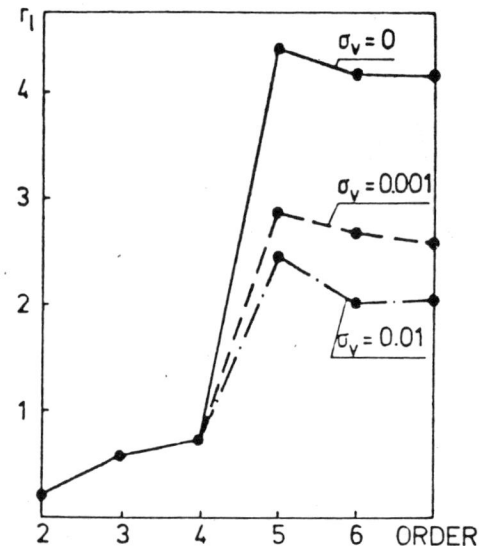

Fig.8.2.   Logarithmic reduction index for the different
             noise levels.

Example 8.3. The ARMA(4) model is given

$$y_i = y_{i-1} - 0.8\ y_{i-2} + y_{i-3} - 0.2\ y_{i-4}$$

$$+ u_{i-1} + 0.7\ u_{i-2} - 0.4\ u_{i-3} - 0.1\ u_{i-4} + v_i$$

and similarly, as in Example 8.2, the data y_i, u_i for
i=1, ...,4000 are available for the system identification.
The process v_i is non-measurable noise with the standard
deviation σ_v. The systems with order 2 to 7 are identified, and for each identified system the reduction index is
determined for the extension order N=50. The plots of r for
the different noise levels are shown in Fig. 8.3. The figure
indicates that the appropriate order of the ARMA model is 4.

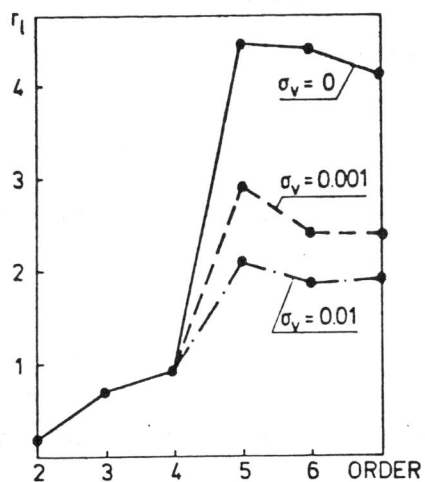

Fig.8.3.   Logarithmic reduction index for the different
              noise levels.

Finally, given the reduced balanced state space representation and its reachability and observabiliy matrices, the ARMA model which relates to this representation should be determined.

From /4/ we have

$$\alpha H(p) = -\beta \qquad (8.12)$$

where α is the row-vector of AR coefficients

$$\alpha = [\ a_{po} \quad a_{po-1} \quad \cdots \quad a_1\ ]\ . \qquad (8.13)$$

$H(p_o)$ is the Hankel matrix of order p_o, and β is the last row of the Hankel matrix $H(p_o+1)$, of order $p+1$. The MA coefficients form the first column of $H(p_o+1)$.

9.   Transfer function realization and identification procedure

The procedure for determining the state space realization (A,B,C) from the finite set of Markov parameters (defined below, cf. (9.3) is presented. This procedure was developed in /6 /.

Here we consider both continuous-time and discrete-time linear systems. Let $\mathcal{C}_n$ and $\mathcal{O}_n$ be the controllability (reachability in the discrete-time case) and observability matrices of order n

$$\mathcal{C}_n = [\, B \ AB \ \ldots \ A^{n-1}B \,], \quad \mathcal{O}_n = [C^T \ (CA)^T \ \ldots \ (CA^{n-1})^T]^T \quad (9.1)$$

and $H_{1,n}$ be its Hankel matrix of order n

$$H_{1,n} = \begin{bmatrix} h_1 & h_2 & \ldots & h_n \\ h_2 & h_3 & \ldots & h_{n+1} \\ \ldots \ldots \ldots \ldots \ldots \\ \ldots \ldots \ldots \ldots \ldots \\ h_n & h_{n+1} & \ldots & h_{2n-1} \end{bmatrix} \quad (9.2)$$

where

$$h_k = C \ A^{k-1} \ B \quad (9.3)$$

are Markov matrices, $p \times q$. Note that

$$H_{1,n} = \mathcal{O}_n \, \mathcal{C}_n \quad (9.4)$$

Let $H_{k,n}$ be the "shifted" Hankel matrix

$$
H_{k,n} = \begin{bmatrix} h_k & h_{k+1} & \cdots & h_{n+k-1} \\ h_{k+1} & h_{k+2} & \cdots & h_{n+k} \\ & & & \\ \cdots\cdots\cdots\cdots\cdots\cdots\cdots \\ \cdots\cdots\cdots\cdots\cdots\cdots\cdots \\ h_{n+k-1} & h_{n+k} & \cdots & h_{2n+k-2} \end{bmatrix} \tag{9.5}
$$

then

$$
H_{k,n} = \sigma_n A^{k-1} \ell_n. \tag{9.6}
$$

Next, we assume that $H_{1,n}$ is of rank m, and that the rank of $H_{1,n+1}$ is also m. Since the matrices $H_{1,n}$ and $H_{1,n+1}$ are obtained from measurement data, it can be difficult to determine n that fulfills this demand. This question is considered later.

Since $H_{1,n}$ is of rank m, it can be decomposed as follows:

$$
H_{1,n} = P Q \tag{9.7}
$$

where P and Q are full rank matrices pn × m and m × qn, m ≤ pn, m ≤ qn, rkP = rkQ = m. The pseudoinverse of $H_{1,n}$ is reflexive, therefore from (9.7) it follows that

$$H^{+}_{1,n} = Q^{+} P^{+} \tag{9.8}$$

where P^{+}, Q^{+} are pseudoinverses of P, Q, such that $P^{+}P = I$, $QQ^{+} = I$.

In /6/ one can find that

$$A = P^{+} H_{2,n} Q^{+} \tag{9.9}$$

$$B = Q E_{q} \tag{9.10}$$

$$C = E^{T}_{p} P \tag{9.11}$$

where $E_{q} = [\underset{q}{I} \ 0 \ \dots \ 0]$. These formulae show that (A,B,C) is fully determined from the Markov parameters.

The realization (A,B,C) depends on the decomposition (9.7) of the Hankel matrix. In particular, one can apply the singular value decomposition, or Cholesky, LU, QR decomposition, if $H_{1,n}$ is nonsingular.

Consider the case of the singular value decomposition of the Hankel matrix, which we denote as follows:

$$H_{1,n} = V_o \Gamma_o^2 U_o^T \qquad (9.12)$$

where V_o, U_o are unitary matrices, and Γ_o is a diagonal positive definite matrix. Comparing the above equation with (9.7) we find

$$P = V_o \Gamma_o , \quad Q = \Gamma_o U_o^T , \quad P^+ = \Gamma_o^{-1} V_o^T, \quad Q^+ = U_o \Gamma_o^{-1} . \qquad (9.13)$$

Finally, consider a symmetric system. A system is symmetric if its transfer function matrix is symmetric, or its Hankel matrix is symmetric (in this case, of course, p=q). For the symmetric system

$$Q = \Sigma \, P^T \qquad (9.14)$$

and for this case the equations (9.9) - (9.11) are

$$A = P^+ H_{2,n} (P^+)^T \Sigma \qquad (9.15)$$

$$B = \Sigma \, P^T E_p = \Sigma \, P_p^T \qquad (9.16)$$

$$C = E_p^T P = P_p \qquad (9.17)$$

where P_p are the first p rows of P, Σ as above.

It is valid to abserve that the system defined by (9.15) - (9.17) is sign-symmetric, since from these equations it follows that

$$A \, \Sigma \; = \; \Sigma \; A^T, \qquad B \; = \; \Sigma \; C^T \; . \qquad\qquad (9.18)$$

Example 9.1. A continuous-time system with the following Markov parameters $h_1 = 0$, $h_2 = 1$, $h_3 = -3.03$, $h_4 = -82.919$, $h_5 = 1102.7$, $h_6 = 9923$, $h_7 = -280780$, and $h_8 = -300660$ is examined. Since the order of the system under investigation is not known a priori, we assume it to be maximally obtainable from 8 Markov parameters, that is p=4.

The matrices $H_{1,4}$ and $H_{2,4}$, according tp (9.2), (9.5) are as follows:

$$
H_{1,4} =
\begin{bmatrix}
0.0000 & 1.0000 & -3.0300 & -82.8190 \\
1.0000 & -3.0300 & -82.8190 & 1102.7 \\
-3.0300 & -82.8190 & 1102.7 & 9923.0 \\
-82.8190 & 1102.7 & 9923.0 & -280780
\end{bmatrix}
$$

$$H_{2,4} = \begin{bmatrix} 1.0000 & -3.0300 & -82.8190 & 1102.7 \\ -3.0300 & -82.8190 & 1102.7 & 9923.0 \\ -82.8190 & 1102.7 & 9923.0 & -280780 \\ 1102.7 & 9923.0 & -280780 & -300660 \end{bmatrix}$$

The singular value decomposition of $H_{1,4}$ is in the form (9.12), with

$$\Gamma_o = \text{diag}(530.2170, \quad 38.1170, \quad 0.6956, \quad 0.7114)$$

$$U_o = \begin{bmatrix} -0.0003 & 0.0041 & 0.7150 & 0.6992 \\ 0.0039 & 0.0303 & 0.6988 & -0.7147 \\ 0.0351 & -0.9989 & 0.0240 & -0.0187 \\ -0.9994 & -0.0350 & 0.0034 & -0.0037 \end{bmatrix}$$

and $V_o = U_o \Sigma$, with $\Sigma = \text{diag}(-1, 1, 1, -1)$.

The following realization is obtained from (9.15) to (9.17)

$$A = \begin{bmatrix} 0.9982 & -14.4051 & 0.1056 & 0.0829 \\ 14.4051 & -7.0132 & 0.0514 & 0.0404 \\ 0.1056 & 0.0514 & 0.2440 & -0.7051 \\ 0.0829 & -0.0404 & 0.7501 & -0.2591 \end{bmatrix}$$

$$B^T = [\ 0.1559 \quad 0.1563 \quad 0.4973 \quad -0.4974\ ],$$

$$C = [-0.1559 \quad 0.1563 \quad 0.4973 \quad 0.4974\ ].$$

When the realization (A,B,C) is determined, the question arises if it is minimal. This question can be answered on the basis of the system balancing theory, and is illustrated in the following example.

Example 9.2. The realization obtained in Example 9.1 is considered and the possibility of its reduction via system balancing is checked.

The balancing coefficients for this system are

$$\Gamma = \text{diag}(4.8875, \quad 4.8363, \quad 0.0594, \quad 0.0482)$$

and its balanced representation is

$$\bar{A} = \begin{bmatrix} -0.0074 & -0.7062 & 0.0032 & 0.0033 \\ 0.7062 & -0.0075 & 0.0033 & 0.0033 \\ 0.0032 & -0.0033 & -2.3257 & -13.8542 \\ -0.0033 & 0.0033 & 13.8542 & -3.6894 \end{bmatrix}$$

$$\bar{B}^T = [-0.5935 \qquad 0.5929 \qquad 0.1282 \qquad -0.1309]$$

$$\bar{C} = [-0.5935 \qquad -0.5929 \qquad 0.1282 \qquad 0.1309]$$

In the matrix $\lceil$ we have $\gamma_2 \gg \gamma_3$, so that the system can be reduced to n=2 state variables by deleting the last two state variables in the balanced representation. In this way the upper left block of $\bar{A}$, as well as the upper block of $\bar{B}$, and the left block of $\bar{C}$ represent the reduced realization of the system

$$\bar{A}_r = \begin{bmatrix} -0.0074 & -0.7062 \\ 0.7062 & -0.0075 \end{bmatrix}, \qquad \bar{B}_r = \begin{bmatrix} -0.5935 \\ 0.5929 \end{bmatrix},$$

$$\bar{C}_r = [-0.5935 \qquad -0.5929].$$

According to (6.6) the reduction error is $i_r \leq 0.0111$.

Since the Markov parameters were taken from the system in Fig. 7.4, for $k_2 = 100$, the reduced realization is almost identical with the reduced one from Example 7.4.

For the discrete-time systems the Markov parameter h_i has the useful physical interpretation - it is the impulse

response at time istant t_i = i Δ t, where Δt is the sampling interval. Therefore, the discrete time system the above
procedure has an additional interpretation - it is an
identification procedure. Given the impulse response, the
realization (A,B,C) is determined. This particular approach,
developed independently in /6/, is presented in /11,12/.

10. Conclusions

The determination of the balanced representation allows
one to evaluate every state variable from the point of view
of its controllability (reachability) and observability.
This, in turn, gives one the possibility of reducing the
variables which are weakly controlled and observed. The
applications have shown that the balancing method is useful
in the determination of the order of different linear
models.

Appendix. Solution of the Lyapunov Equation

This problem is presented in /3/.

A1. <u>Continuous-time case</u>. Consider the matrix A, n x n, with the distinct eigenvalues $\lambda_1, \ldots, \lambda_n$ and the matrix B, n x p. Define also the following matrices

$$\ell = [B \quad AB \quad \ldots \quad A^{n-1} B] \tag{A.1}$$

$$Q_o(t) = Q(t) \otimes I_p \tag{A.2}$$

where $\otimes$ denotes the Kronecker product, and I_p is the unit matrix of order p; further

$$Q(t) = V^{-1} D(t) V^{-T} \tag{A.3}$$

and

$$V = \begin{bmatrix} 1 & \lambda_1 & \lambda_1^2 & \ldots\ldots & \lambda_1^{n-1} \\ & & & & \\ 1 & \lambda_2 & \lambda_2^2 & \ldots\ldots & \lambda_2^{n-1} \\ & & & & \\ & \ldots\ldots\ldots\ldots\ldots\ldots & & \\ & \ldots\ldots\ldots\ldots\ldots\ldots & & \\ & & & & \\ 1 & \lambda_n & \lambda_2^2 & \ldots\ldots & \lambda_n^{n-1} \end{bmatrix} \tag{A.4}$$

$$D(t) = [d_{ij}(t)], \quad i,j=1, \ldots, n \tag{A.5}$$

$$d_{ij}(t) = \frac{1}{\lambda_i + \lambda_j} (e^{(\lambda_i + \lambda_j)t} - 1) \tag{A.6}$$

With this notation we prove

Theorem A.1. The grammian

$$W_c(t) = \int_0^t e^{A\tau} B B^T e^{A^T\tau} d\tau \tag{A.7}$$

is determined from the formula

$$W_c(t) = \mathcal{C} Q_0(t) \mathcal{C}^T. \tag{A.8}$$

Proof. From the Cayley-Hamilton theorem we know

$$e^{At} B = a_0(t) B + a_1(t) AB + \ldots + a_{n-1}(t) A^{n-1}B$$

or

$$e^{At} B = [B \ AB \ \ldots \ A^{n-1} B] \ (a(t) \otimes I_p) \tag{A.9}$$

where

$$a(t) = [a_0(t) \ a_1(t) \ \ldots \ a_{n-1}(t)]^T.$$

The coefficient vector a(t) is determined from

$$a(t) = V^{-1} f(t) \tag{A.10}$$

where V is given by (A.4), and

$$f(t) = [\, e^{\lambda_1 t} \quad e^{\lambda_2 t} \quad \ldots \quad e^{\lambda_n t} \,]^T.$$

Using (A.1) and (A.10) we write (A.9) as

$$e^{At} B = \mathcal{C} \, (V^{-1} f(t) \otimes \underset{p}{I}) \,. \tag{A.11}$$

Now, putting the above to the definition of grammian (A.7) we find

$$W_c(t) = \int_0^t \mathcal{C} (V^{-1} f(\tau) \otimes \underset{p}{I}) (f^T(\tau) V^{-T} \otimes \underset{p}{I}) \mathcal{C}^T d\tau =$$

$$= \mathcal{C} \int_0^t V^{-1} f(\tau) f^T(\tau) V^{-T} \otimes \underset{p}{I} \mathcal{C}^T d\tau = \mathcal{C} (V^{-1} D(t) V^{-T} \otimes \underset{p}{I}) \mathcal{C}^T$$

where

$$D(t) = \int_0^t f(\tau) f^T(\tau) d\tau$$

and the ij-th element of D(t) is

$$d_{ij}(t) = \int_0^t e^{\lambda_i \tau} e^{\lambda_j \tau} d\tau = \frac{1}{\lambda_i + \lambda_j} (e^{(\lambda_i + \lambda_j)t} - 1).$$

Corollary A.1. With C being a p × n matrix, the following grammian

$$W_0(t) = \int_0^t e^{A^T \tau} C^T C e^{A\tau} d\tau \tag{A.12}$$

is determined from

$$W_0(t) = \sigma^T Q(t) \sigma \tag{A.13}$$

with

$$\sigma^T = [C^T \quad (CA)^T \quad \dots \quad (CA^{n-1})^T]. \tag{A.14}$$

Corollary A.2. With the matrices A,B,C is above, and with the real parts of the eigenvalues of A being negative, the solutions of the following Lyapunov equations

$$A W_c + W_c A^T = - B B^T \tag{A.15}$$

$$A^T W_0 + W_0 A = -C^T C \tag{A.16}$$

are given by

$$W_c = \ell \, Q_o \, \ell^T \tag{A.17}$$

$$W_o = \sigma^T \, Q_o \, \sigma \tag{A.18}$$

with

$$Q_o = Q \otimes I_p, \quad Q = V^{-1} D V^{-T} \tag{A.19}$$

V as above, and

$$D = [d_{ij}], \quad i,j=1, \dots, n, \quad d_{ij} = -\frac{1}{\lambda_i + \lambda_j}. \tag{A.20}$$

Example A.1. For the following A and B

$$A = \begin{bmatrix} 0 & 1 \\ -2 & -3 \end{bmatrix}, \quad B = \begin{bmatrix} 1 \\ 0 \end{bmatrix}$$

it is possible to determine the solution of the Lyapunov equation (A.15).

The matrix A has the following eigenvalues: $\lambda_1 = -1$, $\lambda_2 = -2$. Since p=1, $Q_o=Q$ and $Q = V^{-1} D V^{-T}$, where

$$V = \begin{bmatrix} 1 & \lambda_1 \\ 1 & \lambda_2 \end{bmatrix} = \begin{bmatrix} 1 & -1 \\ 1 & -2 \end{bmatrix}, \quad D = -\begin{bmatrix} \frac{1}{2\lambda_1} & \frac{1}{\lambda_1+\lambda_2} \\ \frac{1}{\lambda_1+\lambda_2} & \frac{1}{2\lambda_2} \end{bmatrix} = \begin{bmatrix} 0.5000 & 0.3333 \\ 0.3333 & 0.2500 \end{bmatrix}$$

From these data we obtain

$$Q = V^{-1} D V^{-T} = \begin{bmatrix} 0.9167 & 0.2500 \\ 0.2500 & 0.0833 \end{bmatrix}.$$

Next, the matrix $\mathcal{C}$ is determined

$$\mathcal{C} = [\, B \quad AB \,] = \begin{bmatrix} 1 & 0 \\ 0 & -2 \end{bmatrix}$$

therefore, from (A.17) we find

$$W_c = \mathcal{C} Q_o \mathcal{C}^T = \begin{bmatrix} 0.9167 & -0.5000 \\ -0.5000 & 0.3333 \end{bmatrix}.$$

A2. Discrete time case. Consider the matrix A, n×n, with distinct eigenvalues $\lambda_1 \ldots, \lambda_n$ together with the matrix B, n×p. Let us also define the following matrices

$$Q_o(N) = Q(N) \otimes \underset{p}{I} \tag{A.21}$$

$$Q(N) = V^{-1} D(N) V^{-T} \tag{A.22}$$

where V is given by (A.4) and

$$D(N) = [d_{ij}(N)], \quad i,j=1, \ldots, n \tag{A.23}$$

$$d_{ij}(N) = \frac{(\lambda_i \lambda_j)^N - 1}{\lambda_i \lambda_j - 1} \tag{A.24}$$

With the above notation we prove:

Theorem A.2. The grammian

$$W_c(N) = \sum_{i=0}^{N} A^i BB^T (A^i)^T \tag{A.25}$$

is determined from the formula

$$W_c(N) = \ell Q_o(N) \ell^T . \tag{A.26}$$

where ℓ is given by (A.1).

Proof can be found in /3/.

Corollary A.3. With C being a p x n matrix, the following grammian

$$W_o(N) = \sum_{i=0}^{N} (A^i)^T C^T C A^i \qquad (A.27)$$

is determined from

$$W_o(N) = \sigma^T Q_o(N) \sigma \qquad (A.28)$$

with σ given by (A.14).

Corollary A.4. With matrices A,B,C as above, and A having its eigenvalues within the unitary circle, the solutions of the following Lyapunov equations

$$A W_c A^T - W_c = - B B^T \qquad (A.29)$$

$$A^T W_o A - W_o = - C^T C \qquad (A.30)$$

are given by

$$W_c = \ell Q_o \ell^T \qquad (A.31)$$

$$W_o = \sigma^T Q_o \sigma \qquad (A.32)$$

with

$$Q_o = Q \otimes I_p \, , \quad Q = V^{-1} D V^{-T} \tag{A.33}$$

where V is as above, and

$$D = [d_{ij}] \, , \quad i,j=1,\ldots,n, \quad d_{ij} = \frac{1}{1-\lambda_i \lambda_j} \, . \tag{A.34}$$

Example A.2. The discrete Lyapunov equation (A.29) is considered, with

$$A = \begin{bmatrix} 0 & 1 \\ 0.4 & -0.3 \end{bmatrix}, \quad B = \begin{bmatrix} 0 \\ 1 \end{bmatrix}.$$

The matrix A has the eigenvalues $\lambda_1 = 0.5$, $\lambda_2 = -0.8$. Since p=1, $Q_o = Q$, and $Q = V^{-1} D V^{-T}$, where

$$V = \begin{bmatrix} 1 & \lambda_1 \\ 1 & \lambda_2 \end{bmatrix} = \begin{bmatrix} 1 & 0.5 \\ 1 & -0.8 \end{bmatrix}$$

$$D = \begin{bmatrix} \dfrac{1}{1-\lambda_1^2} & \dfrac{1}{1-\lambda_1 \lambda_2} \\ \dfrac{1}{1-\lambda_1 \lambda_2} & \dfrac{1}{1-\lambda_2^2} \end{bmatrix} = \begin{bmatrix} 1.3333 & 0.7143 \\ 0.7143 & 2.7778 \end{bmatrix}.$$

so that we obtain

$$Q = V^{-1} D V^{-T} = \begin{bmatrix} 1.2539 & -0.3175 \\ -0.3175 & 1.5871 \end{bmatrix}$$

For the above data the matrix $\mathcal{C}$ is

$$\mathcal{C} = [B \quad AB] = \begin{bmatrix} 0 & 1 \\ 1 & -0.3 \end{bmatrix}$$

and from (A.31) we obtain the solution

$$W_c = \mathcal{C} Q_o \mathcal{C}^T = \begin{bmatrix} 1.5872 & -0.7936 \\ -0.7936 & 1.5871 \end{bmatrix}$$

Acknowledgment

Financial support from the Stiftung Volkswagenwerk is gratefully acknowledged.

References

/1/   W. Gawronski, H.G. Natke: Balancing linear systems.
      Int. J. Systems Sci., vol. 18, 1987.

/2/   W. Gawronski, H.G. Natke: On balancing the linear
      symmetric systems. Int. J. Systems Sci., vol. 17, 1986.

/3/   W. Gawronski, H.G. Natke: Gramminans and the solutions
      of the Lyapunov equations. ZAMM (Zeitschrift für ange-
      wandte Mathematik und Mechanik, No. 11, 1986

/4/   W. Gawronski, H.G. Natke: Order estimation of AR and
      ARMA models. Submitted for publication, 1985.

/5/   W. Gawronski, H.G. Natke: Balanced state space repre-
      sentation for vibrating systems. CRI Report,
      CRI-B-6/85.

/6/   W. Gawronski, H.G. Natke: On realizations of the trans-
      fer function matrix. Int. J. Syst. Sci., vol. 18, 1987.

/7/   W. Gawronski, H.G. Natke: On ARMA models for vibrating
      systems. Probabilistic Engineering Mechanics, 1986,
      Vol 1, No. 3.

/8/   C.Z. Gregory, Jr.: Reduction of large flexible space-
      craft models using internal balancing theory.
      J. Guidance, Control and Dynamics, vol. 7, 1984.

/9/   D.C. Hyland, D.S. Bernstein: The optimal projection
      equations for model reduction and relationsips among
      the methods of Wilson, Skelton, and Moore. IEEE Trans.
      Autom. Control, vol. 30, 1985.

/10/ E.A. Jonckheere: Principal component analysis of
      flexible systems, open-loop case. IEEE Trans. Autom.
      Control, vol. 29, 1984.

/11/ J.N. Juang, R.S. Pappa: An eigensystem realization
      algorithm for modal parameter identification and model
      reduction. J. Guidance, Control and Dynamics, vol. 8,
      1985.

/12/ J.N. Juang, R.S. Pappa: Effects of noise on ERA-identi-
      fied modal parameters. AAS/AIAA Astrodynamics Specia-
      list Conf., Vail, 1985.

/13/ R.C.K. Lee: Optimal Estimation, Identification and
      Control, MIT Press, Cambridge, Mass. 1964.

/14/ B.C. Moore: Principal component analysis in linear systems: Controllability, observability and model reduction. IEEE Trans. Autom. Control, vol. 26, 1981.

/15/ L. Parnebo, L. Silverman: Model reduction via balanced state space representations. IEEE Trans. Autom. Control, vol. 27, 1982.

/16/ S. Shokoohi, L.M. Silverman, P.M. Van Dooren: Linear time-variable systems: Balancing and model reduction. IEEE Trans. Autom. Control, vol. 28, 1983.

/17/ R.E. Skelton, P.C. Hughes, H.B. Hablani: Order reduction for models of space structures using modal cost analysis. J. Guidance, Control and Dynamics, vol. 5, 1982.

/18/ C. P. Therapos: On the selection of the reduced order via balanced state representation. IEEE Trans. Autom. Control, vol. 29, 1984.

/19/ S.J. Varoufakis, P.N. Paraskevopulos: A comparative study in model reduction of linear time-invariant systems. Int. J. Modelling & Simulation, vol. 2, 1982.

/20/ A. Yousuff, R.E. Skelton: Covariance equivalent realizations with application to model reduction of large-
scale systems. In _Control_ _and_ _Dynamic_ _Systems_, Academic
Press, N.Y., 1985.

/21/ A. Yousuff, D.A. Wagie, and R.E. Skelton: Linear system
approximation via covariance equivalent realizations.
_J._ _Math._ _Analysis_ and _Appl.,_ vol. 106, 1985.

NON-LINEARITY IN DYNAMICAL SYSTEMS

G.R. Tomlinson
Heriot-Watt University, Riccarton, Edinburgh, U.K.

PART 1. A REVIEW OF DETECTION METHODS IN MODAL TESTING

As a starting point it is worthwhile presenting a global view of the significance of non-linearity and the manner in which this is considered in relation to modal testing and analysis. Figures 1a and 1b show why non-linearity is important and how modal testing inter-relates with non-linearity.

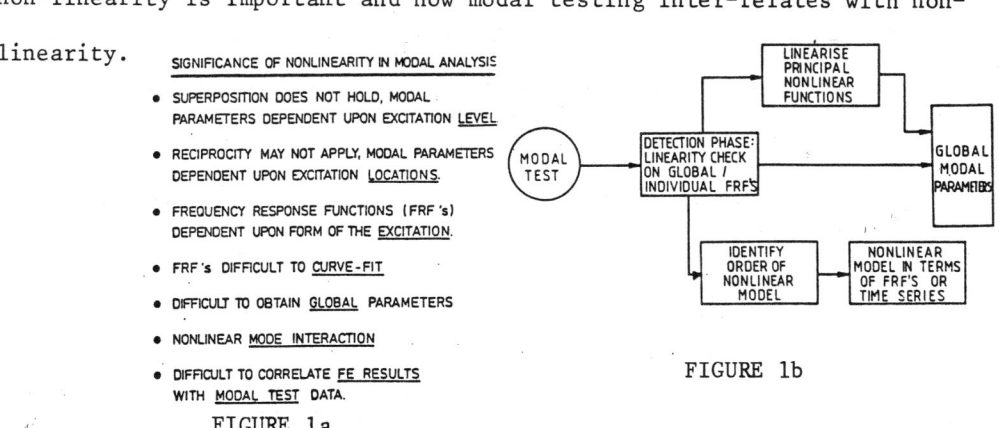

SIGNIFICANCE OF NONLINEARITY IN MODAL ANALYSIS

- SUPERPOSITION DOES NOT HOLD, MODAL PARAMETERS DEPENDENT UPON EXCITATION LEVEL
- RECIPROCITY MAY NOT APPLY, MODAL PARAMETERS DEPENDENT UPON EXCITATION LOCATIONS.
- FREQUENCY RESPONSE FUNCTIONS (FRF's) DEPENDENT UPON FORM OF THE EXCITATION.
- FRF's DIFFICULT TO CURVE-FIT
- DIFFICULT TO OBTAIN GLOBAL PARAMETERS
- NONLINEAR MODE INTERACTION
- DIFFICULT TO CORRELATE FE RESULTS WITH MODAL TEST DATA.

FIGURE 1a

FIGURE 1b

The role of the type of excitation obviously plays an important part in the study of non-linear structures[1] and thus a distinction is drawn between the methods which employ sinusoidal, random and impact excitation.

1.1 Superposition and Linearity

Linearity is defined by the principle of superposition. It is not dependent upon the form of the input excitation. Superposition is simply stated mathematically as:

$$\text{if} \quad x_1(t) \rightarrow y_1(t) \; ; \; x_2(t) \rightarrow y_2(t), \tag{1}$$

$$\text{then} \quad x_1(t) + x_2(t) \rightarrow y_1(t) + y_2(t) \tag{2}$$

However, if this procedure is employed one must check that equation (2) holds for all response and excitation points which makes this a time consuming process. In addition, no obvious identification process can be linked to this approach. An application of this procedure is shown in Figure 2. The total power input from one or more exciters is used to monitor the variation in the resonant frequency of individual modes of vibration of an aircraft. When no variation is evident in the frequency/damping curve with increasing input power then the system is considered linearised.

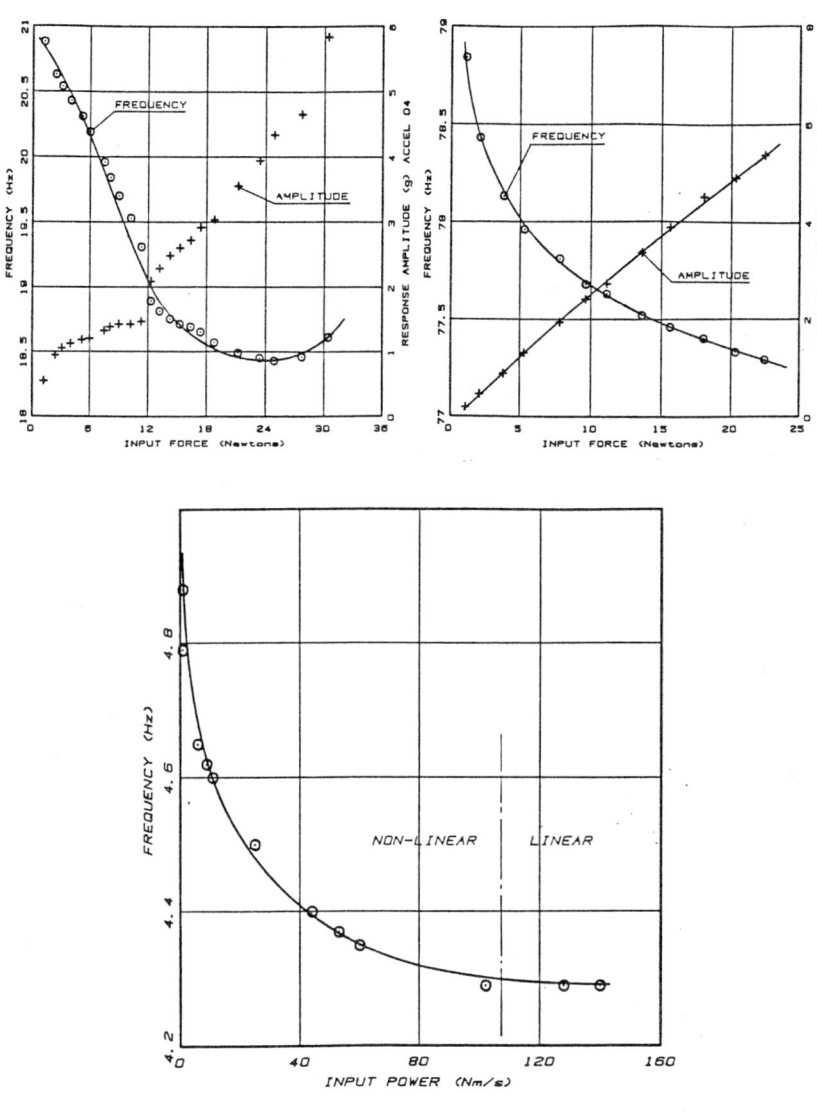

POWER/FREQUENCY CHARACTERISTICS
FOR TANK PITCH MODE

Fig 2   AN EXAMPLE OF 'LINEARITY' CHECK
CARRIED OUT ON AIRCRAFT STRUCTURES

1.2   Nyquist plot distortions/frequency isochrones employing sine excitations.

One of the earliest techniques for detecting nonlinearity during modal testing employed the distortions of the Nyquist plot.[2,3]   If one considers the response of a mode of a single input/single output (SISO) system with a nonlinearity (either damping or stiffness) one can show, using a linearisation procedure such as the method of harmonic balance or similar[4,5] that the response locus in the Nyquist plane due to a harmonic force input is distorted.   This method of analysis is justified since in general the response of the nonlinear system is filtered, leaving only the response due to the fundamental harmonic.   The cause of the distortion is evident from the fact that the linearised equation, represented in terms of its real and imaginary parts can no longer be represented as the equation of a circle.   Consider the equation,

$$m\ddot{x} + c\dot{x} + kx + f(x,\dot{x}) = F\sin(\omega t - \phi) \tag{3}$$

where $f(x,\dot{x})$ are non-linear restoring forces of displacement and velocity.

Assuming $x(t) = A\sin\omega t$ one can show that the complex displacement amplitude is of the form,[3]

$$A^* = A_R + iA_I = A\cos\phi + iA\sin\phi \tag{4}$$

where $A_R, A_I = f(A, A^2 \ldots A^n)$, i.e. the real and imaginary parts are nonlinear functions of the response.

In addition to the distortion of the normally circular vector locus for linear systems, the lines joining points of constant frequency for different excitation levels (isochrones) become distorted in a manner dependent upon the type of the nonlinearity.   These are more sensitive to distortion than the actual vector locus and the pattern of these may be a better method of detection/identification than the actual shape of the diagram.   Figure 3   gives some examples of Nyquist plot distortions.   Techniques for identifying the coefficients of the nonlinear systems based upon the vector plots have been suggested   although one must make the a-priori assumption about the form of the nonlinearity.

The principal limitations of the above methods are that the modes of vibration must be isolated, a constant excitation force within the resonant region must be employed and the nonlinearity must be simple.

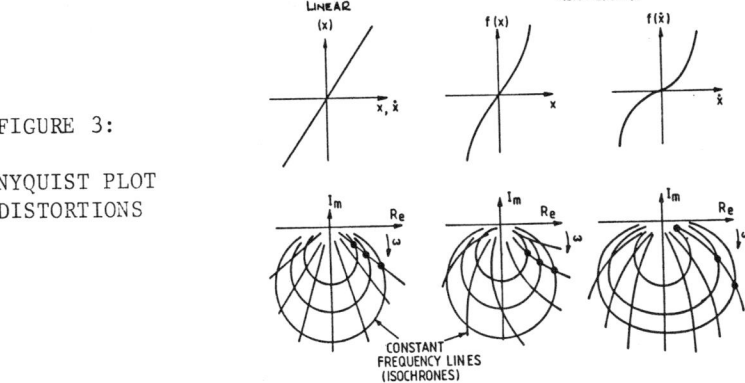

FIGURE 3:

NYQUIST PLOT
DISTORTIONS

1.3   Damping distortion

Reference   6   shows that for a single mode system the damping can be calculated from the formula,

$$\delta_r = \frac{\omega_a^2 - \omega_b^2}{\omega_r^2} \left[\frac{1}{\tan \theta a /_2 + \tan \theta b /_2} \right] \quad ; \delta_r = 2\zeta_r |_{\omega = \omega_r} \qquad (5)$$

By utilising several points both before and after the resonance point on the Nyquist plot one can produce a three-dimensional carpet plot, which, for a linear system is flat. However, in the case of a nonlinear system this becomes distorted. Figure 4 shows an example of this method. This procedure is basically an extension of the Nyquist distortion method described above since it is the phase distortion (variation in the isochrones) which is the sensitive factor. This procedure is easy to implement but has the same limitations as the previous method in that only isolated modes of vibration can successfully be treated.

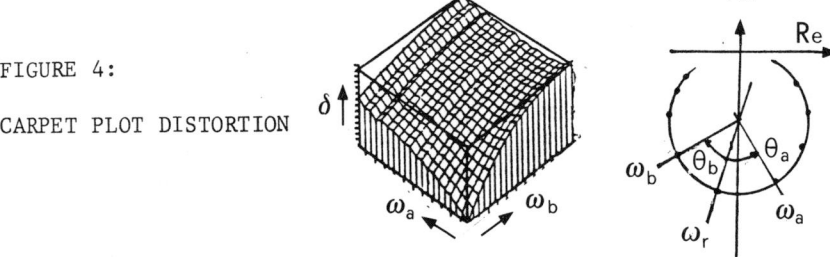

FIGURE 4:

CARPET PLOT DISTORTION

1.4   The "sig-function"

    This procedure basically employs the time data to detect and characterise nonlinearity.[7] This approach utilises the energy present in sub or post harmonics arising from nonlinearities. The "sig-function" is defined for a sinusoidal excitation as,

$$\text{sigf}(\omega) \;=\; \frac{\int_0^T (xx - \text{fund})^2 \, dt}{\int_0^T (xx)^2 \, dt} \tag{6}$$

where xx = original (unfiltered) output time signal

    fund = fundamental (filtered) time signal

Figure 5 shows an application of this method. This technique has not yet

been applied to multi-mode systems and thus it is difficult to comment

on its usefulness for general modal testing. However, it does not offer

any means of linearisation, which has been a major limitation of all the

techniques described so far, and it may be somewhat susceptible to noise

on the response or input signals.

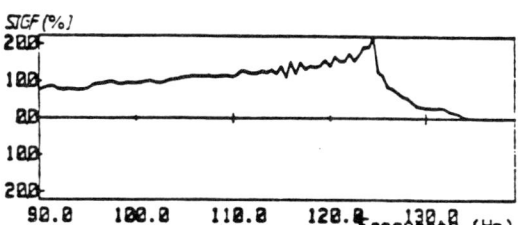

FIGURE 5:

SIG FUNCTION FOR A
CUBIC STIFFNESS NON-
LINEAR SYSTEM.

1.5   The Hilbert Transform

The Hilbert transform relies upon the properties of causal functions.

All physically realisable systems, linear or non-linear, are causal which

means that the impulse response of a frequency response function (FRF)

will satisfy the condition,

$$g(t) = 0, \quad t < 0 \tag{7}$$

where $g(t)$ is due to $\delta(t)$ at $t=0$

A consequence of this is that there is a unique relationship between the

odd and even parts of a causal time function i.e.,

$$g(t) = g(t) \text{ even} + g(t) \text{ odd} \tag{8}$$

where,

$$g(t) \text{ even} = g(t) \text{ odd} \times \text{sgn}(t) \tag{9}$$

$$g(t) \text{ odd} = g(t) \text{ even} \times \text{sgn}(t) \tag{10}$$

and,

$$\text{sgn}(t) = +1, \; t > 0 \tag{11}$$

$$= -1, \; t < 0$$

Assuming that the Fourier transform of $g(t)$ exists, equations (9) and (10) can be written as,

$$\mathcal{F}\{g(t) \text{ even}\} = R_e G(\omega) = \mathcal{F}\{g(t) \text{ odd} \times \text{sgn}(t)\} \tag{12}$$

$$\mathcal{F}\{g(t) \text{ odd}\} = i I_m G(\omega) = \mathcal{F}\{g(t) \text{ even} \times \text{sgn}(t)\} \tag{13}$$

Equations (12) and (13) show that the real part of the frequency response function can be obtained from the imaginary part and vice versa. These are usually written as,

$$H\{R_e G(\omega)\} = I_m G(\omega) \tag{14}$$

$$H\{I_m G(\omega)\} = R_e G(\omega) \tag{15}$$

where $H\{\cdot\}$ represents the Hilbert transform.

Thus for a linear system,

$$H(\omega) = G(\omega) \tag{16}$$

When non-linearity is present in the response of a structure, the inverse Fourier transform of the measured frequency response function is found to be significantly non-causal[8] and as a result of equations (12) and (13), one finds,

$$H(\omega) \neq G(\omega) \tag{17}$$

Equations (16) and (17) are explained graphically in Figure 6a. The method of obtaining Hilbert transforms described above is based on the time domain procedure. However, the formal Hilbert transform is an integral transform[9]

$$H(\omega_c) = -\frac{1}{i\pi} PV\!\!\int_{-\infty}^{+\infty} \frac{G(\omega)\,d\omega}{\omega-\omega_c} \tag{18}$$

where PV = Principal value of the integral.

This can be expressed in terms of the real and imaginary parts of $G(\omega)$ and calculated numerically. Figure 6b shows how the Hilbert transform can be obtained via time and frequency domain procedures. Details of these can be found in references 10, 11. Figures 7a and 7b show how the Hilbert transform detects non-linearity, resulting in the condition for a non-linear system $H(\omega) \neq G(\omega)$.

FREQUENCY DOMAIN – ARRAY CONVOLUTION

$$H_R(\omega_j) = \frac{-2}{\pi} \sum_{\substack{k=1 \\ k \neq j}}^{n} \frac{Im[G(\omega_k)]\, \omega_k\, \Delta\omega}{\omega_k^2 - \omega_j^2}$$

$$H_I(\omega_j) = \frac{2\,\omega_j}{\pi} \sum_{\substack{k=1 \\ k \neq j}}^{n} \frac{Re[G(\omega_k)]\, \Delta\omega}{\omega_k^2 - \omega_j^2}$$

TIME DOMAIN – ARRAY PRODUCT AND FFT

$$\text{g odd }(t)$$

$$H_R(\omega) = \mathscr{F}\left[\mathscr{F}^{-1}\{\,Im[G(\omega)]\,\} \cdot sgn\,(t)\right]$$

$$\text{g even }(t)$$

$$H_I(\omega) = \mathscr{F}\left[\mathscr{F}^{-1}\{\,Re[G(\omega)]\,\} \cdot sgn\,(t)\right]$$

Figure 6b: Realisation of the Hilbert Transform.

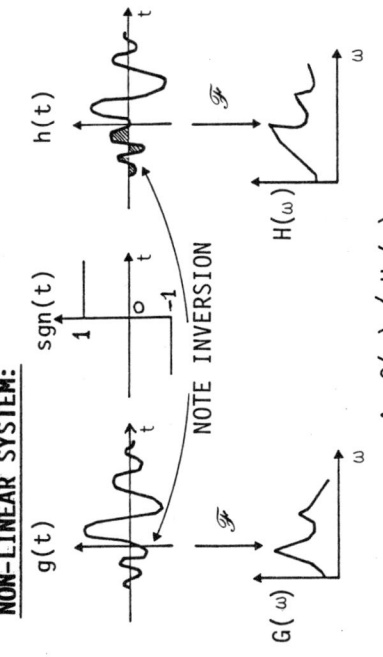
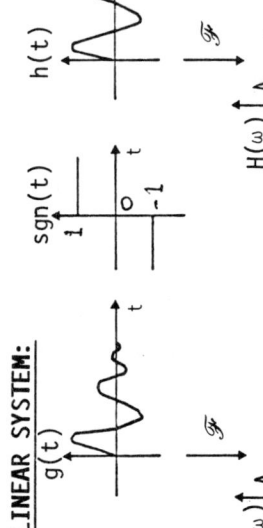
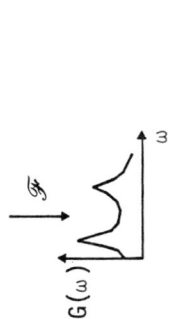

Figure 6a: Graphical explanation of the Hilbert Transform Procedure.

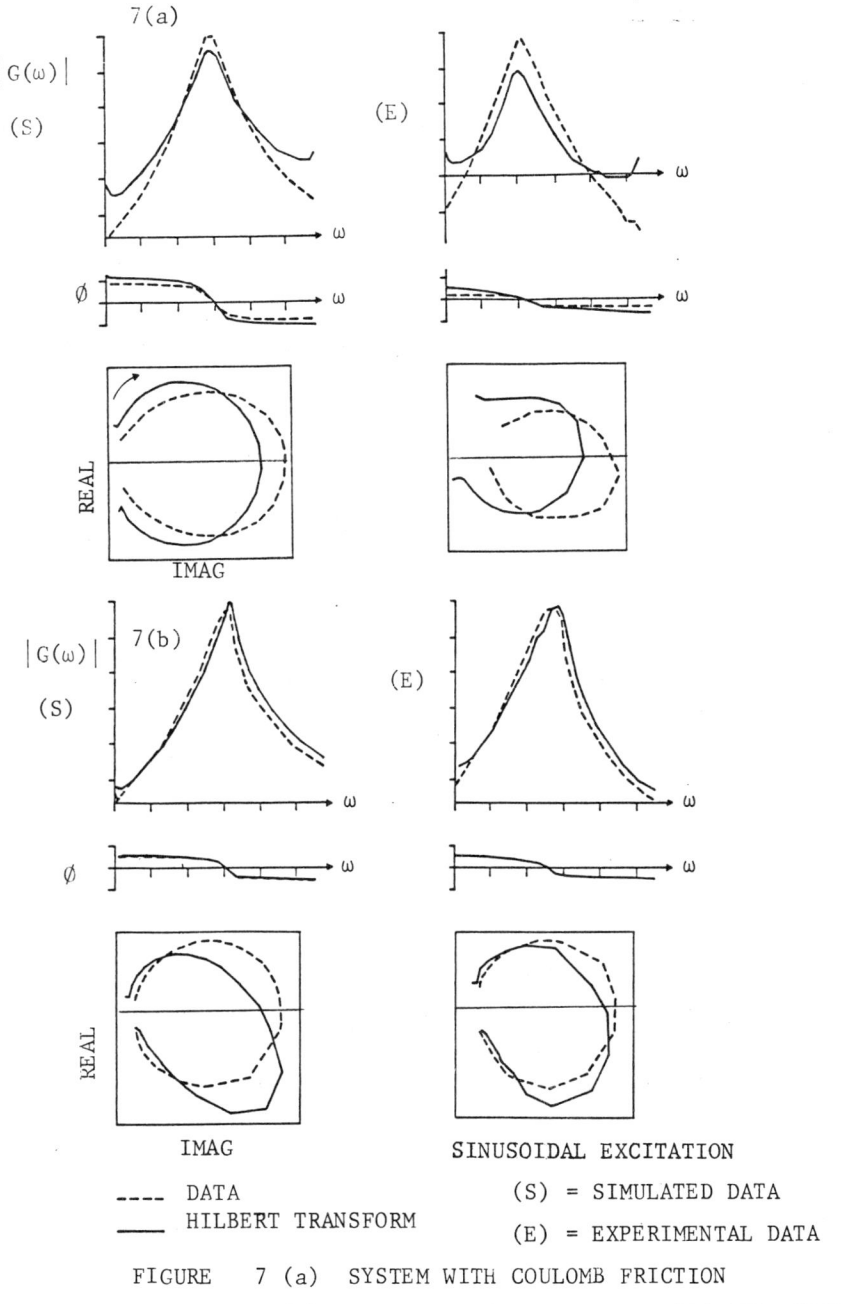

FIGURE   7 (a)   SYSTEM WITH COULOMB FRICTION
          7 (b)   SYSTEM WITH HARDENING CUBIC STIFFNESS

1.6   The use of power spectra

The principal techniques currently employed, and in the process of development, for the detection and identification of nonlinearity when random excitation is used in conjunction with power spectra are discussed in this section.

Optimum Frequency Response Function (FRF)

It has been shown[12] that in the presence of nonlinearity which affects only the output response of a SISO system, the optimum method of obtaining a linearised FRF (linearised in the sense that it corresponds to the best least-squares model of the linear system for a given rms excitation level) is the classical $H_1(\omega)$ function,

$$H_1(\omega) \;=\; \frac{\overline{S}yx}{\overline{S}xx} \;=\; \frac{Y(\omega)X(\omega)}{X^*(\omega)X(\omega)} \tag{19}$$

$\overline{S}yx$ = averaged cross spectrum of the output and input

$\overline{S}xx$ = averaged auto power spectrum of the input

*   = conjugate

In equation (19) the input power spectrum may be polluted by non-linearity arising from the vibration exciter which provides a feedback path.   This can be minimised by modifying equation (19) to,[12]

$$H_3(\omega) \;=\; \frac{\overline{S}yv}{\overline{S}xv} \;=\; \frac{\overline{S}yv}{\overline{S}vv} \Big/ \frac{\overline{S}xv}{\overline{S}vv} \tag{20}$$

where $\overline{S}yv$ = averaged cross spectrum between the output and voltage to the oscillator

$\overline{S}xv$ = averaged cross spectrum between the force input to the struc-

ture and the voltage to the oscillator

$\overline{S}vv$ = auto power spectrum of the voltage to the oscillator

Figure 8 shows the result of applying $H_1(\omega)$ at two very different excita-

tion levels. It can be seen that the FRF's can be curve-fitted success-

fully using a standard algorithm. Characterisation of nonlinearity pro-

ceeds with monitoring the variation of the extracted modal parameters

with rms excitation level. It should also be noted that the use of the

coherence function is not always a good detector of nonlinearity. This

is because the inability to separate leakage and nonlinearity is often

difficult and nonlinearities may be phase-coherent when periodic random

is used.

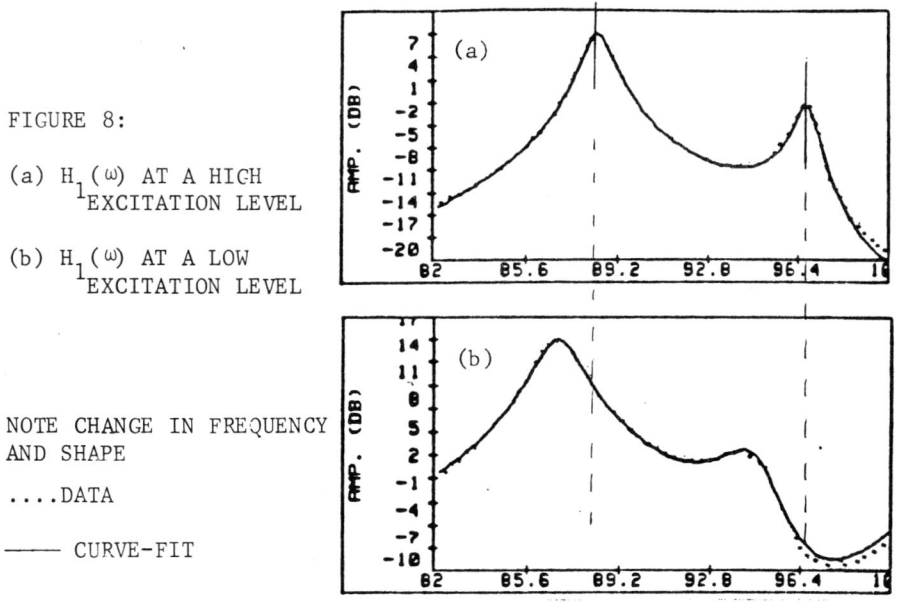

FIGURE 8:

(a) $H_1(\omega)$ AT A HIGH
EXCITATION LEVEL

(b) $H_1(\omega)$ AT A LOW
EXCITATION LEVEL

NOTE CHANGE IN FREQUENCY
AND SHAPE

....DATA

——— CURVE-FIT

The above procedures cannot be used to identify nonlinearity but have the advantage that they generally produce FRF's which are linearised to a given excitation level, thus detection of nonlinearity is possible from a series of tests.

Higher Order Spectra and Time-series Methods

An increasing interest and number of applications is arising in the literature of the use of the Bi-spectrum,[13] which has been shown to be of particular value for problems concerning quadratic type non-linearities when the input is close to Gaussian white noise. This is however, a special case of the Volterra series which is discussed in the next section. The Bi-spectrum is discussed in some detail in references 13, 14, 15, which describe the equations necessary to compute the Bi-spectrum.

Time series methods have been extended from ARMA (Auto Regressive Moving Average) models to NARMAX models[16] (Nonlinear ARMA for Xogeneous). These have been applied prinicpally in to control systems but are currently being considered in relation to general non-linear systems.[17]

These methods appear to have advantages over the functional series (Volterra, Wiener) as they are computationally more efficient, but no application of this approach to structural dynamics problems has been currently reported.

Volterra/Wiener methods have been applied to non-linear structural dynamics problems and these methods are considered in some detail in Part 2.

PART   2.   FUNCTIONAL SERIES ANALYSIS OF SINGLE INPUT, SINGLE

OUTPUT (SISO) NON-LINEAR SYSTEMS

The previous section described several procedures for identifying

and quantifying non-linearity in SISO dynamical systems. However most

of these approaches used either the classical frequency response function,

$$H(\omega) = \frac{Y(\omega)}{X(\omega)} = \frac{\mathscr{F}\{\ y(t)\}}{\mathscr{F}\{\ x(t)\}} \quad \mathscr{F} = \text{Fourier transform} \qquad (21)$$

to characterise the non-linearity by detecting distortion in the response,

or some form of linearisation procedure to describe the behaviour of the

system. Strictly speaking, equation (21) when applied to   linear systems

is based upon the convolution integral, which utilises the principle

$$y(t) = \int_0^\infty h(t)x(t-\tau)d\tau \qquad (22)$$

of superposition to state that the output $y(t)$ is the sum of the systems

individual impluse responses arising from an input $x(t)$ expressed as a sum

of finite impulses.

If we extend this formulation to a non-linear system we begin by

considering the response to two identical impulses applied at different

times t_1 and t_2, here $t_2 > t_1$. These can be expressed as the inputs

$$x_1(t) = a\delta(t-t_1) \tag{23}$$

$$x_2(t) = a\delta(t-t_2), \quad \delta = \text{Dirac delta function}$$

These inputs elicit responses $y_1(t)$ and $y_2(t)$ respectively, which, if the system is time invariant results in,

$$y_1(t) = y_2(t) \tag{24}$$

but with $y_2(t)$ occurring at (t_2-t_1) seconds later in time.

Consider the input excitation as the sum of the two impulses,

$$x(t) = x_1(t) + x_2(t) \tag{25}$$

This produces a response $y(t)$ which in general is,

$$y(t) \neq y_1(t) + y_2(t) \tag{26}$$

However, $y(t)$ only deviates from the sum of $y_1(t)$ and $y_2(t)$ when $t > t_2$. Thus we could express the output as,

$$y(t) = y_1(t) + y_2(t) + f_2(t-t_1,t-t_2)$$
$$\text{with } f_2(t-t_1,t-t_2) = 0 \; ; \; t_1 > t < t_2 \tag{27}$$

where $f(t-t_1, t-t_2)$ represents a correction term for the difference between $y_1(t) + y_2(t)$ to the impulses at t_1 and t_2 respectively.

If we now express a continuous input signal as a summation of impulses i.e.

$$x(t) = \sum_{-\infty}^{\infty} x(\tau) \delta(t-\tau) \Delta\tau \tag{28}$$

then the response of the non-linear system at any time can be considered as the superimposed responses due to each impulse plus the correction terms. Thus we can write,

$$
\begin{aligned}
y_1(t) &= \Sigma\ f_1\ (t-\tau) \\
y_2(t) &= \Sigma\ f_2\ (t-\tau_1,\ t-\tau_2) \\
y_3(t) &= \Sigma\ f_3\ (t-\tau_1,\ t-\tau_2,\ t-\tau_3)
\end{aligned}
\tag{29}
$$

and,

$$y(t) = y_1(t) + y_2(t) + \text{---}\ y_n(t) \tag{30}$$

$$y(t) = \sum_{i=1}^{\infty} y_i(t) \tag{31}$$

Equation (31) can be written as a natural extension of the linear convolution integral,

$$y(t) = \int_{-\infty}^{\infty} h(\tau) x(t-\tau) d\tau$$

$$+ \iint_{-\infty}^{\infty} h_2(\tau_1,\tau_2) x(t-\tau_1) x(t-\tau_2) d\tau_1 d\tau_2 \text{ ----} \tag{32}$$

i.e. $y_n(t) = \underset{\text{n times}}{\int_{-\infty}^{\infty} ,,, \int_{-\infty}^{\infty}} h_n\ (\tau_1, \ \dots\ \tau_n)\ \prod_{i=1}^{n} x(t-\tau_i)\ d\tau_i \tag{33}$

Equation (33) is a functional series expansion and was first credited

to Volterra. As such it is frequently referred to as the Volterra series,[18]

the quantities $h_1(\tau)$, $h_2(\tau_1,\tau_2)$ being referred to as the kernels;

the first order kernel $h_1(\tau)$, representing the manner in which the past

values of the input affect the present value of the response. The nth

order kernel represents the interaction between the 'n' past inputs and

the effect that this interaction has on the system response. The kernels

are zero for any of their arguments being less than zero (i.e. negative

time) since physical systems (linear or non-linear) must obey causality.

Equation (33) can be expressed in the frequency domain by taking

the n dimensional Fourier transforms to give,

$$Y_n(f_1, \ldots f_n) = H_n(f_1,\ldots f_n) \prod_{i=1}^{n} x(f_i) \qquad (34)$$

$$H_n(f_1, \ldots f_n) = \frac{Y_n(f_1 \ldots f_n)}{\prod_{i=1}^{n} x(f_i)} \qquad (35)$$

It should be realised that the nth order frequency response function does

not exclusively describe the system behaviour as the effect of kernels

higher than the nth may also be important.

2.1 Determining the kernels in the functional series

In order to obtain the kernels in the Volterra series as described

above it is necessary to use a series of controlled impulses at times

$t_1, t_2 \ldots t_n$. This has been successfully achieved by Vinh and Chouchai[19]

using two impulses to identify the second order transfer function.

However, there are limitations with this approach since in the general

case it is impractical to consider impulse sets greater than say three

as the control instrumentation requirements become significant. This

implies that it is desirable to measure the kernels individually. This

is possible by using an alternative functional series whose terms are

orthogonal under the condition of a white noise input. A series which

satisfies these conditions is the Wiener series[18]

The Wiener series, relating the output to a Gaussian white noise

input $x(t)$ is,

$$y(t) = ko + \int_0^\infty k_1(\tau)x(t-\tau)d\tau + \{\iint_0^\infty k_2(\tau_1,\tau_2)x(t-\tau_1)x(t-\tau_2)d\tau_1 d\tau_2$$

$$- A \int_0^\infty k_2(\tau_1,\tau_2) d\tau_1\} + \dots \tag{36}$$

i.e. $y_n(t) = k_o + \underset{n \text{ times}}{\int_0^\infty \dots \int} k_n(\tau_1,\dots\tau_n) \prod_{i=1}^{n} x(t-\tau_i)d\tau_i$

$$-(n-1) A \underset{(n-1) \text{ times}}{\int_0^\infty \dots \int} k_n(\tau_1, \dots\tau_n) \prod_{i=1}^{n-1} x(t-\tau_i)d\tau_i \tag{37}$$

where A is the amplitude of the power spectral density of the white noise

input signal.

Here we find the principal difference between the Volterra and

Wiener methods. The Wiener kernels of order higher than the first are

dependent upon the type and level of the input signal whereas the Volterra

kernels are not. In addition, the first order Wiener kernel is, in general, different from the linear response whereas the first order

Volterra kernel represents the linear response. It is also worth noting

that when the system kernels are zero for orders higher than the second,

the Wiener kernels $k_1(\tau)$ and $k_2(\tau_1,\tau_2)$ are the same as the Volterra

kernels $h_1(\tau)$ and $h_2(\tau_1,\tau_2)$.

2.2 Measurement of the first and second order Wiener kernels for

quadratic non-linear systems using cross-correlation.

Lee and Schetzen[18] showed that the Wiener kernels can be obtained

directly using correlation methods. The basic ideas relating to this

will be presented below, reference to 20 gives more complete details.

The nth order Wiener functional is constructed such that it is ortho-

gonal to the n-1 functional,

$$E\left[G_n(k_n \ ; \ x(t))\right] = 0 \ ; \ n > 1 \tag{38}$$

where $E\left[-\right]$ denotes the expected value (or time averaged value) and

$$G_0(k_0 \ ; \ x(t)) = ko, \tag{39}$$

$$G_1(k_1;x(t)) = \int_0^\infty k_1(\tau) \ x \ (t-\tau)d\tau \ etc. \tag{40}$$

$x(t)$ is taken to be a Gaussian white noise zero mean input signal.

Thus the zero order kernel is simply the average value of the out-

put i.e.

$$E[y(t)] = k_o \qquad (41)$$

Determination of the first order kernel is obtained by multiplying both sides of equation (37) by $x(t- \tau_1)$ and taking the expected value. Due to the orthogonality condition all functionals of order higher than one are zero. Also the expected value of the product of the zero order kernel with the time delay is zero i.e.

$$E[k_o \cdot x(t-\tau_1)] = 0, \qquad (42)$$

since $x(t)$ is a signal with zero mean. Thus we find that after multiplying both sides by $x(t- \tau_1)$ and taking the expected values we have,

$$E[y(t) \times (t - \tau)] = \int_0^\infty k_1 (\tau) E[x(t-\tau)x(t-\tau_1)] d\tau \qquad (43)$$

We recognise that the left hand side of equation (43) is the classical cross-correlation function $Ryx(\tau)$. Further, the expected value under the integral sign can be recognised as the auto-correlation function $Rxx (\tau-\tau_1)$.

If we introduce the delta function then we can write,

$$Rxx(\tau-\tau_1) = A \delta(\tau-\tau_1) \qquad (44)$$

thus equation (43) becomes

$$E\left[y(t)x(t-\tau)\right] \;=\; \int_0^\infty k_1(\tau) \; A\delta(\tau-\tau_1) \; d\tau \tag{45}$$

i.e. $Ryx(\tau) = A\,k_1(\tau)$

$$\therefore \; k_1(\tau) = \frac{Ryx(\tau)}{A} = \frac{1}{A}\, E\left[y(t)x(t-\tau)\right] \tag{46}$$

For the second and higher order kernels, additional conditions must be satisfied since the necessity to introduce multi-order delays i.e. $(t-\tau_1)(t-\tau_2)$ --- results in the lower order kernels offering a contribution. This can be seen in the derivation of the second order kernel $k_2(\tau_1,\tau_2)$.

Multiplying both sides of equation [37] by $x(t-\tau_1)\, x\,(t-\tau_2)$ results in all the higher order kernels above two being zero. However, the zero order kernel has an effect as follows;

$$E\left[k_o x(t-\tau_1) \; x \; (t-\tau_2)\right] = A k_o \; \delta(\tau_1-\tau_2) \tag{47}$$

and the complete result is[20],

$$E\left[y(t)x(t-\tau_1)x(t-\tau_2)\right] = A k_o \; \delta(\tau_1-\tau_2) + 2A^2 K_2 \; (\tau_1,\tau_2) \tag{48}$$

In order to remove the effect of the first term on the right-hand side of equation [48], the mean value (k_o) is subtracted from the output time response before the second order correlation is performed, i.e.

$$yo(t) = y(t) - ko = y(t) - E\left[y(t)\right] \tag{49}$$

Thus the second order kernel is obtained from,

$$k_2 \ (\tau_1, \tau_2) \ = \ \frac{1}{2A^2} \ E\left[\ y_o(t) \quad x(t-\tau_1)x(t-\tau_2)\right] \qquad (50)$$

This can be generalised to

$$k_n \ (\tau_1, \ \dots \tau_n) \ = \ \frac{1}{n!A^n} \ E\left[(y(t)-\sum_{m=0}^{m-1} G_m(t))x(t-\tau_1)\dots x(t-\tau_m)\right] \quad (51)$$

and the practical procedure is shown graphically in the figure below.

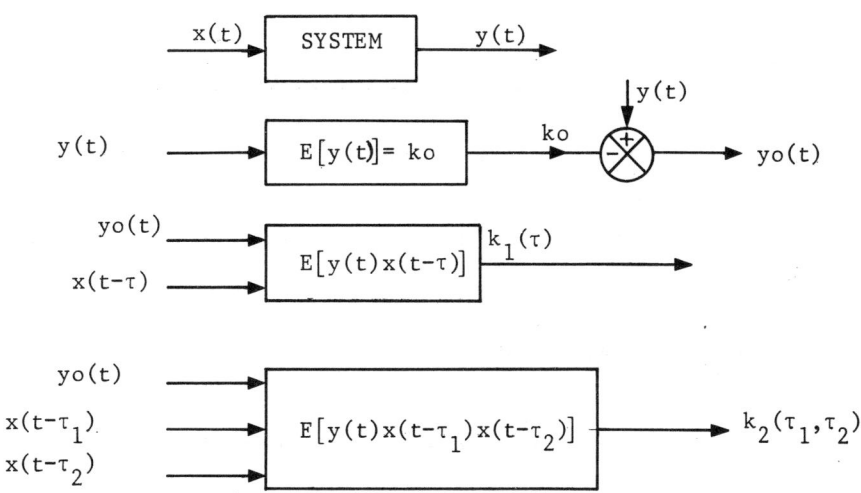

2.3   Application to dynamical systems incorporating non-linearities

The cross-correlation methods described above are applied to the extraction of the first and second order Wiener kernels of the system shown in block diagram form in the figure below.

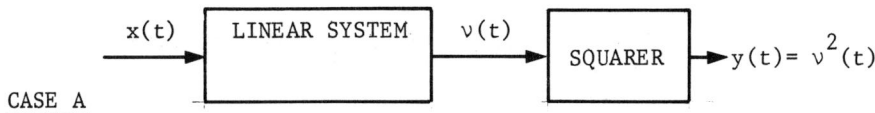

CASE A

CASE B

The dynamics of the linear system are described by the equation,

$$\ddot{y} + 2\zeta\omega_n\dot{y} + \omega_n^2 y = x(t) \qquad\qquad (52)$$

    where ω_n = natural undamped frequency   = 100 rad/s

          ζ = critical damping ratio   = 0.1

        $x(t)$ = band limited, zero mean Gaussian signal whose cut-
                       off frequency is much higher than the system natural
                       frequency.

CASE A :   The output of the system is represented as,

$$y(t) = v^2(t) \tag{53}$$

In this case the Weiner kernels are obtained as follows,

$$v(t) = \int_0^\infty h(\tau) \, x(y-\tau) d\tau \tag{54}$$

$$\therefore \quad y(t) = \int_0^\infty h(\tau_1)h(\tau_2)x(t-\tau_1)x(t-\tau_2) \, d\tau_1 d\tau_2 \tag{55}$$

By inspection the kernels are,

$$
\left.
\begin{aligned}
&k_1(\tau) = 0 \\[4pt]
&k_2(\tau_1,\tau_2) = h(\tau_1)h(\tau_2) \\[4pt]
&k_n(\tau_1,\ldots\tau_n) = 0 \; n > 3
\end{aligned}
\right\} \tag{56}
$$

Equation (52) was simulated on an analogue computer and the output was fed into a squarer. The input $x(t)$ and the output $y(t)$ were simultaneously sampled and stored in time blocks of 512 points. The mean value of the output was removed from each block of data and these data were then used to perform the second order cross-correlation according to equation (56). For the analysis, τ_1 = NΔt and τ_2 = MΔt, where Δt = 5 ms and N,M = 1 → 30. The sampling period was chosen to ensure that the Nyquist frequency criterion was satisfied.

Figures   9a   and   9b   show the results obtained from this test whereby the second order kernel $k_2(\tau_1,\tau_2)$ is plotted as a function of τ_1 and τ_2. For Figure   9a   a total of 2000 averages were used. For Figure 9b   this was increased to 8000 averages. According to equation (56), the exact value of $k_2(\tau_1,\tau_2)$ is simply the product of the linear impulse response functions at τ_1,τ_2. Thus on the leading diagonal we should find,

$$k_2(\tau,\tau) = h^2(\tau) \tag{57}$$

Figure   9c   shows the exact second order kernel $k_2(\tau_1,\tau_2)$ and it can be seen that the results compare very closely with those obtained using cross-correlation.

CASE B : Linear System with quadratic   cubic stiffness non-linearity.

The governing equation of the system is,

$$\ddot{y} + \zeta\omega_n\dot{y} + \omega_n^2 y + ay^2 + by^3 = x(t) \tag{58}$$

In this case the response y(t) will include the first, second and higher order kernels. Equation (58) was simulated on a digital computer with $a = 10^7$ and $b = 5 \times 10^9$ using a Runge-Kutta-Merson 4th order routine with a white noise input signal generated by passing the output of a random number generator through an eigth order Butterworth filter, re-sulting in a band limited Gaussian white noise signal with a cut-off

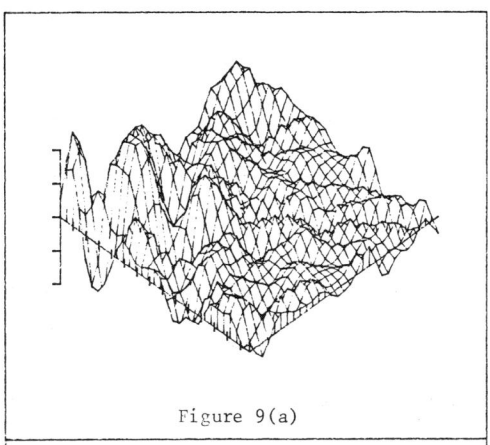

Figure 9(a)

Second Order Wiener Kernel: Linear System Output Squared
4 data blocks averaged

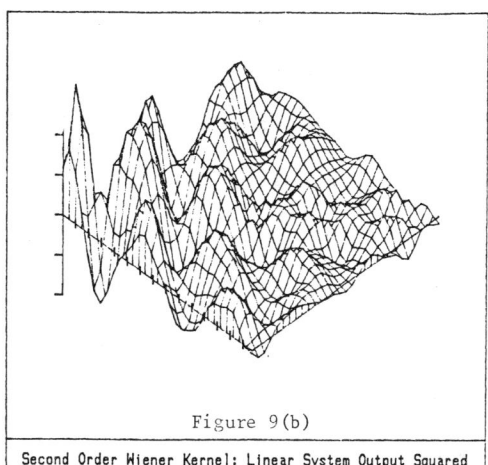

Figure 9(b)

Second Order Wiener Kernel: Linear System Output Squared
16 data blocks averaged

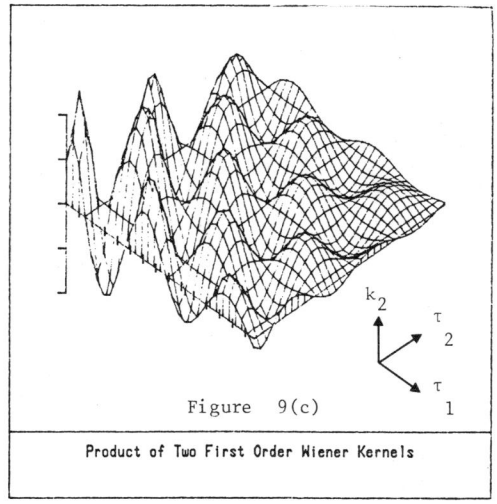

Figure   9(c)

Product of Two First Order Wiener Kernels

FIGURE 9 :   SECOND ORDER KERNELS FOR THE SQUARE

LAW SYSTEM.

frequency of 200 Hz, this being twelve times greater than the natural
frequency of the model and sufficiently rich in harmonic content to excite
the non-linear characteristics. Figure 10 shows the simple frequency response function and coherence for the linear (a,b =0 in equation (58)) and
non-linear models. Equations (46) and (50) were used to generate the
first and second order Wiener kernels by sampling the input and output
of the simulated system. The kernels were then used to predict the actual
output of the non-linear system. Figure 11 shows the actual and predicted
outputs using the first order kernel only and the first and second order
kernels.

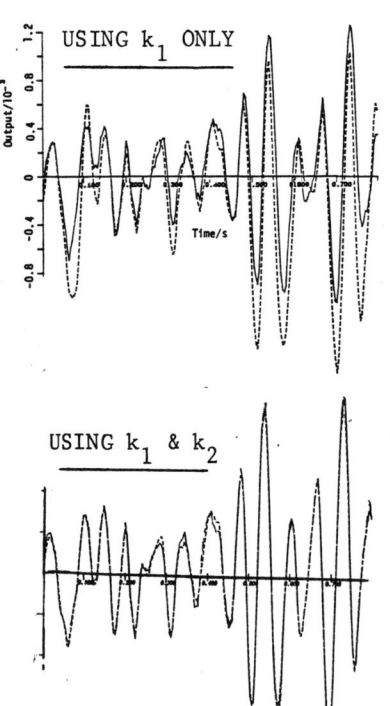

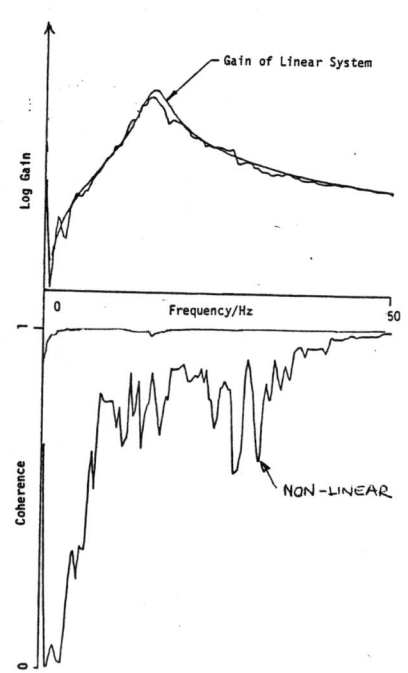

FIGURE 10: FRF AND COHERENCE FOR
LINEAR AND NON-LINEAR
SYSTEM.

FIGURE 11: ACTUAL (----) AND
PREDICTED (———)
OUTPUTS.

The need to extend the analysis to a third order kernel is not necessary

in this case since the combination of the first and second order kernels

is sufficiently accurate. This becomes obvious when one examines the

output of the system to a sinusoidal input whose amplitude produces the

same excursions as the actual input. This is demonstrated by Figure 12

which shows that only the second harmonic is significant for the excita-

tion level employed. Figure 12 was generated by using a fixed amplitude

sine wave whose frequency varied from 0 to 25 Hz.

FIGURE 12:   RESPONSE OF THE
             NON-LINEAR
             SYSTEM TO A
             HARMONIC PROBING
             SIGNAL.

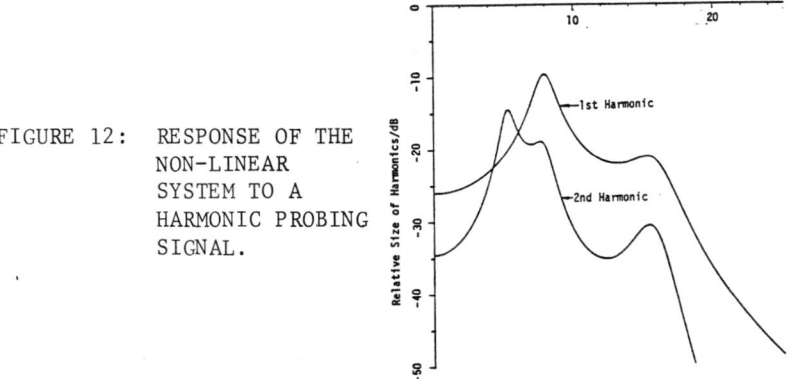

This 'harmonic probing' procedure is a practical way of estimating

the number of Wiener (or Volterra) kernels necessary to define a non-

linear system, assuming that the probing signal produces the maximum

excursions that the non-linear system is likely to experience.

Volterra Representation

The Volterra kernels of equation (58) can also be calculated

using the harmonic probing method.[21] By setting the input to $e^{j\omega t}$ and

equating coefficients of $e^{j\omega t}$, the first order Volterra kernel is found

to be,

$$H_1(j\omega) = \frac{1}{10^4 - \omega^2 + j20\omega} \tag{59}$$

Likewise, putting $x(t) = e^{j(\omega_1 + \omega_2)t}$ and equating coefficients of $e^{j(\omega_1 + \omega_2)t}$ gives the second order frequency response function $H_2(\omega_1, \omega_2)$

i.e.,

$$H_2(\omega_1, \omega_2) = \frac{-a\, H_1(\omega_1)\, H_1(\omega_2)}{\omega^2 - (\omega_1 + \omega_2)^2 + j20(\omega_1 + \omega_2)} \tag{60}$$

Figure 13 shows the modulus of $H_2(\omega_1, \omega_2)$. The two peaks along the leading diagonal where $\omega_1 = \omega_2$ correspond to resonances at $\omega_n/2$ and ω_n. This means that the second order non-linearity will be a maximum at excitation frequencies $\omega = \omega_{1,2/\omega_n}$ and $\omega = \omega_{1,2/2\omega_n}$. The actual kernels can be obtained by taking the nth order inverse Fourier transform of the nth order frequency response function.

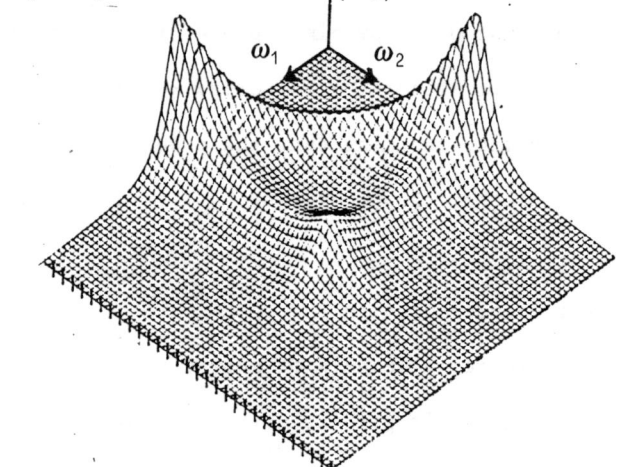

FIGURE 13:

MODULUS OF THE
SECOND ORDER FRF

One of the interesting aspects of the higher order frequency response functions is that they become increasingly complex and smaller in magnitude with increasing spectral number. For example, in the above problem the peak of the $H_2(\omega,\omega)$ function is found to be approximately six times smaller than the peak value of $H_1(\omega)$ and twenty times smaller than $H_3(\omega)$.

The consequences of this are that accurate detection of the higher order spectra using correlation methods may prove to be difficult in practical testing. However, it has been found that by considering only the regions of the kernels close to and inclusive of the leading diagonal, sufficient accuracy can be obtained. The importance of this lies in the fact that a significant computational improvement is gained which makes the above procedures a feasible approach to the detection and prediction of non-linear systems.

References

1. Frachebourg, A. and Gygax, P.E., Comparison of excitation signals: sensitivity to non-linearity and ability to linearise dynamic behaviour; Proc. 10th Int. Sem. on Modal Analysis, K.U. Leuven, Belgium, Pt V, 1985.

2. Tomlinson, G.R., Identification of the dynamics characteristics of a structure with Coulomb friction; Jnl. of Sound and Vibn., 64 (2), 233-242, 1979.

3. Rades, M., Identification of the dynamic characteristics of a simple system with quadratic damping; Serie de Mécanique Appliquée, 28 (4), 439-446, 1983.

4. Breitbach, E., Identification of non-linear systems in: Identification of vibrating structures, CISM courses and lectures No. 272, Springer-Verlag, 1982.

5. Nayfeh, A.H., and Mook, D.T., Nonlinear Oscillations, Wiley Intersciences, 1979.

6. Ewins, D.J., Model Testing - Theory and Practise, Research Studies Press, 1984.

7. Mertens, M. et al., Basic rules of a reliable detection method for non-linear dynamic behaviour, Proc. 10th Int. Sem. on Modal Analysis, K.U. Leuven, Pt IV, 1985.

8. Kirk, N.E., "Modal analysis of non-linear structures employing the Hilbert transform", Ph.D. Thesis, University of Manchester 1985, and in, "Vibn. analysis and identification of non-linear structures", G.R. Tomlinson, Manchester, 1985.

9. Bendat, J.S., The Hilbert transform and applications to correlation measurements, Bruel and Kjaer Publication, 1985.

10. Tomlinson, G.R., Hilbert transform procedures for identifying and quantifying non-linearity in modal testing, To be pusblished in Jnl. of Mechanical Systems and Signal Processing, Vol. 2, 1987.

11. Vinh, T. and Tomlinson, G.R., Vibration Analysis and Identification of Non-linear structures, Short course notes, University of Manchester, England 1985.

12. Goyder, H.G.D., Vibration Analysis and Identification of Non-linear structures, Short course notes, University of Manchester, England, 1985.

13. Linfoot, B.T. and Hall, S.M., Analysis of the motions of scale-model
     sea-cage systems, IFAC Symposium: Aquaculture, Trondheim, Norway,
     1986.

14. Choi, D.W. et al., Application of digital cross-bispectral analy-
     sis to model the non-linear response of a moored vessel in random
     seas, Jnl. of Sound and Vibration, 99 (3), 309-326, 1985.

15. Bendat, J.S., Statistical errors for non-linear system measurements
     involving square-law operations, Jnl of Sound and Vibration, 90 (2),
     275-282, 1983.

16. Billings, S.A., and Voon, W.S.F. Least squares parameter estima-
     tion algorithms for non-linear systems, Int. Jnl. Systems. Sci.,
     15 (6), 601-615, 1984.

17. Billings, S.A. and Tsan, K.A., Estimating Higher Order Spectra, Proc.
     V IMAC, Imperial College, London, 1987.

18. Schetzen, M., The Volterra and Wiener Theories of non-linear systems,
     John Wiley and Sons, New York, 1980.

19. Chouchai, T., Comportement Dynamique Non-Lineaire des Structures,
     Ph.D. Thesis, Institut Supérieur des Materiaux et de la Construction
     Mécanique, Paris, 1986.

20. Marmarellis, P.Z. and Marmarellis, V.Z., Analysis of physiological
     systems, John Wiley Interscience, New York, 1980.

21. Bedrossan, E. and Rice, S.O., The output properties of Volterra
     systems driven by harmonic and Gaussian inputs, Proc. IEEE, 59,
     1688-1707, 1971.

NUMERICAL ACOUSTIC RADIATION MODELS

P. Sas
Katholieke Universiteit Leuven, Belgium

INTRODUCTION

A phenomenon difficult to identify and quantify is the complex relation betweeen structural vibrations and the sound caused by those vibrations (structure borne sound). This relation is important since a large percentage of the noise pollution is of structure borne nature. Whenever a mechanical structure producing vibrations is conceived, measures should be taken to reduce the sound inherent to those vibrations. To do this efficiently the sound radiation mechanism must be first identified and quantified in appropriate models. The theoretical base for such a model is known since more than a century, and is given by the solution of the 3d wave equation, with the surface velocity of the vibrating structure as boundary condition. Several theoretical studies have been devoted to this subject in the past and have resulted in analytical formulas which are only valid for a limited number of relative simple source geometries such as axisymmetric structures (spheres, cylinders) or flat plates. Analytical solutions for the general radiation problem do not exist, numerical solutions are conceivable but rather complex due to numerical instability problems, and hence require considerable computer power.

This paper deals with the implementation and use of a numerical radiation model based on a solution of the Helmholtz integral equation, the model has been integrated into a global sound optimisation philosophy by combining finite element modeling (FEM), modal analysis and the mentioned radiation model. Hereby the vibration patterns generated by the FE model or the experimental modal analysis identification serve as input to the radiation model. Linking FEM results to a radiation model yields an optimal tool for judging a new design with regard to its sound radiation, provided that a dynamic finite element model of the concerned design is available. Linking data from experimental modal analysis identification to a radiation model creates on the other hand the possibility to predict the impact of local structural modifications on the sound radiated by mode shapes at resonance frequencies, which is important for trouble shooting applications or for prototype optimisation. Applying this method, on a real life structure such as a combustion engine the influence of changing locally the stiffness, mass or damping on parameters such as the radiation efficiency or the radiated sound power can be predicted.

THEORY

The general acoustic radiation problem can be formulated as a boundary

value problem (cfr.fig.1): find the solution for the pressure in an

infinite medium (V), with known density and speed of sound, where an

arbitrarily shaped object is immersed.   As boundary condition the normal

component of the velocity at any point on the surface (S₁) of the

immersed object is given. It can be shown that a solution of this

boundary value problem must satisfy the three dimensional wave equation

as well as the Sommerfeld boundary condition. Exact solutions of this

boundary value problem are possible when the surface represents a level

value of a coordinate in one of the few coordinate systems in which the

wave equation can be separated.[1 2 3 4]

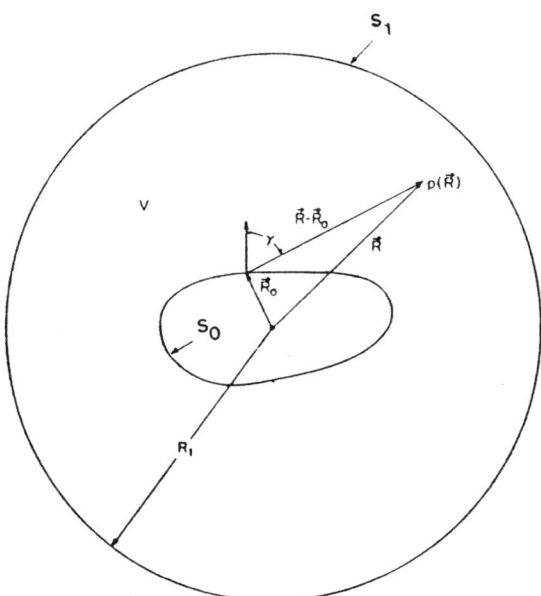

Fig.1 : General field lay-out

The most general of these exact solutions are probably those for axisymmetric surfaces, where the radial factor of the pressure is a combination of spherical Bessel and Neumann functions while the tangential factor of the pressure is expanded in a series of Legendre functions whose coefficients are determined by the boundary conditions. One of the problems inherent to such exact solution is that the boundary conditions must also be expressed in terms of a series of Legendre functions.

An interesting mathematical property is the fact that, for the given boundary condition, the 3D wave equation can be expressed as an integral equation. This implies the introduction of the free-space Green function and the combination of the 3D wave equation with the Gauss-integral theorem which relates a surface to a volume integral. Doing so the pressure field at an arbitrary field point can be formulated as the surface integral of a linear combination of the surface pressure and velocity over the radiating boundary. This integral equation is known as the Helmholtz integral equation.

$$p(\bar{R}) = \int_{S_o} \left[p(\bar{R}_o) \frac{\delta}{\delta n} g(\bar{R}/\bar{R}_o) + j\omega\rho v(\bar{R}_o) g(\bar{R}/\bar{R}_o) \right] dS_o \qquad (1)$$

where $g(\bar{R}/\bar{R}_o) = e^{jkr}/r$ (free space Green function)
$r = /\bar{R} - \bar{R}_o/ \qquad \omega = 2\pi f$
$v(\bar{R}_o) = $ structural velocity $\qquad k = $ wavenumber $(\frac{2\pi f}{c})$.

Since the only prescribed quantity is the surface velocity, the surface pressure being unknown, the pressure in the field can only be determined by allowing the field point $(\bar{R})$ to approach the radiating surface and by

consequently solving the resulting integral equation (Fredhom integral

equation of the second kind) for the unknown surface pressure $p(\bar{R}_0)$. For

arbitrary source configurations, where the solution is not restricted to

a particular frequency range, the surface pressure distribution can only

be obtained by solving the Fredholm integral equation numerically. Once

the surface pressure is obtained, the Helmoltz integral equation becomes

a simple integral representation of the pressure at field points not

located on the radiating surface, which on his turn can be solved

numerically.

The integral equation can be circumvented if a Green function which

satisfies the Neumann boundary conditions can be constructed. In this

case the unknown pressure vanishes in the surface integral, thus reducing

the integral equation to a simple integral representation. Unfortunately

a Green function satisfying the Neumann boundary conditions can only be

constructed if the boundary is completely defined by the value of a

single coordinate, and if the wave equation is separable in this

coordinate system.

$$p(\bar{R}) = - \frac{j\omega\rho}{2\pi} \int_{S_o} \frac{e^{jkr}}{r} v(\bar{R}_o) \, dS_o \qquad (2)$$

Similar to the exact solution those conditions are only met for simple

boundaries such as an infinite plane, cylinder or sphere, if the pressure

can be represented by linear conbinations of wave harmonics. The

solution for infinite plane sources deserves special attention since it yields the well-known Rayleigh's integral (eq.2) from which it can be concluded that a planar source located in an infinite baffle is equivalent to a distribution of point sources. This principle forms the theoretical basis of the point source model.

POINT SOURCE RADIATION MODEL BASED ON RAYLEIGHS INTEGRAL

Since the vibration patterns, originating from finite element models or from experimental modal analysis procedures are discrete, they are only determined in a finite number of points distributed over the surface of interest. As a result it will be necessary to approximate the applied surface integral (eq.2) by a summation over those discrete points. This is the equivalent of substituting the vibrating surface by a finite number of point sources. Each of those elementary sources represents a fraction (S_i) of the original surface. The acoustic strength of those sources is given by the product of the partial surface and the structural velocity of the point where the surface was attributed to. If for example the vibration of a plane surface is given by a pattern of n points, the pressure level for an arbitrary point above the radiation surface is equal to :

$$p_i(\bar{R}_i) = - \sum_{i=1}^{n} (jkcv_i \ \frac{1}{2\pi r_i} \ e^{jkri} S_i)$$
$$\text{where } r_i = |\bar{R}_i - \bar{R}_o|.$$

(3)

Similar expressions can be derived for the air particle velocity.

Consequently the acoustic intensity is easy to derive, since its value is given by the product of particle velocity and pressure. The acoustic power radiated by a surface is given by the average of the intensity normal to the surface, multiplied by the size of the surface. Equation 3 is only valid if the acoustic environment is reflection free. The influence of eventual reflecting surfaces close to the radiating surface can therefore be simulated by simply adding image sources at the opposite side of the reflecting surface. The reflection influence is taken into account by adding a supplementary pattern of noise sources to the source lay-out, this supplementary pattern is the image of the original source pattern.

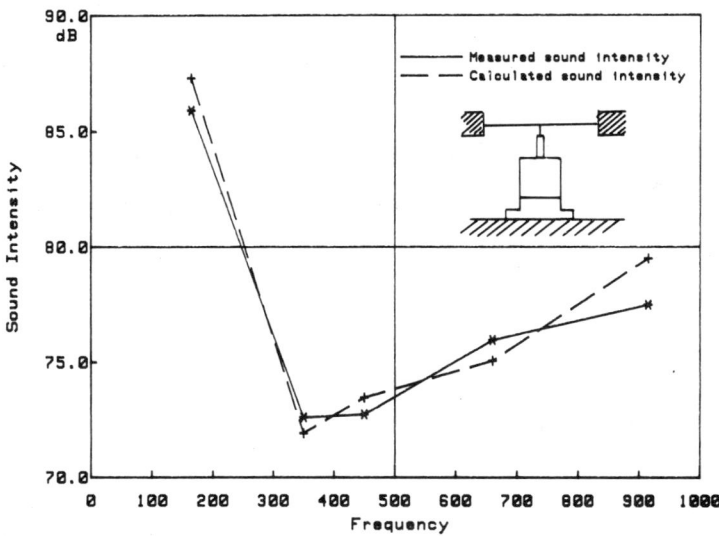

Fig.2 : Predicted and measured sound intensity for a clamped
plate, grid element size (1.5x1.5cm).

The accuracy of the radiation model has been verified by comparing the predictions of the radiation model with experimental results. The result of such a verification experiment is given in fig.2. This diagram shows the average near field sound intensity levels for the first five resonance frequencies of a constrained plate. The plate (50x50x.2cm) was clamped in a concrete block and excited with white noise (0-2000Hz). The near field intensity patterns was measured together with the modal deformation patterns, which served as input for the acoustic radiation model. The intensity measurements were conducted with the aid of the intensity measurement robot using a two microphone intensity probe. As can be noticed from those results, the agreement between experiment and prediction is fairly good. Up to the fourth resonance frequency the accuracy is better than 1dB. The larger error of the fifth resonant frequency is probably due to an insufficient number of source points.

The main advantage of the point source model is its simplicity, it requires up to 100 times less computing time than the general solution treated hereafter, it is however limited to 2D-sources.

NUMERICAL SOLUTION OF THE HELHOLTZ INTEGRAL EQUATION

Searching through literature one will find several papers treating the numerical solution of Rayleighs integral formula (eq.2), but publications treating the numerical solution of the 3D Helmholtz integral equation (eq.3) are rather scarce. Schenck[5] was the first to mention the problems caused by the non-uniqueness of such solutions at frequencies corresponding to the eigenfrequencies of the associated interior problem. To overcome this problem he introduced the Combined Helmholtz Integral Equation Formulation (CHIEF) where an overdetermined system of algebraic equations is obtained by combining the surface Helmholtz integral formulation with additional equations generated from the interior Helmholtz integral formulation. The difficulty of this approach is the determination of the optimal number and position of the interior points which are used to generate the additional equations. Except for a recent paper of Koopman[6] no application of the combined Helmholtz integral equation formulation has been reported so far.

Another method for overcoming the uniqueness problem was introduced by Burton and has been refined by Meyer[8] et al, and later by Filippi[9]. This method is based upon the fact that a unique solution for the acoustic pressure can be obtained by solving a modified integral equation consisting of the original integral equation and its differentiated form. This combination yields a unique solution for all frequency values. Unfortunately, the differentiated form of the integral equation contains a strongly singular integral, which cannot be directly integrated. Burton[7] and Mayer[8] approach this problem by using a transformation to interpret the singular integral. This method is efficient but results in a more complicated integral equation which requires more computer power. For a more detailed review of numerical radiation models we refer to Vandeponseeele[11]. The majority of those publications however do not include an experimental verification of the accuracy of the proposed numerical solution. They limit themselves to a comparison with analytically predicted results for simple configurations such as spheres. In a few cases correlation with measurements is verified but only for simple source geometries.

For our applications the principle of the CHIEF method was retained since it is a reasonable compromise between accuracy, computational effort and versatility, at least for the frequency range where the density of resonance frequencies is low, which for most mechanical noise sources is the frequency range where most noise problems occur. The difference with previous investigations is the requirement for the computational scheme to be compatible with those used in the existing experimental modal analysis (GMAP) and finite element methods.

PRINCIPLE OF THE NUMERICAL SOLUTION

The pressure at a point R on a closed surface (fig.1) is given by the Helmholtz integral equation (eq.1). By taking the derivative of the free space Green function the integral equation can be rewritten as follows :

$$p(\bar{R}_o) = -\frac{k}{2\pi} \int_{S_o} p(\bar{R}_o)\, \frac{e^{jkr}}{r} (\frac{1}{kr} - j) \cos\gamma\; dS_o \qquad (4)$$

$$-\frac{j\rho ck}{2\pi} \int_{S_o} v(\bar{R}_o)\, \frac{e^{jkr}}{r}\, dS_o$$

Equation (4) can be solved by direct numerical integration, but even for simple sources the computations are too time consuming. An approximate solution is possible by dividing the source surface S into N planar surface elements. Planar surfaces are far more interesting since integrations over two dimensional surfaces are more tractable. Often closed form expressions can be found, thereby avoiding numerical integration. The number of surface elements is chosen such that one can assume that the velocity and pressure distribution over the elements is uniform. Equation (4) can then be written as :

$$p(R_i) = \sum_{j=1}^{n} [p(R_j) D_{ij} + v(R_j) M_{ij}] \qquad (5)$$

where $p(R)$: pressure at the center of element i

$$D_{ij} = -\frac{k}{2\pi} \int_S \frac{e^{jkr_{ij}}}{r_{ij}} \left(\frac{1}{k\,r_{ij}} - j \right) \cos\gamma_{ij}\, dS$$

$$M_{ij} = -j\rho c \frac{k}{2\pi} \int_S \frac{e^{jkr_{ij}}}{r_{ij}}\, dS$$

The dipole coefficient D_{ij} relates the contribution of the pressure at element j to the pressure at element i. Similarly, the monopole coefficient M_{ij} relates the contribution of the normal velocity of element j to the pressure at element i. If equation 5 is repeated for each centerpoint of all N elements, one obtains a NxN system of linear inhomogeneous equations.

$$[[\delta_{ij}] - [D_{ij}]] [P_j] = [M_{ij}] [V_{ij}]^T \qquad (6)$$

This NxN matrix system can be solved for the element pressures by one the many appropriate methods. We applied the Householder method since it handles also overdetermined systems. Once the pressure on the source surface known, it is easy to derive the radiated acoustic power. By definition the radiated acoustic power is given by :

$$W = \text{Re} \int_S pu_r \, dS \qquad\qquad (7)$$

where   W : Acoustic power
               u_r : Particle velocity

Since on the radiating surface the particle velocity is equal to the
surface velocity, the radiated power can be calculated once the surface
pressure distribution is known. The integration algorithm to calculate
the D_{ij} and M_{ij} terms of eq.5 is a specially developed analytical
sectorial integration algorithm which is far more accurate and
numerically stable than the usually applied algorithms (cfr.
Vandeponseele [12]).

OVERDETERMINATION OF THE SYSTEM OF EQUATIONS

    It is well known that the Helmholtz integral equation system will
fail to yield the unique solution of the acoustic radiation problem at
certain characteristic frequencies. Copley[10] and Schenck[5] have shown that
these characteristic frequencies are identical to the eigenfrequencies of
the corresponding homogeneous interior (Dirichlet) problem. Since the
uniqueness problem occurs only at certain frequencies the problem can be
avoided by considering only frequencies which are not close to internal
eigenvalues. This is not feasible since the internal eigenvalues are not
known a priori and since at higher frequencies the eigenvalue density is
so dense that it is almost impossible to avoid the internal eigenvalues.
Moreover since the integral equation is discretized into a system of

algebraic equations, there is no longer a specific eigenvalue but a range
of values at which the matrix is ill-conditioned.

Schenck[5] suggested to overcome this non-uniqueness problem by solving
an overdetermined sytem of algebraic equations, obtained by combining the
system of algebraic equations generated from the standard integral
equation (eq.1) with additional equations originating from the internal
Helmholtz integral equation. These additional equations are necessarily
associated with points laying inside the source surface. It can be shown
that the overdetermined system of equations yields always a unique
solution. But in practice the position of those internal points is
important. Indeed if the internal points are chosen near the nodal
points of the interior eigenfunctions, the additional equations fail to
provide the necessary constraints to ensure uniqueness. In general, both
the eigenvalues and nodal lines of the internal problem are unknown so
that one has no guidance in selecting the additional internal points. It
is therefore suggested to use a sufficient number of internal points.

The Householder method which was chosen for solving the matrix
system is well suited for handling overdetermined systems, and was
therefore preferred over other methods. The condition number of the
matrix is used as an indicator, warning the user at which frequencies
singularities occur.

ACCURACY OF THE GENERAL RADIATION MODEL

To verify the accuracy of the radiation model, the acoustic power radiated by a pulsating sphere has been predicted by the model and compared to the levels resulting from analytical formulas which can be found in most textbooks on theoretical acoustics. Actually radiation efficiencies have been compared, since they are independant of size and amplitude of the source (sphere). The pulsating sphere has been modeled using 72 triangular elements. To illustrate the influence of the overdetermination the radiation efficiencies have been calculated twice, once without overdetermination, and once with one additional internal point situated at r=0.3R (where R is the radius of the sphere). The results are represented in fig.3 and show close agreement between the predicted and analytical efficiencies.

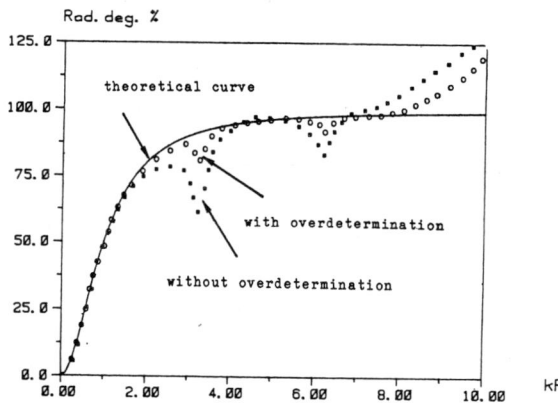

Fig.3 : Radiation efficiencies of a pulsating sphere

The pronounced discrepancies between predicted and analytical results at kr=3.4 and kr=6.8 are due to the non-uniqueness problem. Indeed those wavenumbers correspond to the first and the second internal resonance frequency of the sphere. This confirms the fact that the method, we selected for solving the Helmholtz integral equation, yields erroneous results for frequencies around the internal resonances. Overdetermination of the system using additional internal points improves the accuracy, but to our judgment further research is required to optimise the position and number of overdetermination points.

ACOUSTIC RADIATION OF THE OIL SUMP OF A DIESEL ENGINE

A first real life verification test of the radiation model was conducted on the oil sump of a diesel engine. The oil sump is, because of its size and location one of the noisiest components of combustion engines. The experimentally measured vibration amplitudes of the oil sump have been used as input for the radiation model, while the predicted sound power levels have been compared with the measured power levels. Test object was a bare engine block with only the oil sump connected to it. An electromagnetic shaker was used to excite the engine with white noise in the frequency range (0-2000Hz). During excitation the vibration levels as well as the radiated sound were measured. The vibration levels were derived from acceleration measurements in 82 points all over the structure. Some of those vibration patterns are shown in fig.4. The experimental sound power levels are based on surface scanned near field acoustic intensity measurements. All surfaces of the oil sump have been scanned with a two microphone acoustic intensity probe, yielding the average intensity for each surface. The sound power radiated by each surface is given by the product of the averaged intensity levels with the size of the considered surface. Summing up the sound power for all the surfaces of the oil sump yields the power radiated by the oil sump.

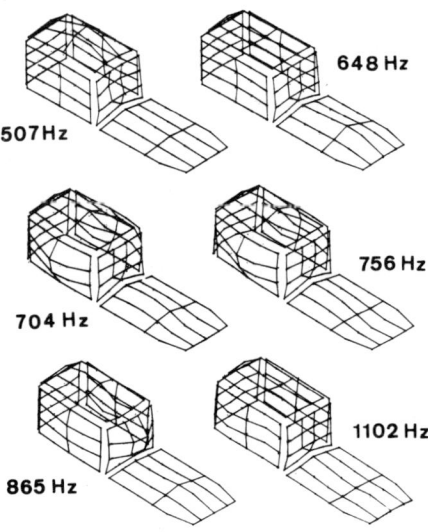

Fig.4 Vibration patterns of an oil sump at resonance frequency

The predicted and measured power levels are summarised in the following

table. Only the most pronounced resonance frequencies have been listed.

Freq.	507.1	605	648	670	704	755.2	795	865	1005	1102
Exper.	65.1	62.4	65.8	59.5	66.0	66.5	62.7	68.8	60.4	65.0
Model	64.3	63.1	66.7	60.2	64.8	67.2	62.1	70.2	61.3	63.5

The average difference between experiments and model is +/- 1dB. However

by monitoring   the condition   number of   the matrix   system, the

non-uniqueness problem did not occur for the listed frequencies. Much larger discrepancies between measurement and model predictions might occur for those frequencies where the matrix system is ill-conditioned.

CONCLUSION

Two computational procedures to model the relation between structural vibrations and radiated sound have been introduced. They have been conceived such that structural deformation patterns originating from experimental modal analyis or from finite element calculations can be used as input data. The first model is based on a reduction of the Green's function in discrete points, and is computationally much simpler than the second which is based on a general 3D solution. Both models yield accurate predictions compared to experimental results but the method based on the Greens's function is limited to 2D sources.

REFERENCES

1. Junger M.C., D.Feit, 'Sound, Structures and their interaction', MIT, Lincoln Laboratory, Lexington, 1972, pp.61-117.

2. Morse P. ,Ungard K., 'Theoretical acoustics', Mc.Graw-Hill NY, 1977, pp332-366

3. Pierce A.D., 'Acoustics an introduction to its physical principles and applications' Mc. Graw-Hill NY, 1981, pp.153-203.

4. Atkinson F.V., 'On Sommerfelds radiation condition' Phil. Mag, 1949,

Vol40, deel 7, pp.645-651

5. Schenk H.A., 'Improved formulation for acoustic radiation problems' J.A.S.A., 1968, Vol.44, deel 2, pp.41-51

6. Koopman G.H., 'Application of sound intensity computation based on the Helmholtz equation', Proc. 11th Int. Congress on Acoustics, 1983, Vol.10, pp.83-87.

7. Burton A.J., Miller G.F., 'The application of integral equation methods to the numericak solution of some exterior boundary value problems, Proc. Roy. Soc. London, 1976, Vol A323, pp.201-210

8. Meyer W.L, Bell W.A., Zinn B.T., Stallybrass M.P., 'Boundary integral solutions of the three dimensional acoustic radiation problem', J.S.V, 1978, Vol.

9. Filippi P.J.T., 'Layer potentials and acoustic diffraction' J.S.V., 1977, Vol.54, deel 4, pp.473-500

10. Copley L.G., 'Fundamental results concerning integral representations in acoustic radiation', J.A.S.A., 1968, Vol.44, deel 1, pp.28-32.

11. Vandeponseele P., Sas P., 'Three-dimensional sound radiation models: an overview', Proc. 11th Int. Modal Analysis Seminar, K.U.Leuven, sept.1986, Vol.2, pp.B1.1-36.

12. Sas P., Vandeponseele P., 'A numerical solution of the 3-D Helmholtz integral equation for modeling the relation between structural vibration and related sound', Proc. 8th. Int. Modal Analysis Seminar, K.U.Leuven, 1983, Vol.3B, pp.322-344.

DIGITAL ACOUSTIC INTENSITY MEASUREMENTS

P. Sas
Katholieke Universiteit Leuven, Belgium

ABSTRACT

A measurement method (digital acoustic intensity measurements) is proposed, which, in contrast to the classical methods, enables the measurement of the acoustic intensity in the near field of a vibrating structure. As will be demonstrated the grafic representation of the near field intensity vector is a useful resource in the experimental study of the sound radiation mechanisms, in identifying noise sources or in estimating the sound contribution of different subcomponents of a complex source. As such it is a unique tool to identify the radiation mechanism of vibrating structures and to study the complex relation between structural vibrations and radiated sound.

INTRODUCTION

The acoustic intensity (I) of a sound wave is defined as the average acoustic power transmitted per unit area in the direction (r) of the wave propagation (u) and is therefore equal to the average over a period (T) of the the product of pressure (p) and air particle velocity ($\bar{u}$),

$$\bar{I} = \frac{1}{T} \int_0^T p(t)\bar{u}(t)dt \qquad (1)$$

Since harmonic signals are commonly expressed in terms of their rms values, equation (1) can be written as

$$\hat{\bar{I}} = \hat{p}\,\hat{u}\,\cos\varphi \qquad (2)$$

where
$\bar{I}$: rms value of I(t)
$\hat{p}$: rms value of p(t)
$\hat{u}$: rms value of $\bar{u}(t)$
φ : phase angle between p(t) and u(t)

The impact of the phase angle (φ) is most apparent marked in strong reactive fields, for example standing waves, where sound pressure and particle velocity oscillate in opposite phase. As a result no acoustic energy is transmitted and the resulting intensity becomes zero. In a free progressing sound wave on the contrary, pressure and particle velocity are in phase and a maximum amount of acoustic energy is transmitted. This situation is characteristic for the far field where the intensity is given by :

$$|\bar{I}| = p^2/\rho c \qquad (3)$$

where
c : velocity of sound
ρ : air density

Since in this expression the acoustic pressure is the only unknown quantity, most sound power estimation standards are based upon it. However, it is only valid in the far field and yields considerable errors when applied to the near field. The error can be as large as 10dB close to the radiating surface even for a simple source such as a 'pulsating' sphere. Near field sound intensity measurements as dealt with in this paper do not suffer from these constraints.

REVIEW OF ACOUSTIC INTENSITY TECHNIQUES

The first attempts to estimate the sound intensity in the near field were based on the physical measurement of pressure and particle velocity. Already in 1932 a patent was granted to H.F. Olson[1] for an acoustic watt-meter which was based on a velocity-ribbon microphone and two crystal pressure microphones. Experiments of Clapp et al[2], Baker[3], Zyl et al[4], with similar equipment, illustrated the limitations of this technique. Indeed, due to extraneous air movement, insufficient damping of the wire at the wire resonance frequencies, and the need for a a steady air supply the technique is limited to laboratory conditions.

Since the particle velocity itself is difficult to measure, it is obvious to derive it from the pressure gradient which, within certain frequency limits, can be measured easily. Schultz[5] was the first to apply this principle, he used two pressure-sensitive condensor microphones, mounted back to back to approximate the pressure gradient. Due to the limitations of the electronics of those days, the subsequent integration and multiplication of the microphone signals suffered from innacuracy. This is probably the reason why one has to wait until the seventies to see this principle applied by others.

H.P.Lambrich and W.A. Stahel[6] and F. Friundi[7] reported in 1976 on an acoustic intensity measuring instrument commercially available and manufactured by the Interkeller AG. The instrument was conceived for the rapid and reliable localization of acoustic sources and sinks on vehicles.

Around the same period of time similar equipment was developed at some university laboratories and research on the subject was started, for example the accuracy study of Fahy,[8] and Hamanns[9] applications on machine tools.

Fahy[10] formulated the acoustic intensity in terms of the cross-spectrum between two closely spaced microphones. This finding permits the use of commercially available equipment to derive the acoustic intensity, since the cross-spectrum is a standard function on most Fast Fourier Analysers (FFT). As a result the special electronic circuitry of the past is no longer necessary and post processing of the measurements becomes possible since digital equipment is used. This principle was first applied and refined by Chung and Pope[11] for the analysis of engine noise. Their studies stressed the need for accurate phase matching of the microphones. They also investigated the effects of microphone spacing, number of averages and the size of the microphones.

Several researchers, both at universities and industry, have since

then successfully improved and applied the method. This culminated in an
international congress on acoustic intensity in 1981 organised by the
CETIM. During the last years however one sees an unbridled expansion of
acoustic intensity measurements, which are applied in season and out of
season, emphasizing a definite need for standardisation of the
measurement procedure and of the application field. This does not alter
the fact that the intensity measurement technique is an highly efficient
tool for visualising the relation between structural vibrations and
radiated sound, as will be illustrated in this paper.

PRINCIPLE AND MEASUREMENT PROCEDURE

Digital acoustic intensity measurements are based on the estimation
of the pressure gradient using two closely spaced microphones. The
applied procedure is similar to the one presented by Chung and Pope[11], but
has been refined integrated in a dedicated software package. Applying
Newtons second law to a small volume of air yields an equation for the
air particle velocity as a function of the pressure gradient :

$$- \text{grad } p = \rho \, \frac{\delta \bar{u}}{\delta t} \qquad (4)$$

or if only one direction is concerned

$$- \frac{\delta p}{\delta r} = \rho \, \frac{\delta \bar{u}_r}{\delta t} \qquad (5)$$

solving for u yields

$$\bar{u}_r = - \frac{1}{\rho} \int \frac{\delta p}{\delta r} \, dt \qquad (6)$$

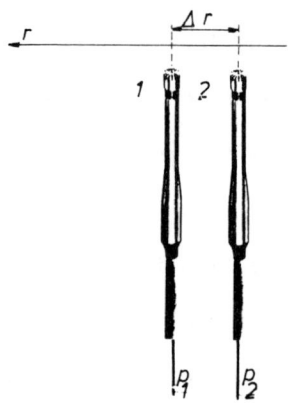

Fig.1 : Two microphone measurement set-up

Using a measurement probe consisting of two closely spaced pressure microphones (fig.1), the pressure gradient in the direction (r) can be approximated by

$$\frac{\delta p}{\delta r} = \frac{p_1 - p_2}{\Delta r} \qquad (7)$$

Δr : microphone separation distance Similarily the pressure at the point midway between the microphones can be approximated as

$$p = \frac{p_1 + p_2}{2} \qquad (8)$$

Substituting (8),(7),(6) in (1) yields following equation for the acoustic intensity in the direction r

$$I_r = \frac{1}{T} \int_0^T \left\{ \frac{p_1 + p_i}{2} \left(\frac{1}{\rho \Delta r} \right) \int (p_2 - p_1) dt \right\} dt \qquad (9)$$

Transforming to the frequency domain yields following equation for the real part of the acoustic intensity

$$I_{r \text{ (reel)}} = \frac{1}{4\pi\rho\Delta r} \int_{-\infty}^{\infty} \frac{\text{Im}(P_1 P_2^*)}{f} \, df \qquad (10)$$

where   Im $(P_1 P_2^*)$:   imaginary part of the cross spectrum

between both pressure signals.

f :   frequency

Since the cross spectrum is a standard function, available on most FFT
analysers, acoustic intensity measurements are easy to implement and no
longer require elaborate electronic equipment.

The vector character of the intensity levels implied by equation (10)
should be emphasized.   One measures only the intensity vector component
parallel to the orientation of the measurement probe. Measuring the
intensity component in three perpendicular orientations yields the
spatial intensity vector.   Since the spatial intensity vectors form the
actual data base for the visualisation of the intensity field, a three
dimensional probe has been developed enabling the determination of three
components simultaneously.

HARDWARE

Quarter inch B&K condensor microphones have been selected as
transducers for both the one dimensional probe as well as for the three
dimensional one.   Their small size,their stability and their excellent
phase match are the main reasons for this choice. Those microphones were
used in combination with the B&K 2618 preamplifiers to boost the

microphone output signals with 20 dB, which enables the use of long
cables. The microphone spacing is adjustable in accordance with the
frequency range of interest. The measurement set up is shown in fig 2.

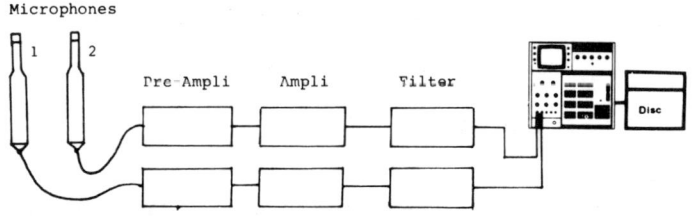

Fig.2 : Measurement set-up

The three dimensional probe is schematically represented in fig.3,
and consists of four identical 1/4 inch microphones, geometrically
arranged in such a way to yield three perpendicular intensity components.

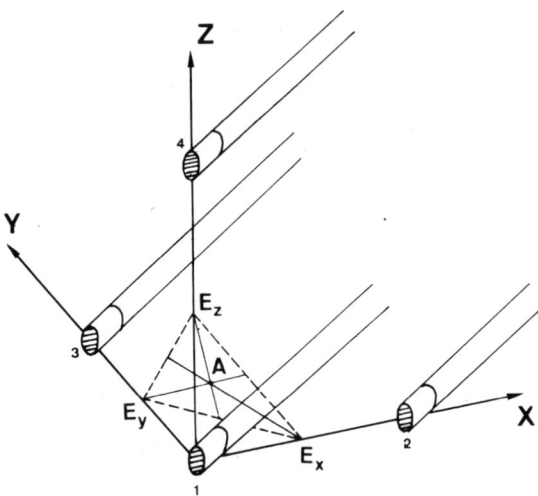

Fig.3 : Schematical view of the 3D measurement probe

The X-component of the intensity is given by microphone combination (1-2), the Y-component by combination (1-3) and the Z-component by combination (1-4). The resulting intensity components do not relate to one single point (A), but only approximate the intensity vector in that point. The related approximation error, which is function of the frequency and of the complexity of the acoustic field, is of the same order of magnitude as the approximation error due to the pressure gradient approach. The actual realisation is shown in figure 4.

Fig.4 : Three dimensional acoustic intensity measurement probe

AUTOMATED 3-D ACOUSTIC INTENSITY MEASUREMENTS

To facilitate the intensity measurements necessary for visualising the relation between structural vibrations and radiated sound, two automatic microphone positioning system have been developed.   The first

positioning system is a so called rectangular coordinate robot with four

degrees of freedom: x,y,z translation and one rotation of the wrist

(fig.5). The robot is driven by the processor of the FFT analyser, as a

result robot control and measurements are managed by the same processing

unit.

Fig.5 : View of the positioning robot

Due to its limited x,y,z range (100x80x50cm), this robot is mainly

intended for research applications. For example the resonant near field

patterns of various vibrating plates, randomly excited with white noise,

and other small sources have been measured with this robot.

The second robot (fig.6) was specially designed for the analysis of

larger sources. It is a rectangular coordinate robot with three degrees
of freedom (x,y,z translation) intended to scan large surfaces such as
the body of a car, or a machine tool. It was designed for easy
assembling and disassembling in site.

Fig.6 : View of the surface scan positioning robot

SOFTWARE

The cross spectrum calculations and the other calculations related
to equation (10) are carried out on a four channel Fast Fourier Analyser
with post processing capability (HP 5451B). The software implemented
consists of two main packages, the measurement package and the analysis
package. Each package is designed around a monitor, which adresses
several modules (input, output, display, clculations, robot control,

initialisation...).   An octave band option has been included in the

intensity calculation module to determine the intensity levels in octave

or third octave bands.   The software can handle up to 750 cross spectra

contineously, allowing the calculation of the spatial intensity vector in

250 measurement points, for up to 30 single frequencies.   Dedicated

graphic software has been developed for representing those intensity

vectors, yielding animated intensity patterns or intensity contour plots.

ERRORS INVOLVED IN ACOUSTIC INTENSITY MEASUREMENTS

The frequency band suited for unbiased intensity estimations depends

upon the microphone spacing. As pointed out by several authors this is

by far the most stringent limitation of the acoustic intensity method.

For a given phase mismatch (θ) between both measurement channels the

normalized bias error (ε_r) can be estimated by means of the following

equation :

$$\varepsilon_b = \frac{E(I_r) - I_r}{I_r} = \frac{\theta c}{2\pi f \Delta r} \qquad (11)$$

where   E(I_r) : intensity estimation

For example if one allows a bias error of .5 dB with a microphone spacing

of 20.5 mm and with a phase mismatch of .5 degrees, the lower frequency

limit will be 360Hz. To minimize this bias error a correction procedure

has been introduced which consists of a multiplication of every measured

cross-spectrum with a previously recorded calibration spectrum.   The calibration spectrum is the inverse of the frequency response function between the two microphones. This is a far more elegant and less time consuming procedure than the earlier proposed measurement circuit switch as practiced by Chung[11]

The higher limit of the unbiased frequency range is also dependent upon the microphone spacing due to the approximation of the sound pressure midpoint both microphones. The related bias error can be written in terms of frequency and microphone spacing if one expresses equation (10) in Taylor-series, supressing the higher order terms yields the following equation :

$$\varepsilon_b = \frac{E(I_r) - I_r}{I_r} = \frac{2}{3} \left(\frac{\pi \Delta r f}{c} \right)^2 \tag{12}$$

For example if one allows a bias error of .5 dB for a microphone spacing of 20.5 mm, the maximum frequency will be 2342Hz.

Other important bias errors which may not be neglected, are those caused by the digital signal processing, the most important are aliasing and leakage.

VERIFICATION EXPERIMENTS

Verification experiments were conducted to verify the accuracy of the measurement procedure and to demonstrate it's feasibilty. The measurements were conducted in an anechoic room using a reference source. The reference source was fed with white noise ranging from 0 to 5000Hz

and the microphone separation distance was 20mm. The difference between
the intensity levels based on pressure readings, and the intensity levels
recorded with the pressure gradient method, is represented in fig.7.

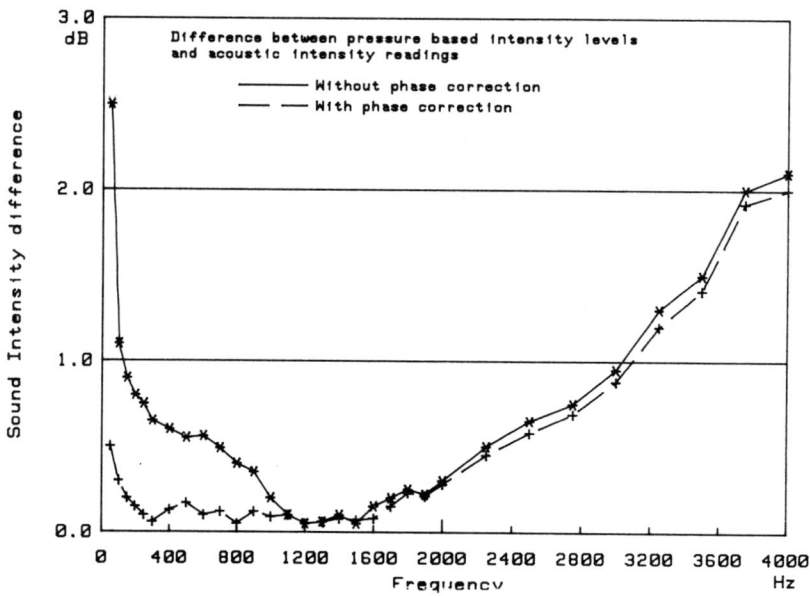

Fig.7 : Difference between standard far field measurements and
        acoustic intensity measurements

The measurements have been conducted   with and without phase
correction.   The resulting   error levels   clearly illustrate   the
limitations imposed by by the instrument phase mismatch (low frequency
error), and by the finite difference approximation (high frequency
error). The proposed phase correction procedure proves to be efficient
since the low frequency error is drastically reduced when applying phase
correction.

APPLICATIONS

The most straightforward applications of the intensity measurement technique are those were one uses the technique as a diagnosis tool to determine the sound power radiated by the various components of a larger noise source (source ranking). Such applications are beyond the scope of this paper where we will concentrate on the grafic representation of the intensity patterns.

Since long, attempts have been made to visualize the sound patterns in order to gain some insight in the noise radiation mechanism. Ingenious equipment resulted from those attempts, such as the one described by W.E.Kock.[14] It basically consists of a pressure microphone which is continuously moved through the area of interest. An electric lamp mounted on top of the microphone is fed with the amplified microphone output, and translates the pressure variations into brightness variations. The brightness pattern is photographically recorded. Unfortunately this technique does not present a quantitative result. It is time consuming and requires a special dark room and yields only two dimensional patterns.

INTENSITY PATTERN OF A SIMULATED DIPOLE

To illustrate the power of the proposes intensity measurement technique, the radiation pattern of a simulated dipole has been measured. The acoustic dipole has been simulated by a pair of loudspeakers fed with signals of equal amplitude but opposite sign.

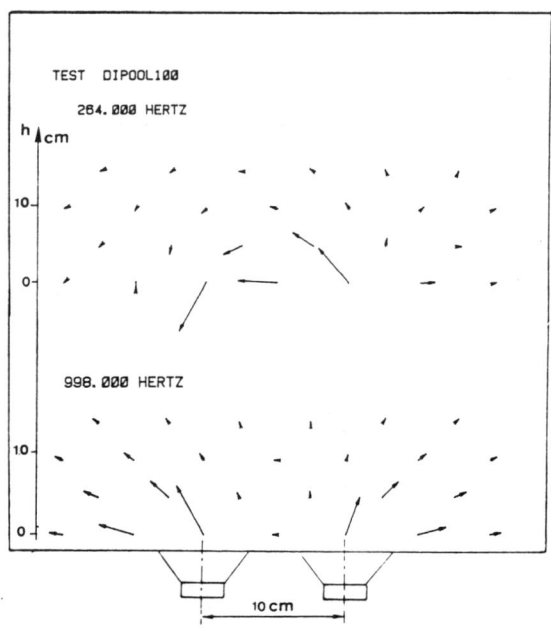

Fig.8 : Measured intensity patterns for opposite phased loudspeakers

The intensity vectors have been measured in planes parallel and perpendicular to the baffle plane. The signal fed to the loudspeakers was a saw-tooth like signal to generate sufficient harmonic components. For each of the harmonic components the intensity pattern was determined. The measurements were carried out twice, once with both loudspeakers in phase, and once with both loudspeakers in opposite phase.

Some of the results are shown in fig.8 for the plane perpendicular to the baffle. The separation distance between both loudspeakers was 10 cm. Those intensity patterns clearly illustrate the hydrodynamical short-circuit. At 264 Hz the short-circuit is most pronounced, showing that, for this frequency, the air particles do escape compression by moving out of the way laterally to the second loudspeaker, which is vibrating in opposite phase. At higher frequencies, for example 998Hz the sound wave is not fast enough to interfere with the other pole. No lateral sound energy leakage occurs and all acoustic energy is transmitted into the far field.

It has to be remarked that the intensity amplitudes are plotted on a linear scale. Moreover the concerned intensity levels represent by definition the time avarages of the actual intensity over one period, and not the momentary amplitudes. For a perfect dipole the average intensity over one period should be zero at those points where hydrodynamical short circuit occurs, since the lateral leakage of acoustic energy is equal in both directions. The reason why the intensity pattern of fig.8 still shows a clear leakage of acoustic energy from one loudspeaker to the other, is the difference between both loudspeakers. Indeed, they were not, as was falsely assumed, identical and radiating equal acoustic power. Their respective internal resistance was 10.7 and 11.8 Ohm, yielding a difference of 10% for the radiated sound power. Consequently, the leakage of acoustic energy no longer is symmetrical and a residual energy transfer from the largest pole to the smaller one can be visualized.

INTENSITY PATTERNS OF VIBRATING PLATES

The acoustic radiation behavior of vibrating plates is of special interest since plates are a current component in mechanical structures.

In platelike structures it is important to know the impact, on the radiated sound power, of parameters such as the plate geometry, the plate material, additional damping or stiffning ribs. Although several theoretical studies can be found in literature on this subject, few real configurations are treated and one has to fall back on experimental results. For such measurements the intensity technique is the ideal tool.

When dealing with the radiation of platelike structures, one often introduces the notion radiation efficiency. This parameter, which varies between 0 and 1, is defined as the ratio between the sound power radiated by a noise source and its potential source strength.

$$\sigma = \frac{W}{\rho c S v^2} \qquad\qquad (13)$$

where   S : Size of the vibrating surface
           v : rms value of the average surface velocity
           W: radiated sound power
           σ : radiation efficiency

The potential source strength is equal to the sound power radiated by a rigid piston of the same surface size and average velocity as the original source. The measurement of the radiation efficiency for a vibrating platelike structure is based on a subsequent or simultaneaous acquisition of both surface velocity and acoustic intensity. The acoustic intensity is measured using the surface scanning method, while the surface velocity is measured with an acceleration transducer (accelerometer). The accelerometer is positioned in a limited number of points randomly chosen on the surface.

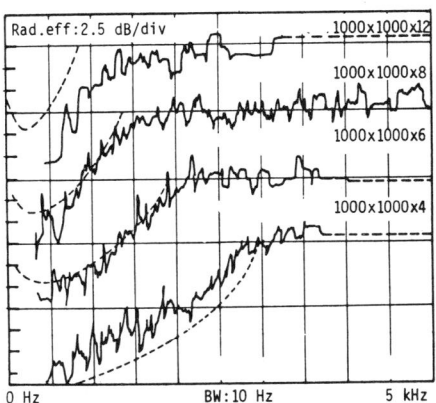

Fig.9 : Influence of thickness on the radiation efficiency of plates

A series of measurements has been carried out in order to classify

the radiation efficiencies for a number of typical plate configurations.

The influences of plate thickness and various stiffening ribs on the

radiation efficiency have been experimentally verified   using the

described measurement procedure and have been summarized in a catalog,

which can be consulted by designers.

Some of the results are represented in fig.9. They clearly show

that stiffening the plate increases the radiation efficiency.   The

estimation of the radiation efficiency alone is not sufficient to explain

the radiation mechanism, it has to be interpreted together with the

corresponding intensity patterns. For that purpose the intensity pattern

of various free/free plates, randomly excited with white noise, has been

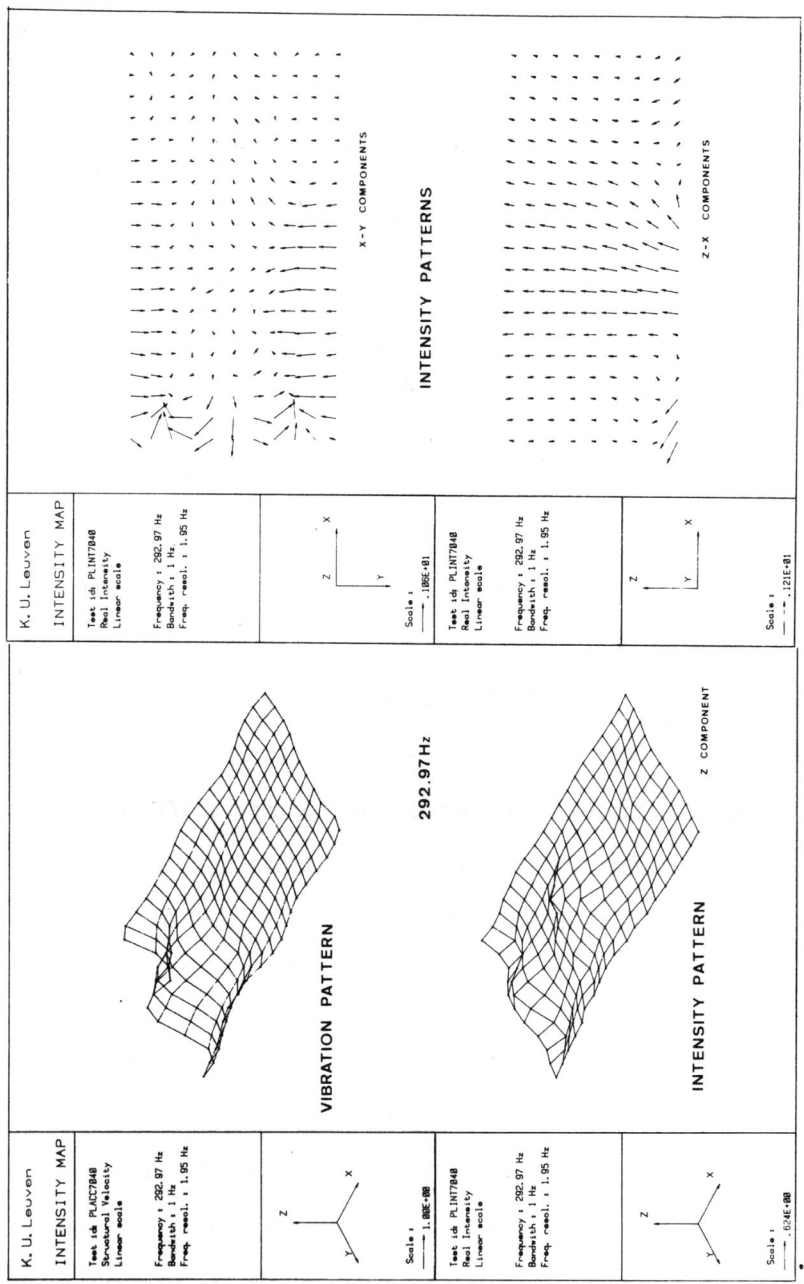

Fig.10 :Acceleration and measured intensity pattern of a free/free suspended and randomly excited plate (292Hz)

measured using the intensity measurement robot (fig.5). One of the resulting intensity patterns, namely those of a rectangular, free-free suspended plate at 292Hz is shown in fig.10. On this representation the intensity patterns, normal and parallel to the plate, as well as the deformation pattern are shown. The intensities have been measured in two planes: one parallel, and one normal to the plate. Especially on the intensity patterns, parallel to the plate, the acoustic short-circuit is clearly marked by the leakage of acoustic energy from and to areas of the plate which are vibrating in opposite phase. The acoustic leakage is concentrated around the nodal lines. This was expected since they form the boundary between opposite phase areas.

Fig.11 : Measurement set-up for the diesel engine oil-sump

The close correlation between the modal deformations and the near field intensity component normal to the plate is remarkable and clearly illustrates the fact that some areas of the plate act as sources while others are sinks. This agrees with theory since, taking into account the dimensions of the plate (700x500mm), the considered resonance frequencies are far below the critical frequency. Moreover since the plate is free-free suspended, acoustic short-circuit occurs around the edges of the plate. This is the reason why, as can be noticed on fig.10 the normal intensity is much smaller on the edges than one might guess taking into account the modal deformations of the edges.

INTENSITY PATTERNS OF AN OIL SUMP

The near field intensity patterns of the oil sump of a truck diesel engine have been measured, together with the vibration pattern, in order to determine the relation between structural vibrations and radiated sound for such a structure. The oil sump was free/free suspended and excited with white noise, the measurements were conducted with the intensity measurement robot (fig.11).

Two of the resulting intensity and velocity patterns are shown in fig.12 and 13. The velocity pattern of 380.86 is characterized by a symmetric first bending of both sidewalls, together with the longitudinal bending of the rear plane. Although the amplitudes of both bendings are of the same order of magnitude, the radiated intensities are not. The most pronounced radiation is due to the sidewalls and only minor radiation activity is noticed at the rear plane.

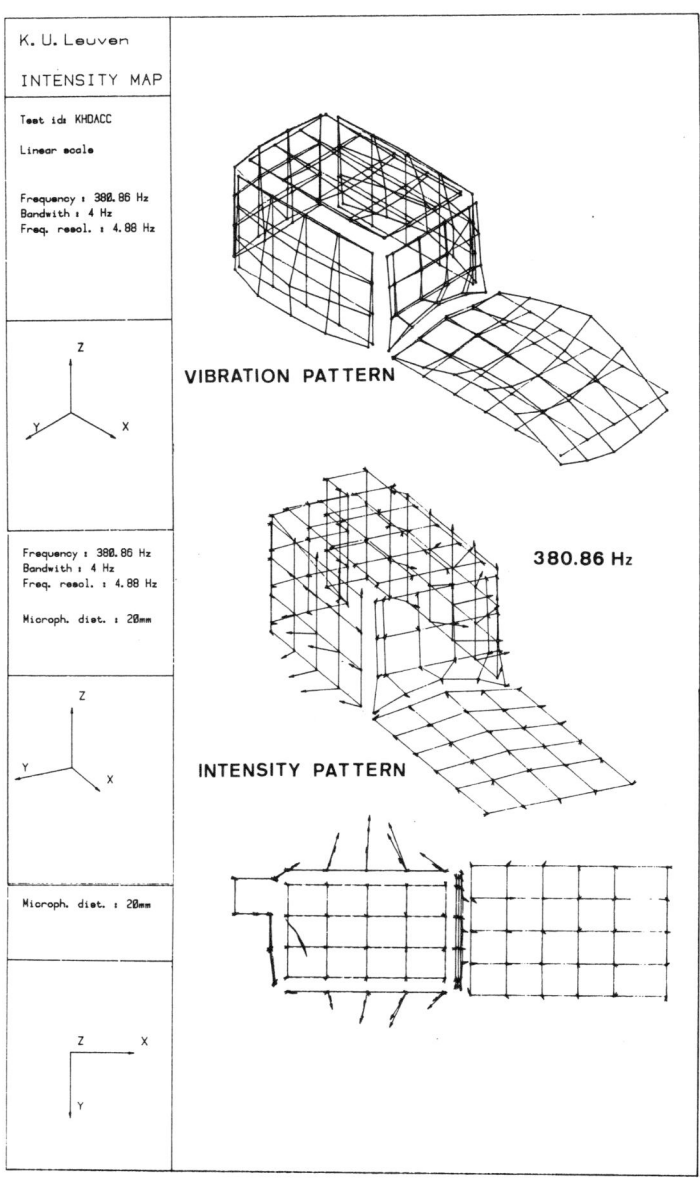

Fig.12 : Acceleration and measured intensity patterns of an oil sump

(freq = 380.6Hz)

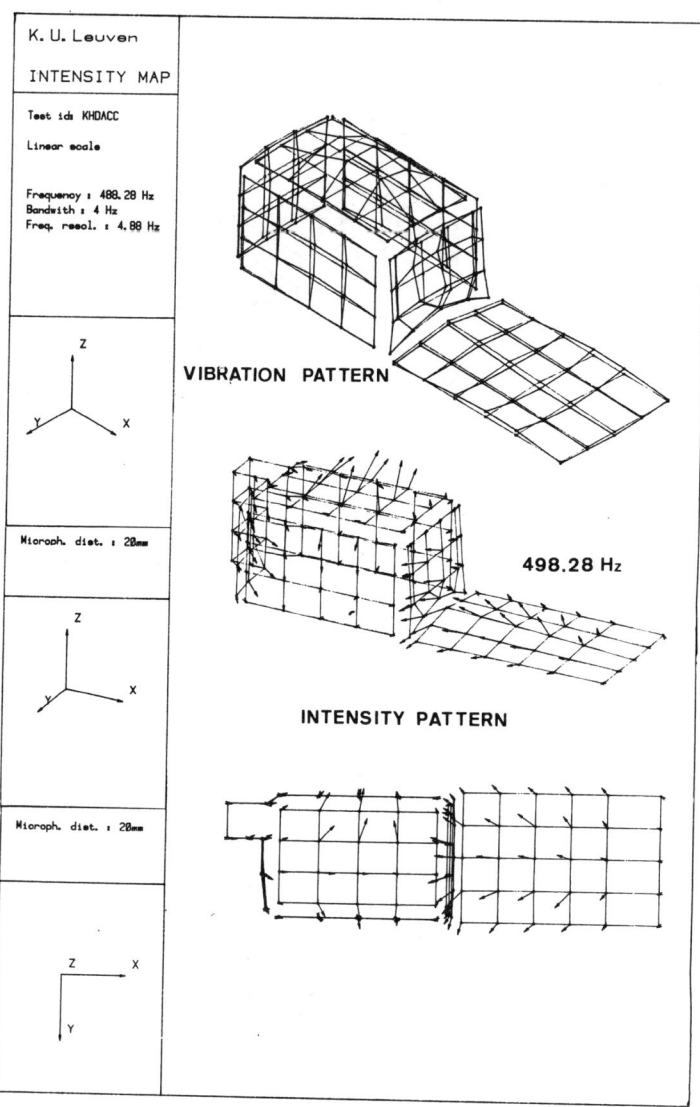

Fig.13 : Acceleration and measured intensity patterns of an oil sump

(freq = 498.3Hz)

The resonance frequency of 498.28Hz clearly shows an acoustic short-circuit between the back plane and the rear plane. The acoustic energy radiated by the rear plane is partly neutralized by the back plane.

Those results clearly illustrate the use of the intensity measurement technique since conclusions based on the vibration pattern only would have been false.

INTENSITY PATTERN OF A LOUDSPEAKER BAFFLE

Another near field radiation pattern which was analysed using the measurement robot was a basreflex loudspeaker box. The result, fig.14, shows the intensity pattern normal to the front of the loudspeaker box at 71, 129, and 214Hz. It has to be remarked that in this representation the intensity vectors have been plotted in opposite sense to their propagation direction, so that they point to the source. Those patterns clearly illustrate that the basreflex effect is most pronounced at the low frequencies. Indeed at 71Hz the intensity seems to originate from the reflex slit, while at 214Hz the intensity vectors indicate the loudspeaker as being the most dominant source.

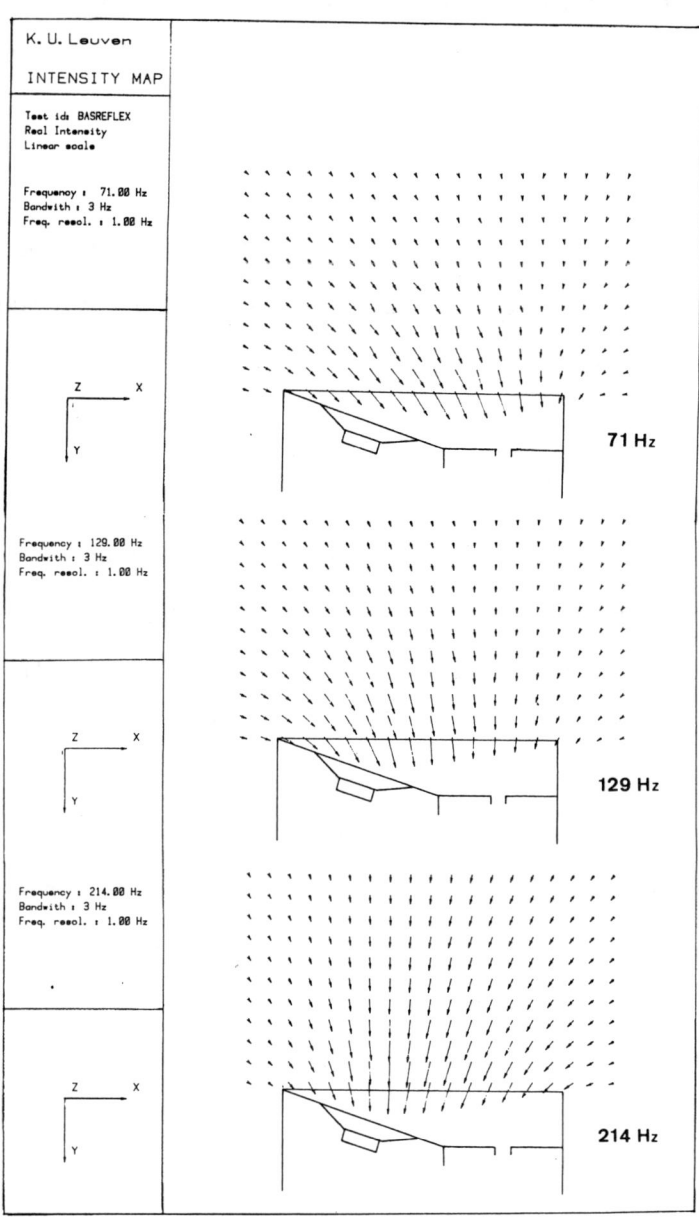

Fig.14 : Intensity pattern of a bas-reflex loudspeaker box

REFERENCES

1. OLSON H.F., 'System response to the energy flow of sound waves', U.S.
    Patent No 1.892.644, 1932

2. CLAPP C.W., FIRESTONE F.A., 'Acoustic Wattmeter', Journal of the
    Acoustical Society of America (J.A.S.A), 1941, Vol.28, pp 693-713.

3. BAKER S., 'An Acoustic Intensity Meter', J.A.S.A., 1955, Vol.27,

4. VAN ZYL B.G., ANDERSON F., 'Evaluation of the intensity method of
    sound power determination', J.A.S.A., 1975, vol.57, deel 3,

5. SCHULTZ T.J., 'Acoustic Wattmeter', J.A.S.A, 1956, vol.28, deel 4, pp
    693-697.

6. STAHEL W., LAMBRICH H.P., 'Development of an instrument for the
    measurement of Sound Intensity and its application in Car Acoustics',
    Proceedings of the 53th AES Convention, 1976

7. FRIUNDI F., 'The utilisation of the intensity meter for the
    investigation of the sound radiation of surfaces', External
    Publication, Interkeller A.G., Zurich 1977.

8. FAHY F.J., 'A technique for measuring sound intensity with a sound
    level meter', Noise Control Engineering, 1977, vol.9, deel 3.

9. HAMANN M., 'Untersuchungen zur Schalleistungsbestimmung in der
    Larmbekampfung', Messen & Prufen/Automatik, 1979, vol.5

10. CHUNG J.Y., 'Cross-spectral Method of Measuring Acousital
    Intensity', General Motors Research Laboratories, 1977, Publ.
    GMR-2717

11. FAHY F.J., 'Measurement of Acoustic Intensity using the
    cross-spectral density of two Microphone Signals', J.A.S.A., 1977,

vol.62, deel 4

12. International Congress on Recent Developments in Acoustic Intensity
     Measurements, Senlis (France), 1981, CETIM

13. SAS P., 'The use of digital signal processing techniques in acoustic
     noise source localisation, including acoustic intensity
     measurements', doct. thesis K.U.leuven, 1982

MULTIPLE INPUT/OUTPUT ANALYSIS:
AN IDENTIFICATION TOOL IN ACOUSTICS

P. Sas
Katholieke Universiteit Leuven, Belgium

SUMMARY

An overview is given of digital techniques which can be applied to analyse acoustical multiple input/output systems. The overview includes frequency domain as well as time domain techniques. The merits and limits of techniques such as correlation, coherence, and transfer function analysis when applied to source localization and transmission path identification problems will be discussed. Recent techniques such as partial and virtual coherence or structural inverse filtering (force identification) have been included in this survey.

INTRODUCTION

A large number of noise problems, especially those where structure borne sound is involved can be viewed as multiple input/output problems. Indeed when analysing noise problems one is often confronted with the situation where several sources contribute to the sound power that has to be reduced. In such cases it is essential to determine the relations between every source and the output, even when the sources are mutually dependant. To enable efficient noise control actions the contributions from the acting sources must be ranked and the noise and vibration transmission paths from every input to the output identified. Multiple input/output analysis can be a valuable aid for such an analysis but should be used in combination with other techniques.

An example will illustrate the principles of multiple input/output analysis. consider the situation where one intends to analyse, for a rear wheel driven car, the contribution of the rear axe to the noise in the passenger compartment. Besides the wheels, which are important vibration sources connected to the rear axe, the rear axe itself contains several vibration sources (wheel bearings, gears and bearings of the differential...) which are potential sound contributors. Identification and ranking of those vibration sources is the first task of the multiple input/output analysis. The rear axe, excited by those vibration sources will radiate noise that, depending on the sound barriers between source and receiver, will be transmitted inside the passenger compartment. The rear axe vibrations itself have a wide choice of transmission paths to reach the body and generate noise at the receiver point (springs, dampers, transmission axe, torsion bar....). Identification and ranking of the transmission paths is therefore the second task of the multiple input/output analysis.

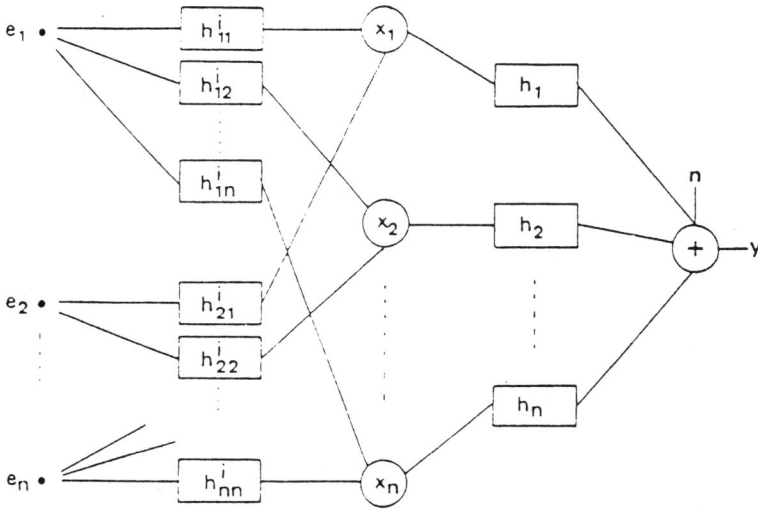

Fig. 1 : Schematical representation of a multiple input-output problem

A grafical representation of this multiple input/output problem is given by the scheme of fig.1, which is characteristic for most multiple input/output problems. In this diagram H_{ij} represents the frequency response function that characterises the transmission path between source (i) and receiver (j). Remark that a series of intermediate source/receivers x_i has been defined, such that the original problem is splitted up in one multiple input/multiple output problem (sources/connection points) where the connection points act as receivers, and in on one multiple input/single output problem (connection points/inside noise) where the connection points act as sources.

The state of the art approach to analyse such a multiple input/output problem consists of applying the 'Decoupling (striptease) Method', where the possible noise sources are physically eliminated one at a time. The car is therefore installed on a special test bench where each of the rear wheels can be driven independantly, one would then systematically disconnect or neutralize (cladding or encapsulating) all sources and transmission paths except one, drive one wheel and conduct a sound measurement at the receiver point to determine the contribution of the active transmission path or source.

This method is time consuming and tedious, requiring sophisticated test set-ups and rig-fixtures to simulate the real world drive conditions as close as possible, which is not easy and often impossible to realise. Even then the results may be misleading since the underlying assumption of linear superposition is not fulfilled. Especially in the low frequency range the phase relation between the sources and/or transmission paths is important, different source inputs may indeed cancel or add up.

Therefore a need exists for methods, allowing the identification of noise sources and transmission paths, without physically interfering with the the normal function or construction of the vehicle, and analysing all sources simulteneously. The following paragraphs give an overview of multiple input/output analysis techniques which attempt to realise this goal

REVIEW OF COHERENCE ANALYSIS

Coherence functions are potentially very attractive for energy source identification.[1] They are able to attribute percentage of output energy to individual measured system inputs enabling for example the ranking of the sources. Practically, however, coherence techniques are limited by several factors. In most real machinery, for example, either the real sources are unknown or cannot be measured directly. Thus, measurements usually occur on the transmission path, as close as possible to the original source, consequently contamination by the energy from all the sources cannot always be avoided. Secondly the classic coherence techniques are derived for non-deterministic (non- coherent signals) such that for cases where coherence exists between the inputs improved coherence techniques such as partial or virtual coherence must be used. Partial coherence is based on the calculation of conditioned power spectra, this conditioning however assumes a specific system model, which is not known a priori, this often limits the applications of the partial coherence technique (ref.2,3,6). More information on the system model can be gained from a linear transformation technique based on a singular value decomposition yielding the principal sources[4] of the system or the virtual coherence.[5] The mentioned coherence techniques will be briefly discussed in this paper, for more details and examples, we refer to the references.[2,3,6]

Ordinary Coherence Function [1]

The coherence function indicates which fraction of the output power is due to source (i). The coherence function has wide applications in the field of digital signal processing, as indicated by its applications in the error analysis of frequency response measurements. Another important application related to source identification is the calculation of coherent output power spectra, and the ranking of sources.

The ordinary coherence function, when applied to source identification is only valid for independant (non-coherent) sources. This is generally the case when dealing with mechanical systems that are structurally not connected. However, when dealing with sources in a structurally interconn ected system, such as combustion engines, the dynamic response at a given point is likely to be related to a number of sources. In that type of problem teh ordinary coherence fucmtion will give erroneous results.

Multiple Coherence Function [4]

The multiple coherence function is defined as the coefficient describing the causal relationship between the output and all considered inputs. In source identification applications, it is used in combination with ordinary and partial coherence functions to to judge the consistence of the input signals. A poor multiple coherence indicates that some important sources have been overlooked when defining the inputs. The multiple coherence is therefore the function that must be evaluated first, before proceeding to any further analysis.

Partial coherence function [1],[2]

   The partial coherence is defined as the ordinary coherence between a
conditioned input and the conditioned output. The conditioning consists
of removing from the considered input and the output the potential
contributions from other inputs.   According to Bendat&Piersol the
conditioning can be formulated on a linear least square basis where the
order of the coherent signals has to be specified by the investigator.
This implies that one has to define which sources are the contaminators
and which the contaminated. The problem which arises when interpreting
partial coherences is the fact that they do not relate to physical
sources, but to conditioned inputs.   It is therefore impossible, using
partial coherence functions, to determine the absolute contribution of
each physical source to the input, it is only possible to attribute
energy from conditioned inputs to the conditioned output. One of the
limitations of the partial coherence technique is the fact that numerical
problems occur when two or more inputs are fully coherent.

Virtual coherence functions [4],[5]

   Since the numerical problems, which arise in the case of coherent
sources, are due to the singularity of the input matrix, it has been
proposed to use mathematical techniques that detect and remove linear
dependancy between elements without requiring an initial ordering of the
inputs. The transformation of the input matrix G   to its principal

values is such a technique, originating from statistical analysis techniques and also known as principal component or singular value decomposition.[7] Applying the singular value decomposition to the input matrix $[G_{xx}]$ (Hermetian and positive semi-definite) yields.

$$[G_{xx}]' = [V]^H [G_{xx}] [V]$$

where superscipt H denotes the Hermetian transpose of the vector or matrix it modifies. $[V]$ is a unitary matrix composed of the eigenvectors of $[G_{xx}]$, and $[G_{xx}]$ is diagonal matrix with as diagonal elements the eigenvalues of $[G_{xx}]$ in descending order. Those eigenvalues are the principal or singular values of the input matrix $[G_{xx}]$. If (r) is the rank of the (nxn) input matrix $[G_{xx}]$, then (r) eigenvalues and hence (r) singular values will be found. If $r < n$ then (n-r) of the singular values are zero, or the (n) original inputs are reduced to (r) 'virtual' uncorrelated inputs. The auto-power spectrum of those inputs is given by the diagonal elements of the $[G_{xx}]'$ matrix, which can be calculated for all frequency points. This implies that for each frequency point an eigenvalue problem must be solved, fortunately the eigenvalue solution is real since the input matrix $[G_{xx}]$ is Hermetian.

Both the number of virtual inputs and the auto-power spectrum of those virtual inputs are valuable information when analysing a multiple input problem. The concept of virtual components can be extended to the cross spectrum vector and the frequency response matrix thereby enabling the calculation of virtual coherences.[5] The virtual coherence indicates the contribution of the virtual source mechanism (i) to the signal measured at the output.

Value and use of the coherence functions

In conclusion then the principal value decomposition is usefull for reducing a system of n inputs (coherent or not) to a system of (r) incoherent (virtual) inputs. In contrast to the partial coherence technique the calculation of the virtual coherence requires no ordering of the inputs and still generates a valid solution if two or more inputs are 100% coherent. However, as was the case for partial coherences, the interpretation of the virtual coherence is not always straightforward, since the link between the monitored sources and the virtual sources is not known; nor is the information resulting from the virtual coherences always relevant.

An example will illustrate those problems. Assume, that on a passenger car, one wants to determine the vibration energy transmitted from engine to body through each of the engine mounts. This can be done by measuring and analysing the vibration of the mounts simultaneously with the interior noise. It is obvious that the only independant vibration source is the engine itself, consequently the input vibration signals will be higly coherent (up to 100%). This can easily be be verified by calculating the ordinary coherence functions. In such a situation partial coherences will numerically fail and virtual coherence analysis will detect one obvious virtual source, the engine, but will not yield more insight in the vibration energy transmission mechanism.

FORCE IDENTIFICATION (STRUCTURAL INVERSE FILTERING)

A technique that potentially could solve the energy transmission problem
mentioned in the preceding paragraph, is the so called force
identificaion or structural inverse filtering technique.[8,9,10] Knowledge of
the acting forces and input loads on machinery components would indeed be
useful when solving structural dynamic problems, related to source and
transmission path identification or fatigue life prediction.

The direct measurement of those forces or loads is in principle
conceivable using force transducers and/or strain gauges, but in practice
the mounting of those sensors will be often very cumbrous or even
impossible. It is therefore much more attractive to measure the
vibrations resulting from these dynamic forces, and use this information
to extrapolate the acting forces. However, vibrations measured in one
point are necessarily influenced by all acting forces which are
conditioned by the respective transmission paths. It is therefore
necessary to use a transmission path model to derive the forces from the
measured vibrations. In the frequency domain the frequency response
matrix between input (forces) and output (vibrations) can be used as
model since they characterise the transmission paths. In matrix notation
the input/output relations for a linear system between the input vector
(forces, $\{F\}$) and the response vector $\{Y\}$ can be written as :

$$\{ Y \}=[H]\{ F \} \qquad\qquad (1)$$

where $[H]$ is the frequency response matrix. Since the frequency response is a structural characteristic independant from the excitation mode, the

matrix [H] can can be measured, in laboratory conditions, by articially

excitating the structure in each input point, and measuring all the

responses. Once [H] is determined, the acting forces can be calculated

on the basis of vibrations measured during field operation, by inverting

the [H] matrix.

$$\{F\} = [H]^{-1} \{Y\} \tag{2}$$

In principle this procedure is an attractive solution to determine the

acting forces, but in practice some problems limit the applicability of

this method :

1) The measurement of the [H] matrix requires the excitation of the

   structure in three directions for each point where the acting forces

   must be defined.   This is often a difficult, if nor impossible,

   procedure which requires a partial disassembling of the structure. In

   that case the boundary conditions are different from the real life

   situations, a fact that must be taken in account when interpreting the

   results.

   The number of measurements, necessary to determine the H matrix can

   be considerably reduced if one uses modal synthesis techniques.

   Indeed using the modal parameters (eigenvalues and eigenvectors),

   which can be derived from one column or row of the H matrix, it is

   possible to synthesize all the other rows or colums of the H matrix.

2) The second problem relates to the numerical restriction inherent to

the inversion of the $\lfloor H \rfloor$ matrix. Indeed at the structural resonance frequencies the $[H]$ matrix tends to be singular, especially for sharply pronounced (lightly damped) resonance frequencies. For those frequencies no inverse matrix exists.

The numerical problems related to the singularity of the $[H]$ matrix at particular frequencies can be overcome in two ways

1) Avoiding the matrix inversion by describing the system in dynamic stiffness, rather than in dynamic flexibility form. This flexiblity matrix $[K]$ can be directly measured by inverting the system in- and output signals.

$$\{F\} = [K] \{X\} \tag{3}$$

This evidently simple technique has the drawback not to allow the modal transformation, such that synthesis of the flexibility matrix based on one row or column is not possible, hence the flexibility matrix must be determined completely by measurements which is a highly laboriuos procedure.

3) The singularity can also be overcome by using the pseudo-inverse technique, this means that instead of solving (n) unknown variables (forces) out of a system of (n) equations (vibration measurements), one overdetermines the system by including redundant response points. This system can be solved using a complex least squares approximation. The pseudo-inverse matrix that is calculated using this method is a full rank pseudo inverse, and its accuracy is dependant on the choice

of the redundant response points. A more robust technique to calculate the pseudo-inverse, which yields at the same time an estimation of the rank of the [H]matrix, is the singular value decomposition which has been described in the preceding section.

CONCLUSIONS

An overview has been given of signal processing techniques which can be applied to analyse acoustic multiple input/output problems. Many acoustic problems, especially those related to structure borne noise, can be reduced to multiple input-output problems, it is therefore worthwhile to consider the application of multiple input-output signal processing techniques when one is faced with an acoustic problem. Especially techniques such as virtual coherence and structural inverse filtering are promessing, but still in their research phase, such that more research and field applications are still necessary before those techniques can be widely applied.

REFERENCES

1) A.BENDAT,G.PIERSOL, 'Engineering applications of correlation and spectral analysis' Wiley&Sons, New-York, 1980

2) R.J.ALFREDSON, 'The partial coherence techniaque for source identification on a diesel engine', J. Sound and Vibration, 55(9), pp.487-494, 1977

3) P.R.WAGSTAFF, J.C.HENRIO, 'Identification of noise sources in

steam turbine plant using multiple input analysis on a digital computer' Proc. of Noise Con 85, pp.293-297

4) J.LEURIDAN, M.MERGEAY, U.VANDEURZEN, G.DESANGHERE, 'Multiple input estimation of frequency response functions for experimental modal analysis : currently used method and some new developments' Proc. 9th Int. Seminar on Modal Analysis, 1984, L.M.S., Leuven, Belgium

5) S.M.PRICE, R.J.BERNHARD, 'Virtual coherence: a 'digital signal processing technique for incoherent source identification' Proc. IMAC4, pp.1256-1262, 1986, Union College, Schenectady, USA

6) P.VANDEPONSEELE, 'Applications of coherence techniques, Proc. 8th Int. Seminar on Modal Analysis, part3b, 1983, K.U.Leuven, Belgium

7) B.NOBEL, J.DANIEL, 'Applied linear algebra', Second Edition, Prentice Hall, 1977

8) H.BATHELT, 'Analyse der korperschall-ubertragungswege in kraftfahrzeugen ' Automobil-Industrie, 1/81, pp.27-34.

9) D.ROESEMS, 'Analysis of structure-borne transmission paths' Proc.6th Int. Seminar on Modal Analysis, 1981, K.U.Leuven, Belgium

10) G.DESANGHERE, 'Identification of external forces based on frequency response measurements' Proc.8th Int. Seminar on Modal Analysis, part3b, K.U.Leuven, Belgium.

11) R.E.POWELL, W.SEERING, 'Multichannel structural inverse filtering' Transactions of the ASME, Vol.106, pp.22-28, 1984

12) P.SAS 'The use of digital signal processing techniques in acoustic noise source localisation, including acoustic intensity measurements' Phd. thesis, K.U.Leuven, 1982, Belgium.

STRUCTURAL IDENTIFICATION OF NONLINEAR SYSTEMS
SUBJECTED TO QUASISTATIC LOADING

A. Nappi
Politecnico di Milano, Milan, Italy

Introduction

During the last decade the mathematical methods developed within the context of System Identification [1-2] have been applied to the solution of 'inverse problems' in structural engineering. Most of the research activity in this field has primarily concerned dynamic systems, with the purpose of identifying, on the basis of experimental data, parameters hardly susceptible to be measured directly, such as damping characteristics and local stiffness [3-14]. In certain civil engineering situations, however, measurements have to be made in static conditions on systems which exhibit non-linear behaviour under external actions. An interesting example is provided by models concerned with the flexural behaviour of reinforced concrete beams under cyclic loading [15]. Another case is represented by the parameters related to the local strength of rock masses (e.g.: cohesion and friction angle). Such parameters are hardly measurable 'in situ' and sometimes may be estimated through nonlinear response measurements [16-17]. Indeed, the application of system identification to geotechnical problems seems to be very promising and parameter estimation techniques have been the object of remarkable research activity [16-22]. This is mainly due to the importance of 'in situ' measurements, which can often provide much more information about large rock masses than laboratory tests.

In these notes, attention is given to some aspects of 'inverse' problems concerned with systems under quasi-static loading conditions. These problems can be described as follows. We consider a discrete model developed for analysis purposes, and we assume that some parameters related to the geometry and/or to the mechanical behaviour be unknown. A prototype of the system is subjected to convenient quasi-static loading conditions and some measurements are taken of meaningful displacements and/or rotations. The purpose is now to determine the unknown or uncertain parameters included in the mathematical model, on the basis of the available experimental data.

Two general engineering motivations can be found: (i) the 'calibration' of models with respect to uncertain parameters in order to obtain an improved model for design purposes; (ii) the indirect assessment of possible local structural changes or damage which might not be detectable by direct inspections.

In what follows we will focus mainly on piecewise-linear discrete models, which result from reasonable idealizations of structures entailing plastic yielding. We shall deal mostly with reversible ('holonomic') elastic plastic laws. However irreversible phenomena, such as local unstressing will be also accounted for by adopting an incremental, 'non-holonomic' theory of plasticity and a Kalman Filtering technique. Namely, the topics presented in these notes can be summarized as follows:

. Bayesian estimation of: (i) local strength parameters which characterize a nonlinear elastic structure; (ii) elastic moduli and geometrical parameters related to a bi-dimensional, geotechnical model. As typical of the Bayesian approach, optimum estimates of the parameters are derived not only from mean values and convariance matrices of measured displacements, but also from initial values of the parameters assessed by 'a priori' engineering judgement [23].

. Estimation of parameters related to the local plastic deformability by Kalman Filtering. This technique has been applied in the context of the incremental (reversible, 'non-holonomic') theory of plasticity with the aim of exploiting sequentially the experimental information gathered along the evolution of a system under a history of external actions. As such, it appears to fit ideally the path-dependent, irreversible nature of plastic structural response and the sequential character of the measurements usually performed in a statical test [24].

The study briefly presented here has been part of a research project carried out at the Department of Structural Engineering (Politecnico of Milan). Specifically, the theoretical background and the numerical examples which follow have been taken without significant changes from Refs. 23 and 24, where the reader can find further details. Other results from the same project can be found in Refs. 16, 17, 21, 22, 25 - 29.

A mathematical model in the presence of holonomic piecewise-linear local plastic deformability law

We shall consider only simple elastic-plastic structural models such as trusses or beams in bending. The peculiar feature is to exhibit a

local (sectional) behaviour with a single generalized stress and strain
component. However, a much broader class of aggregates of multicomponent
elements with piecewise-linearized yield surfaces, is implicitly covered
simply by re-interpretation of symbols [30].

The element behaviour is assumed as elastic-plastic with linear
hardening in the spirit of the holonomic (path-independent, reversible)
plasticity theory. As well known, in this case only local unstressing may
cause discrepancy from truly irreversible elastic-plastic constitutions.
For the present (identification) purposes, the holonomic assumption
appears to be acceptable and realistic, since the loading process may be
chosen as proportional. Then, irreversibility manifestations (local
unstressing) are unlikely to occur or to have significant effects on
measured displacements. Anyway their occurrence can be checked by means
of the analysis through the mathematical model.

The assumed behaviour of the i-th element (say beam section or truss
member) is shown in Fig.1, where: Q^i is the generalized stress; q^i the
generalized strain; S^i denotes elastic stiffness, $r^i_{1,2}$ absolute values of
yield limits, and $H^i_{1,2}$ hardening moduli; λ^i_1 and λ^i_2 are measures of the
plastic strains according to each of the yielding modes separately, such
that:

$$\lambda^i_1, \lambda^i_2 \geq 0; \quad \phi^i_1 \lambda^i_1 = \phi^i_2 \lambda^i_2 = 0 \qquad (1a,b)$$

having defined two yield functions ϕ^i_1 and ϕ^i_2 as:

$$\phi^i_1 = - (r^i_1 + H^i_1 \lambda^i_1) + Q^i \leq 0 \qquad (1c)$$

$$\phi^i_2 = - (r^i_1 + H^i_2 \lambda^i_2) - Q^i \leq 0 \qquad (1d)$$

Setting: $\underline{H}^i \equiv \mathrm{diag} \, [H^i_1, H^i_2]$, $\underline{r}^T_i = [r^i_1, r^i_2]$, $\underline{\phi}^T_i = [\phi^i_1, \phi^i_2]$, $\underline{N}^i = [1, -1]$,
$\underline{\lambda}^T_i = [\lambda^i_1, \lambda^i_2]$, Eqs. (1) can be re-written in the following matrix form:

$$\underline{\phi}^i = \underline{N}^T_i Q^i - \underline{r}^i - \underline{H}^i \underline{\lambda}^i \leq \underline{0} \qquad (2a)$$

$$\underline{\lambda}^i \geq \underline{0}, \quad \underline{\phi}^T_i \underline{\lambda}^i = 0 \qquad (2b)$$

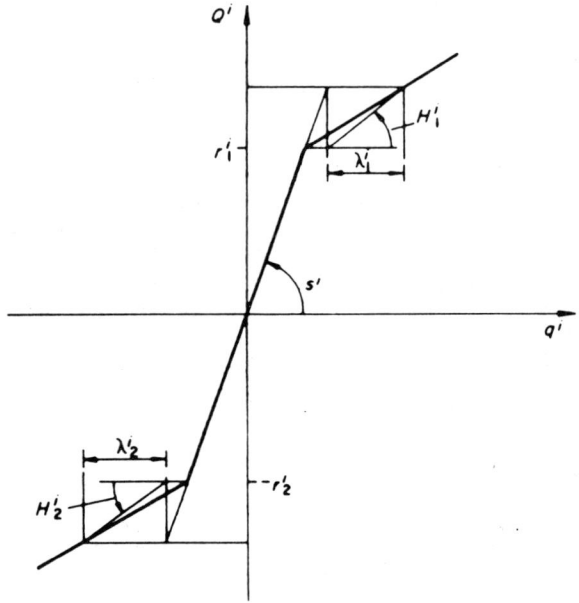

Fig. 1 - Elastic-hardening law for a truss member: specification of symbols

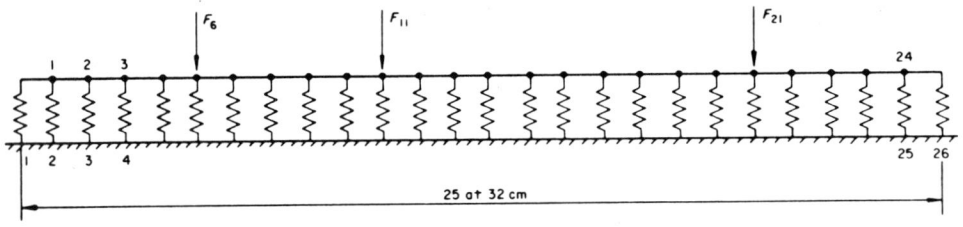

Fig. 2 - Model of elastic-plastic beam on elastic-plastic foundation

This representation of the plastic behaviour has to be supplemented by the description of the elastic behaviour ($p^i = \underline{N}^i \underline{\lambda}^i$ being the plastic strain):

$$Q^i = S^i (q^i - \underline{N}^i \underline{\lambda}^i) \tag{2c}$$

For $\underline{H} = \underline{0}$ the element behaviour represented would become (holonomic) elastic-perfectly plastic.

Eqs. (2) can also be written for all m elements ($i = 1, \ldots, m$) simultaneously as follows:

$$Q = \underline{S} \, \underline{g} - \underline{S} \, \underline{N} \, \underline{\lambda} \; ; \quad \underline{\lambda} \geq \underline{0} \tag{3a,b}$$

$$\underline{\phi} = \underline{N}^T Q - \underline{r} - \underline{H} \, \underline{\lambda} \leq \underline{0} \tag{3c}$$

$$\underline{\phi}^T \underline{\lambda} = 0 \tag{3d}$$

For a kinematically determinate trusslike structural model with n degrees-of-freedom the compatibility and equilibrium equations read:

$$\underline{g} = \underline{C} \, \underline{u} \; ; \quad \underline{C}^T Q = \alpha \underline{F} \tag{4a,b}$$

where $\underline{u}$ and $\underline{F}$ denote the n-vectors of nodal displacementes and loads, respectively; α is a load factor; $\underline{C}$ is the full-column rank matrix depending on the undeformed geometry only (if the "small" deformations hypothesis applies).

The relation set (3)-(4) governs the nonlinear response of the structure to external loads and represents the mathematical model referred to in most of the identification problems to discuss later.

By using Eqs. (3a) and (4a), Eq. (4b) becomes:

$$\underline{C}^T \underline{S} \, \underline{C} \, u - \underline{C}^T \underline{S} \, \underline{N} \, \lambda = \alpha \underline{F} \tag{5}$$

Hence, we can express vector $\underline{u}$ as the sum of a linear elastic contribution $\underline{u}^E$ and of a term due to plastic deformations:

$$\underline{u} = \alpha \, \underline{u}^E + \underline{G} \, \underline{\lambda} \tag{6a}$$

having set:

$$\underline{u}^E \equiv \underline{K}^{-1} \underline{F} \quad ; \quad \underline{G} \equiv \underline{K}^{-1} \underline{C}^T \underline{S} \underline{N} \tag{6b,c}$$

where $\underline{K} \equiv \underline{C}^T \underline{S} \underline{C}$ is the (symmetric, positive definite) elastic stiffness matrix of the system. By further substitutions, the following alternative set of governing relations in ϕ, λ only is generated:

$$\underline{\phi} = \underline{N}^T \underline{Q}^E \underline{\alpha} - \underline{A} \underline{\lambda} - \underline{r} \le \underline{0} \tag{7a}$$

$$\underline{\lambda} \ge \underline{0} \quad ; \quad \underline{\phi}^T \underline{\lambda} = 0 \tag{7b,c}$$

having set:

$$\underline{Q}^E \equiv \underline{S} \underline{C} \underline{K}^{-1} \underline{F} \quad ; \quad \underline{A} \equiv \underline{H} + \underline{N}^T \underline{S} \underline{N} - \underline{G}^T \underline{K} \underline{G} \tag{7d,e}$$

Eqs. (7) represent a 'Linear Complementarity Problem'. By virtue of the Karush-Kuhn-Tucker theorem of mathematical optimization [31] it turns out to be equivalent to a convex quadratic program [32], as matrix $\underline{A}$ is symmetric and positive semidefinite if $H^i_{1,2} \ge 0$ for all i's (positive definite if all hardening parameters are positive).

Bayesian estimation of yield limits

We assume that a test is carried out on an elastic plastic truss-like structure modelled as specified in the preceding Section and that d displacements are measured at a single load level $\bar{\alpha}$.

Measurements are considered as affected by (Gaussian) statistically independent random errors with null mean values and known variances. In other terms, they are defined by a d-vector $\underline{u}^M$ of mean values and by a covariance matrix $\underline{C}_u$.

The independent parameters to identify, collected in vector $\underline{P}$, are those which govern the yield limit distribution defined by vector $\underline{r}$ through a linear relation:

$$\underline{r} = \underline{R} \underline{P} \tag{8}$$

where $\underline{R}$ is a binary 'selective' matrix. Namely, the i-th column of $\underline{R}$ contains unit entries in correspondence to yield limits which are equal (to the same value P_i) for technological reasons and contains zero entries elsewhere.

Let the 'a priori' knowledge on the parameters be expressed by a p-vector P_o of mean values and by a diagonal covariance matrix C_P^o. Vector P_o is thought of as generated by 'best guesses' resting on engineering judgement and previous experience or on any source of information which would be used in the absence of system identification; C_P^o reflects the confidence levels attributed to these values.

For a hypothetical mathematical model linear in the parameters, we would express the computed displacements as follows:

$$u^C = L \ P \tag{9}$$

L being a constant matrix depending on the external actions and on the system characteristics different from the parametrs P. If Eq. (9) were applicable, it could be shown [33] that the optimal (minimum variance) estimator of P on the basis of the above experimental and 'a priori' stochastic information, is given by:

$$\hat{P} = P_o + M \ (u^M - L \ P_o) \tag{10}$$

Matrix M in Eq. (10) is a 'mapping matrix' which combines the uncertainties of measures C_u and of initial estimates (C_P^o):

$$M = C_P^o \ L^T \ [L \ C_P^o \ L^T + C_u]^{-1} \tag{11}$$

The covariance matrix associated to the optimal estimates (10) turns out to be [33]:

$$\hat{C}_P = C_P^o - M \ L \ C_P^o = [C_P^{o^{-1}} + L^T \ C_u^{-1} \ L]^{-1} \tag{12}$$

Actually, in the model under consideration, the dependence of the displacements from the parameters is not linear. However, for small perturbations around a vector P' of current values for the parameters, a Taylor series expansion truncated after the linear terms reads:

$$u^C(P) \simeq u^C(P') + L(P') \ \{P - P'\} \equiv u^C(P,P') \tag{13}$$

Here $\underline{L}(\underline{P}')$ represents the 'sensitivity matrix'. It contains the derivatives of u^C taken with respect to the parameters $\underline{P}$ and calculated at $\underline{P} = \underline{P}'$. Vector u^C depends linearly through Eq. (6a) on vector λ, which in turn depends on $\underline{P}$ through (7) and (8). For a given assigned parameter vector $\underline{P}'$ the complementarity problem (7) can be solved by some standard quadratic programming algorithm. In the numerical solution of (7) for $\underline{P} = \underline{P}'$, let ϕ' indicate the subvector of all zero ϕ variables, and λ' the subvector of the corresponding λ variables (the only ones which may be nonzero). Marking by a prime the corresponding submatrices of $\underline{N}$, $\underline{A}$, $\underline{R}$, one can write:

$$\phi = \underline{N}'^T Q^E \alpha - \underline{A}' \underline{\lambda}' - \underline{R}' \underline{P}' = \underline{0} \tag{14}$$

This equation provides the gradients of the plastic multipliers λ with respect to the parameters in the form:

$$\left[\frac{\partial \underline{\lambda}'}{\partial \underline{P}^T} \right]_{\underline{P}'} = - [\underline{A}']^{-1} \underline{R}' \tag{15}$$

For hardening models (H>0 for all modes) problem (7) has always a unique solution and $\underline{A}'$ is positive definite, so that the matrix (15) exists. For the perfectly-plastic idealization ($\underline{H} = \underline{0}$), submatrix $\underline{A}'$ may be singular in the following two circumstances:
(i) the test loading attains the carrying capacity exhibited by the structural model for $\underline{P} = \underline{P}'$: then Eq. (14) admits an unbounded set of solutions and λ' corresponds to a collapse mechanism; if the carrying capacity is exceeded, problem (7) has no solution;
(ii) below collapse there may be a bounded set of solutions, corresponding to configurations in equilibrium under the same loads (limited or pseudo-mechanism).
By introducing into the model a fictitious hardening of small magnitude (say, moduli H 2-3 orders of magnitude smaller than the member elastic stiffness S) Eq. (15) can still be used in both above circumstances. Computational experience showed that the inaccuracy thus generated in the sensitivity matrix generally does not prevent convergence nor does it reduce the speed of convergence to the true value of $\underline{P}$ in the iterative procedure which will be adopted later.

Let us now introduce a binary matrix $\underline{B}$ which selects the displacements susceptible to measurement among all calculated displacements:

$$\underline{u}^C = \underline{B}\ \underline{u} \tag{16}$$

vector $\underline{u}$ being given by Eq. (6a). Hence, in view of Eqs. (6a) and (16), the sensitivity matrix $\underline{L}(P')$ in Eq. (13) acquires the simple closed form:

$$\underline{L}(P') \equiv \left[\frac{\partial \underline{u}^C}{\partial \underline{P}^T}\right]_{P'} = \left[\frac{\partial \underline{u}^C}{\partial \underline{\lambda}'^T}\right]\left[\frac{\partial \underline{\lambda}'}{\partial \underline{P}'^T}\right]_{P'} = -\ \underline{B}\ \underline{G}'\ \underline{A}'^{-1}\ \underline{R}' \tag{17}$$

where, the submatrix $\underline{G}'$ of $\underline{G}$, Eq. (6c), corresponds to the zero-components of ϕ in the solution of problem (7) for the current parameter vector $\underline{P}'$.

For the present nonlinear model an iterative parameter estimation procedure can now be devised on the basis of the above linearizations. At each iteration the optimal estimators (10) - (12) valid for the model (9) linear in the parameters are applied to the linearized model (13). A similar procedure has been used in linear dynamics (see e.g. [4] and [5]). The iteration scheme is based on the estimator (10) and (11) in the following recurrent form:

$$\hat{\underline{P}} = \underline{P}_o + \underline{M}(\underline{P}')\ [\underline{u}^M - \underline{u}^C(\underline{P}_o,\ \underline{P}')] \tag{18a}$$

$$\underline{M}(\underline{P}') = \underline{C}_p^o\ \underline{L}^T(\underline{P}')\ [\underline{L}(\underline{P}')\ \underline{C}_p^o\ \underline{L}^T(\underline{P}') + \underline{C}_u]^{-1} \tag{18b}$$

$\underline{u}^C(\underline{P}_o,\ \underline{P}')$ being defined by Eq.(13).

The iterative procedure can be described by the following sequence.

(1) On the basis of the initial vector $\underline{P}_o$ of guessed values for the parameters, solve the analysis problem (7). This provides the calculated measurable displacements $\underline{u}^C(\underline{P}_o)$ through (16) and the sensitivity matrix $\underline{L}(\underline{P}_o)$.

(2) Generate the mapping matrix $\underline{M}(\underline{P}_o)$ by means of (18b) setting $\underline{L} = \underline{L}(\underline{P}_o)$ and evaluate the estimates $\hat{\underline{P}}$ by means of Eq.(18a), for $\underline{P}' = \underline{P}_o$.

(3) Setting $\underline{P}' = \hat{\underline{P}}$, solve again the LCP (7) for $\underline{P} = \underline{P}'$, and thus obtain $\underline{L}(\underline{P}')$ and $\underline{u}^C(\underline{P}_o, \underline{P}')$, through Eq.(13).

(4) Determine $\underline{M}(\underline{P}')$ by (18b) with $\underline{L}(\underline{P}')$ and compute new estimates $\hat{\underline{P}}$.

(5) Carry out a convergence test: if there are significant changes between the last and the preceding estimates, return to step 3.

After termination of the above procedure, say at the estimate $\underline{P}^*$, the corresponding covariance matrix $\underline{C}_p^*$ is supplied by Eq.(12):

$$\underline{C}_p^* = \underline{C}_p^o - \underline{M}(\underline{P}^*)\ \underline{L}(\underline{P}^*)\ \underline{C}_p^o = [\underline{C}_p^{o^{-1}} + \underline{L}^T(\underline{P}^*)\ \underline{C}_u^{-1}\ \underline{L}(\underline{P}^*)]^{-1} \qquad (19)$$

This matrix $\underline{C}_p^*$ associated to the achieved estimates $\underline{P}^*$, provides a quantification of the uncertainties affecting them: its main diagonal entries represent the new variances of the parameters at the end of the iterative process [4].

Statistical estimation without prior information - Large variances with respect to mean values (i.e. large diagonal entries in the covariance matrix $\underline{C}_p^o$) reflect scarcity of 'a priori' information (prior to the experiments) about the parameters to identify. When the only available statistical data are those provided by the measurements ($\underline{u}^M$, $\underline{C}_u$), one can specialize the preceding method by a limiting process which reduces the inverse of the initial covariance matrix for the parameters to a matrix with zero entries: $[\underline{C}_p^o]^{-1} \to \underline{0}$. Transforming the r.h.s. of Eq. (18b) by a matrix identity [2] and passing to the limit, the mapping matrix becomes:

$$\underline{M}(\underline{P}') = [\underline{C}_p^{o^{-1}} + \underline{L}^T(\underline{P}')\ \underline{C}_u^{-1}\ \underline{L}^T(\underline{P}')]^{-1}\ \underline{L}^T(\underline{P}')\ \underline{C}_u^{-1} =$$

$$= [\underline{L}^T(\underline{P}')\ \underline{C}_u^{-1}\ \underline{L}(\underline{P}')]^{-1}\ \underline{L}^T(\underline{P}')\ \underline{C}_u^{-1} \tag{20}$$

The covariance matrix of the final estimates, given by Eq. (19), acquires the form:

$$\underline{C}_p^* = [\underline{L}^T(\underline{P}^*)\ \underline{C}_u^{-1}\ \underline{L}(\underline{P}^*)]^{-1} \tag{21}$$

In this situation of uncertain a priori knowledge the initial guessed vector $\underline{P}_o$ plays the role of initialization vector and has, at least in principle, no influence on the resulting end estimates $\underline{P}^*$. In fact, substitute into Eq.(18a) the latter of the expressions (20) for the mapping matrix and Eq. (13) for $\underline{u}^C(\underline{P}_o, \underline{P}')$. Thus Eqs. (18), to be used recurrently in the iterative procedure, reduce to:

$$\hat{\underline{P}} = [\underline{L}^T(\underline{P}')\ \underline{C}_u^{-1}\ \underline{L}(\underline{P}')]^{-1}\ \underline{L}^T(\underline{P}')\ \underline{C}_u^{-1}\ \{\underline{u}^M - \underline{u}^C(\underline{P}')\} + \underline{P}' \tag{22}$$

which is equivalent to the result provided by the generalized least square method [2].

Numerical test with a piecewise-linear model - The above approach has been tested by considering the model of an elastic-plastic beam on elastic-plastic foundation shown in Fig. 2. The beam model is made of 24 elastic-perfectly plastic hinge elements, while the foundations are represented by 26 elastic-perfectly plastic springs with no-tension capacity (unilateral support). The lower yield limit of the springs P_1 and the common absolute value P_2 of the flexural yield limits of the hinges, are regarded as unknown parameters.
     We consider simulated experiments. These are carried out by computing a set of displacements through a given structural model and by using such displacements as fictitious 'measured data' for parameter estimation. Clearly, in this way modelling errors are ruled out at all. The 'measured' displacements have been generated by assuming as 'true' values of the parameters $P_1 = 1.6$ KN, $P_2 = 1.25$ KN m and setting $F_6 = F_{11} = 6$ KN, $F_{21} = 8$ KN for the external loads. As shown in Tab. 1, several sets of fictitious measurements have been considered; a single variance has been attributed to all measures in each case. Tab. 1 specifies the initial and final mean values and variances of the parameters given by the Bayesian identification procedure described above. The number of iterations reported in Tab. 1 is referred to the steps after which the variations of the estimates have been less than 10^{-4} KN and 10^{-4} KN m, respectively. Details of the 16th and of the 17th case are illustrated in Figs. 3 and 4, where the updated values after each iteration are plotted

CASE	MEASURED DISPLACEMENT LOCATIONS	VARIANCE OF MEASURES	INITIAL VALUE	ESTIMATED VALUE	INITIAL VARIANCE	ESTIMATED VARIANCE
1	5,10,15,20,25	.00001	2.0	1.676	0.01	.00442
2	5,10,15,20,25	.000001	2.0	1.630	0.01	.00166
3	5,10,15,20,25	.0000001	2.0	1.607	0.01	.000229
4	5,10,15,20,25	.00001	1.3	1.589	0.01	.00442
5	5,10,15,20,25	.000001	1.3	1.598	0.01	.00166
6	5,10,15,20,25	.0000001	1.3	1.602	0.01	.000229
7	17 - 21	.00001	2.0	1.663	0.01	.00405
8	17 - 21	.000001	2.0	1.624	0.01	.00145
9	17 - 21	.0000001	2.0	1.605	0.01	.000196
10	17 - 21	.00001	1.3	1.599	0.01	.00393
11	17 - 21	.000001	1.3	1.603	0.01	.00144
12	17 - 21	.0000001	1.3	1.603	0.01	.000196
13	19,20,22,23	.00001	2.0	1.662	0.01	.00429
14	19,20,22,23	.000001	2.0	1.631	0.01	.00218
15	19,20,22,23	.0000001	2.0	1.607	0.01	.000372
16	19,20,22,23	.00001	1.3	1.610	0.01	.00411
17	19,20,22,23	.000001	1.3	1.605	0.01	.00218
18	19,20,22,23	.0000001	1.3	1.603	0.01	.000372

a)

CASE	MEASURED DISPLACEMENT LOCATIONS	VARIANCE OF MEASURES	INITIAL VALUE	ESTIMATED VALUE	INITIAL VARIANCE	ESTIMATED VARIANCE
1	5,10,15,20,25	.00001	1.5	1.183	0.01	.00380
2	5,10,15,20,25	.000001	1.5	1.223	0.01	.00142
3	5,10,15,20,25	.0000001	1.5	1.224	0.01	.000196
4	5,10,15,20,25	.00001	.95	1.257	0.01	.00380
5	5,10,15,20,25	.000001	.95	1.251	0.01	.00142
6	5,10,15,20,25	.0000001	.95	1.248	0.01	.000196
7	17 - 21	.00001	1.5	1.191	0.01	.00404
8	17 - 21	.000001	1.5	1.227	0.01	.00145
9	17 - 21	.0000001	1.5	1.245	0.01	.000212
10	17 - 21	.00001	.95	1.248	0.01	.00426
11	17 - 21	.000001	.95	1.245	0.01	.00155
12	17 - 21	.0000001	.95	1.245	0.01	.000212
13	19,20,22,23	.00001	1.5	1.193	0.01	.00477
14	19,20,22,23	.000001	1.5	1.218	0.01	.00243
15	19,20,22,23	.0000001	1.5	1.243	0.01	.000415
16	19,20,22,23	.00001	.95	1.230	0.01	.00505
17	19,20,22,23	.000001	.95	1.245	0.01	.00243
18	19,20,22,23	.0000001	.95	1.247	0.01	.000415

b)

Tab. 1 - Results obtained with elastic-plastic beam: (a) estimation of P_1 (limit compression expressed in KN); (b) estimation of P_2 (yield limit expressed in KN m).

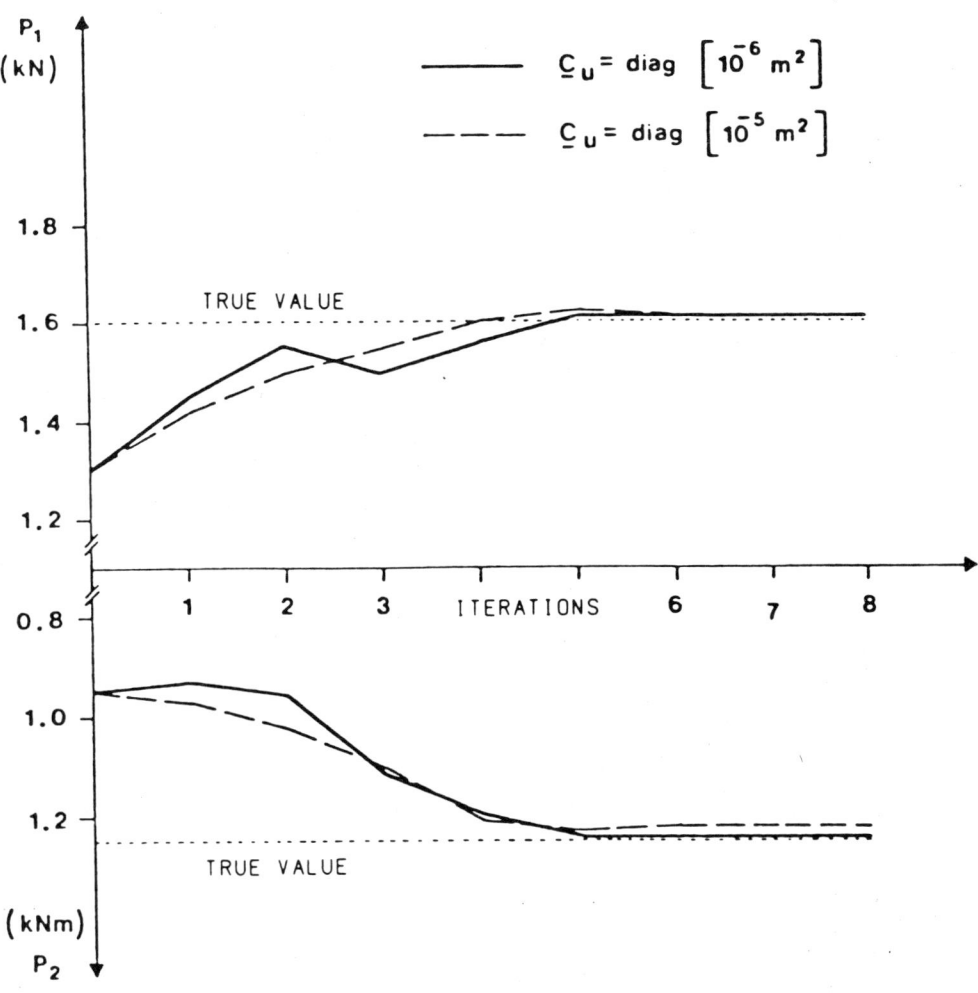

Fig. 3 - Parameter estimates for case 16 (dashed lines) and case 17
        (solid lines) of Tab. 1

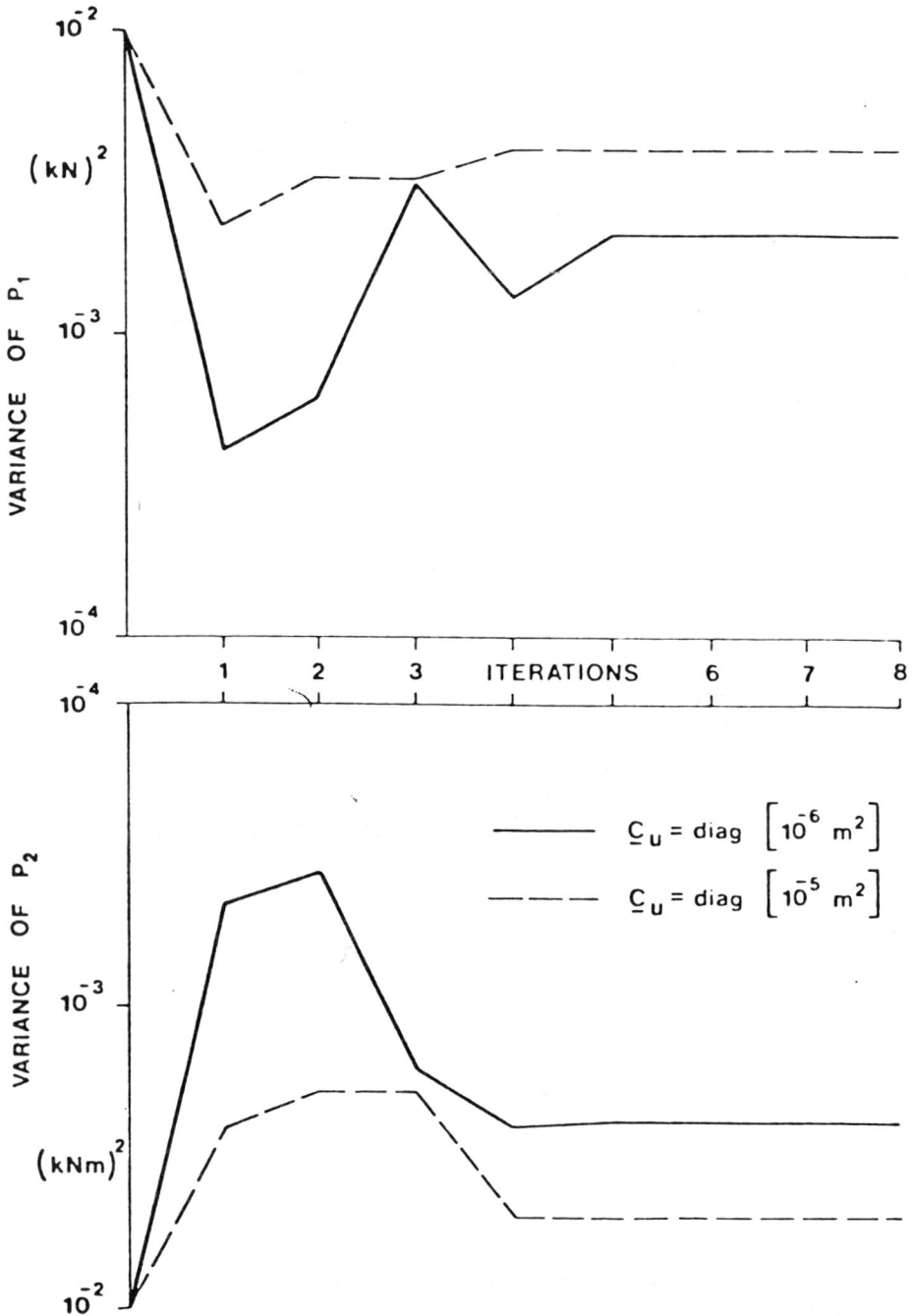

Fig. 4 - Variances of estimates for case 16 (dashed lines) and case 17
         (solid lines) of Tab. 1

along with the relevant variances; in the figures, dashed lines are referred to the 16th case (C_u = diag $[10^{-5} m^2]$). As well expected in a Bayesian inference context, one observes in Tab. 1 that the discrepancy between actual and estimated parameters decreases, at equal prior information, as the variance of the measures (i.e. the uncertainty of the experimental information) decreases; in fact, the influence of the analyst's initial assessment on the final results becomes less and less significant. Tab. 2 illustrates limit situations of lacking prior knowledge on the parameters; these cases have been dealt with by the specialized iterative procedure described in the preceding Section. As expected, in such situations the identification process always converges to the true values of the parameters. The number of iterations reported in Tab. 2 has been obtained by using the same convergence criterion (or tolerance) given above for the Bayesian estimation procedure. Numerical tests have shown that by increasing standard deviations of the initial parameter estimates, we obtain final values which tend to reproduce the measured response. Conversely, if more credit is given to the 'a priori' guess by associating small variances to the initial parameters, final estimates do not differ much from the initial values. For details see Ref. 23.

Bayesian estimation of parameters concerned with an elastic bi-dimensional system - The same concepts discussed in the previous Sections have been applied also to a geotechnical system and the following identification problems have been solved:
(i) the finite element model of a lens included in a layer of homogeneous rock is considered (Fig. 5). Under the hypothesis of linear elastic behaviour, different values of Young's moduli are assumed for the layer and the inclusion. A test is simulated by imposing a parabolic load applied along the surface and vertical displacements are calculated at the 20 points shown in Fig. 5. These displacements have been assumed as 'in situ' measurements, with the aim of estimating the elastic moduli E_1 and E_2 of the layer and of the inclusion, respectively. The identification process has been exactly the same as before. The only remarkable difference consists in the fact that the sensitivity matrix has been evaluated numerically. Results are reported in nondimensional form in Tab. 3, where starred symbols denote true values.
(ii) In a second set of tests the inclusion length and its distance from a fixed vertical line (parameters b^* and d^* in Fig. 5) have been estimated, while elastic moduli have been assumed as known. Again, derivatives have been computed numerically. Some results are given in Tab. 3.
    It is worth noting that the same parameters have been estimated on the basis of a deterministic approach [22] by generating random ficti-tious measures with given mean values (corresponding to the 'true' displacements, as computed on the basis of the actual parameters E_1^*, E_2^*, b^*, d^*). It is interesting to observe that the scattering of the

CASE	MEASURED DISPLACEMENT LOCATIONS	VARIANCE OF MEASURES	INITIAL VALUE	ESTIMATED VALUE	ESTIMATED VARIANCE
1	5,10,15,20,25	.00001	2.0	1.603	.02389
2	17 - 21	.00001	2.0	1.603	.02047
3	19,20,22,23	.00001	2.0	1.602	.04034

a)

CASE	MEASURED DISPLACEMENT LOCATIONS	VARIANCE OF MEASURES	INITIAL VALUE	ESTIMATED VALUE	ESTIMATED VARIANCE
1	5,10,15,20,25	.00001	1.5	1.247	.02047
2	17 - 21	.00001	1.5	1.247	.02216
3	19,20,22,23	.00001	1.5	1.248	.04501

b)

Tab. 2 - Results obtained with elastic-plastic beam without 'a priori' information on parameters: (a) estimation of P_1 (limit compression expressed in KN); (b) estimation of P_2 (yield limit expressed in KN m).

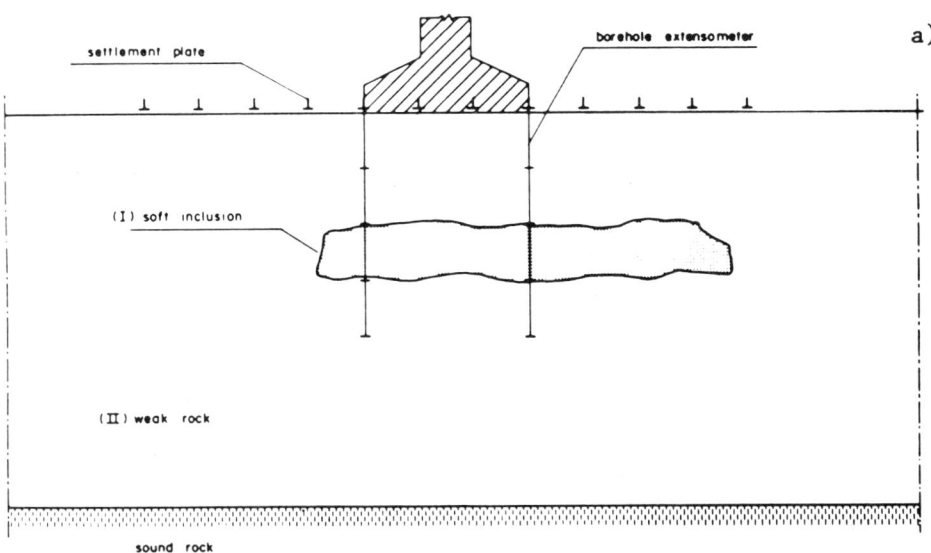

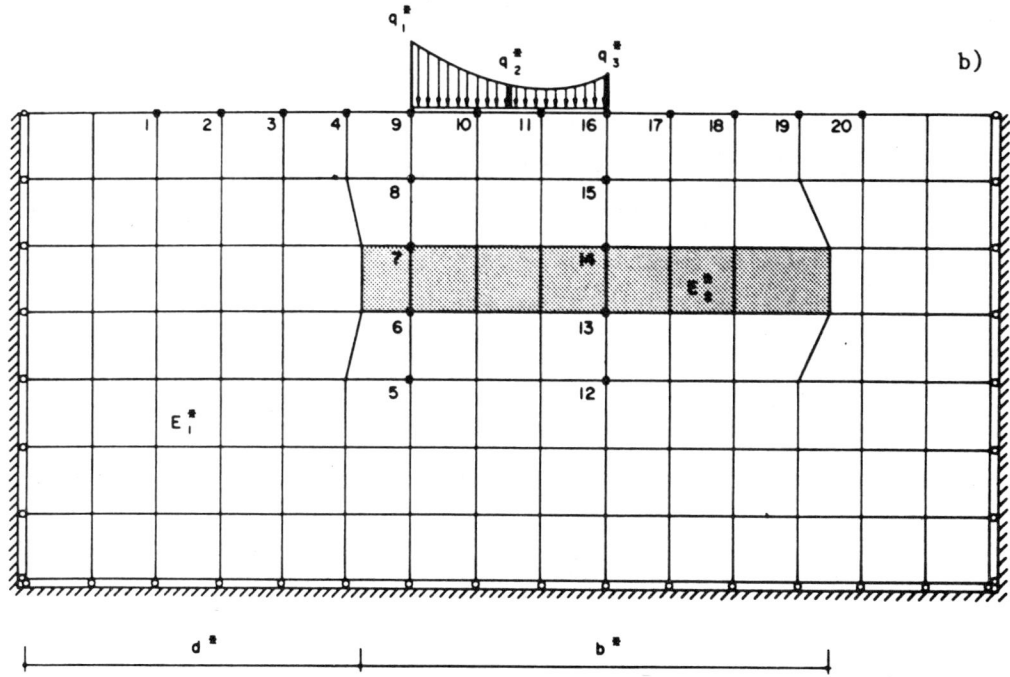

Fig. 5 - Schematic representation of an inclusion (a) and relevant finite element model (b)

	INITIAL VALUES		INITIAL VARIANCES		ESTIMATES		FINAL VARIANCES	
CASE	E_1^o/E_1^*	E_2^o/E_2^*	$c_{E_1}^o/E_1^{*2}$	$c_{E_2}^o/E_2^{*2}$	E_1/E_1^*	E_2/E_2^*	c_{E_1}/E_1^{*2}	c_{E_2}/E_2^{*2}
1	0.917	1.500	.00698	.14063	0.985	1.024	.00091	.00282
2	0.917	1.500	.00310	.06250	0.972	1.048	.00075	.00264
3	0.917	1.500	.00078	.01563	0.937	1.124	.00039	.00209
4	0.917	1.500	.00310	.06250	0.937	1.124	.00157	.00835
5	1.083	0.500	.00310	.06250	1.029	0.956	.00084	.00203

	INITIAL VALUES		INITIAL VARIANCES		ESTIMATES		FINAL VARIANCES	
CASE	b^o/b^*	d^o/d^*	c_b^o/b^{*2}	c_d^o/d^{*2}	b/b^*	d/d^*	c_b/b^{*2}	c_d/d^{*2}
1	0.690	0.952	.02778	.02778	0.948	0.998	0.00498	0.00009
2	0.690	0.952	.02778	.02778	0.899	0.995	0.00832	0.00043
3	1.280	0.720	.02778	.02778	1.241	1.005	0.02436	0.00010

Tab. 3 - Results of Bayesian parameter estimation for the model of Fig. 5
(upper part: Young's moduli; lower part: geometrical parameters)

estimated parameters is about the same (with the Bayesian and the deterministic approach), as shown in Fig. 6. Clearly, however, the computational effort required by the deterministic estimation procedure is much higher and appears to be non-feasible for large problems. As supplementary information, the histograms of the estimated elastic moduli and the corresponding log-normal probability density functions obtained after 1000 deterministic estimation processes [22] are shown in Fig. 7.

Yield limit estimation by Kalman filtering

In this Section we consider elastic-plastic structural models fully allowing for the path-dependent, irreversible nature of plastic behaviour. The problem of estimating parameters related to the local 'plastic' deformability will be investigated in a stochastic context, employing the methodology of Kalman filtering [34, 35, 36]. This methodology is intended to exploit sequentially the experimental information gathered along the evaluation of a system under a history of external actions. As such, it appears to fit ideally both the path-dependent nature of plastic structural response and the sequential character of the measurements usually performed in a statical test.

Since the material model is non-reversible, the elastic-plastic, quasi-static response of a structure will be computed by considering small load increments, say from 'instant' t to t+1 of an ordering variable or 'time' t. We assume that the unit time-change corresponds to a loading step small enough to verify the holonomic (path-independence) hypothesis for each individual step. If we again consider piecewise linear yield surfaces, it is possible to formulate a Linear Complementarity Problem which is formally similar to (7) and is referred to a single time step:

$$\phi_{t+1} = N^T Q^E_{t+1} - A(\lambda_t + \Delta\lambda_{t+1}) - r \leq 0 \tag{23a}$$

$$\phi^T_{t+1} \ \Delta\lambda_{t+1} = 0 \quad ; \quad \Delta\lambda_{t+1} \geq 0 \tag{23b,c}$$

where λ_t represents the plastic multipliers at the end of step t. Vectors $\Delta\lambda_{t+1}$ and ϕ_{t+1} are the sign constrained variables which solve problem (23). We can also write an expression analogous to (6a):

$$u_{t+1} = u^E_{t+1} + G(\lambda_t + \Delta\lambda_{t+1}) \tag{24}$$

The parameters to be estimated in the following on the basis of experimental information on these displacements, will be yield limits (the independent ones) included in vector P.

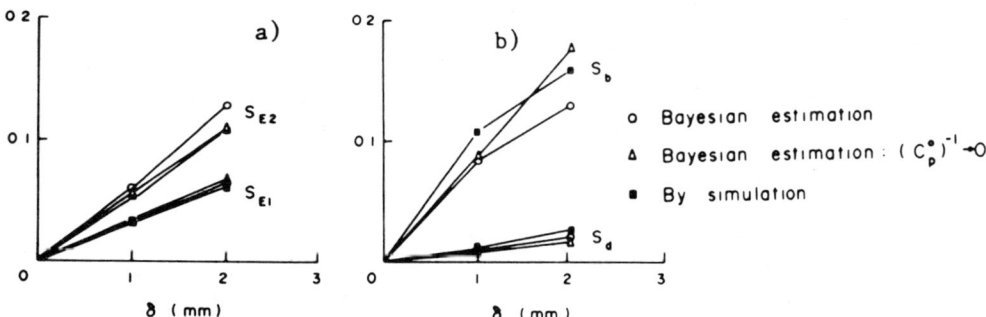

Fig. 6 - Scattering of estimated values vs. measurement resolution δ = 3σ
         for: (a) estimation of elastic moduli; (b) estimation of
         geometrical parameters

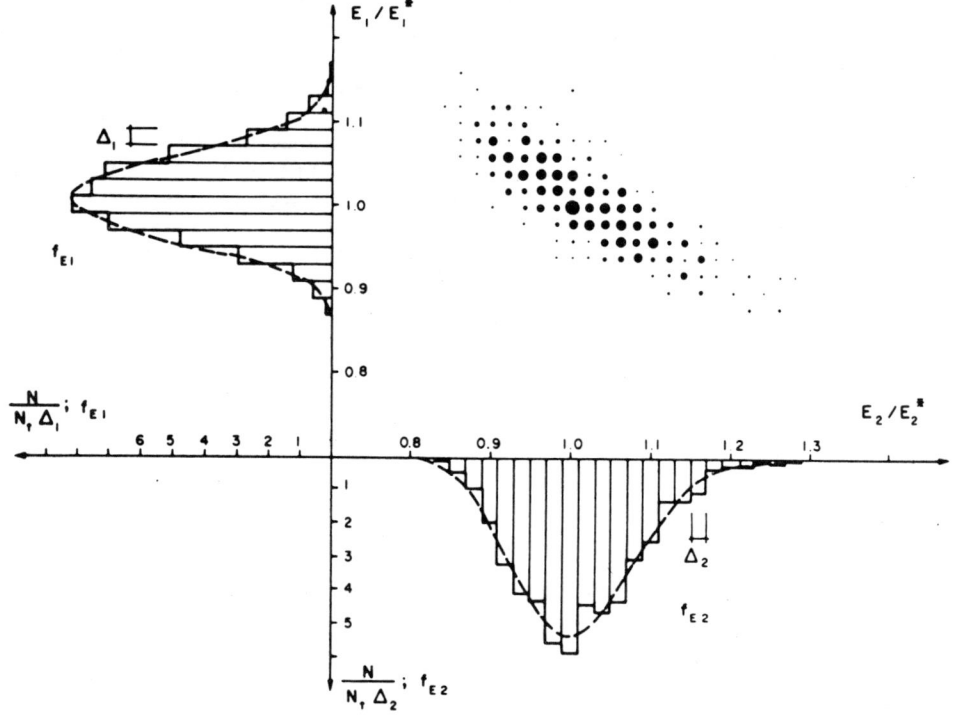

Fig. 7 - Histograms of the optimal values of the elastic moduli and
         related log-normal probability density functions obtained on the
         basis of 1000 deterministic estimates [22]

The natural 'state variables' of the system are the plastic multipliers gathered in vector λ_t. The 'state equations' which describe the system evolution, relate their values at instant t+1 to those at t, to the parameter vector P_t at t and (though not explicitly indicated) to the change ΔQ^E_{t+1} of the elastic stress response to external actions. The (time-independent) parameters P_t will be considered as additional state variables, thus reducing their identification to a state estimation, in the spirit of the 'extended' Kalman filter methodology [34,37]. Therefore, the 'state equations' read:

$$\lambda_{t+1} = \lambda_t + \Delta\lambda_{t+1} = \lambda_t + f(\lambda_t, P_t) \tag{25a}$$

$$P_{t+1} = P_t \tag{25b}$$

Here f represents the functional dependence implicit in the structural model. Let μ_t be a vector of experimental data, i.e. resulting from measurements at t of meaningful nodal displacements (or rotations) u_t. Vector μ_t is conceived as a random variable:

$$\mu_t = u_t(\lambda_t, P_t) + n_t \tag{26}$$

In this 'output equation' the 'noise' term n_t represents the measurement error. As usual, we assume (denoting by E [·] the expected value):

$$E[n_t] = 0; \quad E[n_t n_t^T] = V \text{ for } t = \tau; \quad 0 \text{ for } t \neq \tau \tag{27a,b}$$

In other terms, n_t has zero expected value and zero cross-correlation between n_t and n_τ for any $\tau \neq t$ ('white' noise). The covariance matrix V will be a diagonal matrix of (positive) variances, since the measures are conceived here as uncorrelated random variables.

The nonlinear dependence (25a) is linearized around a given vector $(\bar{\lambda}_t, \bar{P}_t)$ by a truncated Taylor expansion of function f:

$$\Delta\lambda_{t+1} = f(\bar{\lambda}_t, \bar{P}_t) + \left[\begin{array}{c|c} \dfrac{\partial f}{\partial \lambda_t^T} & \dfrac{\partial f}{\partial P_t^T} \end{array}\right]_{\bar{\lambda}_t, \bar{P}_t} \left\{\begin{array}{c} \lambda_t - \bar{\lambda}_t \\ P_t - \bar{P}_t \end{array}\right\} \tag{28}$$

The first term on the r.h.s. represents the distribution of plastic strains at instant t+1 if the system state at t is identified by $(\bar{\lambda}_t, \bar{P}_t)$.

Let us turn now to the 'output equation' (26). Measurable displacements u_{t+1} do not explicitly depend on the parameters P_{t+1}, but only on λ_{t+1} through (24). Hence, in view of (25a), we obtain:

$$u_{t+1}(\lambda_{t+1}, P_{t+1}) = u_{t+1}^E + G \lambda_t + G \Delta\lambda_{t+1}(\lambda_t, P_t) \tag{29}$$

This can also be linearized around vector $(\bar{\lambda}_t, \bar{P}_t)$ and the 'sensitivity matrix' becomes:

$$L \equiv \left[\frac{\partial u_{t+1}}{\partial P_t^T} \right] \tag{30}$$

The derivatives of Eq.(28) and the sensitivity matrix (30) can be computed in closed form, by a procedure expounded in Ref. [24] and omitted here for brevity.

We outline below an iterated (Bayesian) estimation scheme which is typical of the Kalman filtering technique when applied to nonlinear models.

The iteration procedure for each loading step (or time step) is basically the scheme illustrated in the previous section. The conceptually new aspect is that the 'a priori' information at each loading step consists of the predicted 'state variables' (plastic multipliers and parameters) and of their predicted covariance matrix, as computed at the end of the previous step. In other terms the final estimates will depend on a sequence of measured data μ_t; at each time

step new measurements are processed and an iteration scheme is employed in order to account for nonlinearities.

(1a) In view of the 'state equations' (25), we define the predicted state variables at time t making use of the information at time t-1 (i.e., after estimating their values on the basis of measurements μ_{t-1}):

$$\left\{ \begin{array}{c} P_{t/t-1} \\ \lambda_{t/t-1} \end{array} \right\} = \left\{ \begin{array}{c} P_{t-1/t-1} \\ \lambda_{t-1/t-1} + f(\lambda_{t-1/t-1}, P_{t-1/t-1}) \end{array} \right\} \tag{31}$$

(1b) The predicted ('a priori') covariance matrix of the state variables (at time t based on observations up to time t-1) is given by:

$$W_{t/t-1} = D \; W_{t-1/t-1} \; D^T \tag{32}$$

where D is defined as follows, account taken of Eq. (28):

$$D \equiv \begin{bmatrix} I & \vdots & 0 \\ \text{---} & \text{---} & \text{---} \\ \dfrac{\partial f}{\partial P_t^T} & \vdots & \dfrac{\partial f}{\partial \lambda_t^T} \end{bmatrix}_{\lambda_{t-1/t-1}, \; P_{t-1/t-1}} \tag{33}$$

Note that $W_{t-1/t-1}$ is function of $\lambda_{t-2/t-2}$, $P_{t-2/t-2}$, $P_{t-1/t-2}$, as it will be clear below, at point (2b). On the contrary, $W_{t/t-1}$ is function of $\lambda_{t-1/t-1}$, $P_{t-1/t-1}$ through matrix (33).

(2a) Optimal values for the state variables at each time step are given by an iterative procedure similar to the scheme considered in the Bayesian context:

$$\begin{Bmatrix} P_{t/t}^i \\ \lambda_{t/t}^i \end{Bmatrix} = \begin{Bmatrix} P_{t/t-1} \\ P_{t/t-1} \end{Bmatrix} + M_{t/t-1} \; \{ \mu_t - u \, (P_{t/t}^{i-1}) - L(P_{t/t}^{i-1}) \cdot$$

$$\cdot \; (P_{t/t-1} - P_{t/t}^{i-1}) \} \tag{34a}$$

Where $i = 1, \ldots, \nu_t$ if ν_t is the number of subiterations required to account for the nonlinearities at step t. As in the previous section ν_t depends upon the tolerance set on the difference between corresponding terms of vectors $P_{t/t}^{i-1}$. $M_{t/t-1}$ in Eq.(34a) is the mapping matrix and is defined as follows:

$$M_{t/t-1} = W_{t/t-1} \; S^T \; [S \; W_{t/t-1} \; S^T + V]^{-1} \tag{34b}$$

where V is the covariance matrix of the measurement errors and S is the sensitivity matrix. This can be partitioned by separating the derivatives with respect to the parameters from the derivatives with respect to the plastic multipliers:

$$S = S(P_{t/t}^{i-1}) = [L(P_{t/t}^{i-1}) \mid 0] \tag{34c}$$

Hence, $M_{t/t-1}$ is function of $P_{t/t}$ through matrix S and is function of $\lambda_{t-1/t-1}$, $P_{t-1/t-1}$ through matrix $W_{t/t-1}$ as shown by Eq.(32). As pointed out before, Eqs.(34a) and (34b) are analogous to (18a) and (18b).

(2b) Finally, we must define the 'a posteriori' covariance matrix $W_{t/t}$, which expresses the uncertainty of the estimated state variables:

$$W_{t/t} = W_{t/t-1} - M_{t/t-1} \, S(P_{t/t-1}) \, W_{t/t-1} \tag{35}$$

Since $M_{t/t-1}$ appears in Eq.(35), also $W_{t/t}$ is function of $\lambda_{t-1/t-1}$, $P_{t-1/t-1}$, $P_{t/t}$.

Numerical tests based on Kalman Filtering approach - The frame and the loading condition shown in Fig. 8a, have been chosen in order to make some numerical tests. The critical sections at the column ends are marked by dashes. The plastic behaviour is depicted in Fig. 8b and is assumed to be symmetric. A very small fictitious hardening H is introduced to improve the numerical performance of the procedure. The only sectional properties accounted for in the analysis are elastic bending stiffness EI, and yield limits r. The following values (in KN m^2 and KN m, respectively) have been used for the analysis: left column: EI_1 = 2100; r_1 = 40; mid column: EI_2 = 1050; r_2 = 30; right column: EI_3 = 2100, r_3 = 40; horizontal girders EI = 8400; r = 80. The load has been increased from zero up to F_{max} = 72 KN (F_E = 61.14 being the elastic limit load) and then decreased to F_{min} = -72 KN. The displacements and rotations calculated through the analysis have been considered as measured quantities μ. Two displacements are plotted in Fig. 9, while data and main results of the identification process are presented in Tab. 4.

The following cases have been considered:

(i) Two measured displacements (u_1, u_2) and variance $\sigma^2 = 10^{-7} \, m^2$ corresponding to a standard deviation σ = 0·316 mm have been assumed. Roughly speaking, this corresponds to a maximum instrument error of about ± 1 mm; in fact, a Gaussian distribution implies a probability over 99% that the error takes a value within the range ± 3σ $\simeq$ ± 1 mm. The limit bending moments P_1, P_2, P_3 of the left, central and right columns, respectively, are assumed as uncertain parameters. The 'a priori' information (initial estimated values and standard deviations) are reported in Tab. 4. A typical estimation path is illustrated in Fig. 10. Each triangle shows the end of a recursive step, when the best estimate is found on the basis of the available information at a given load

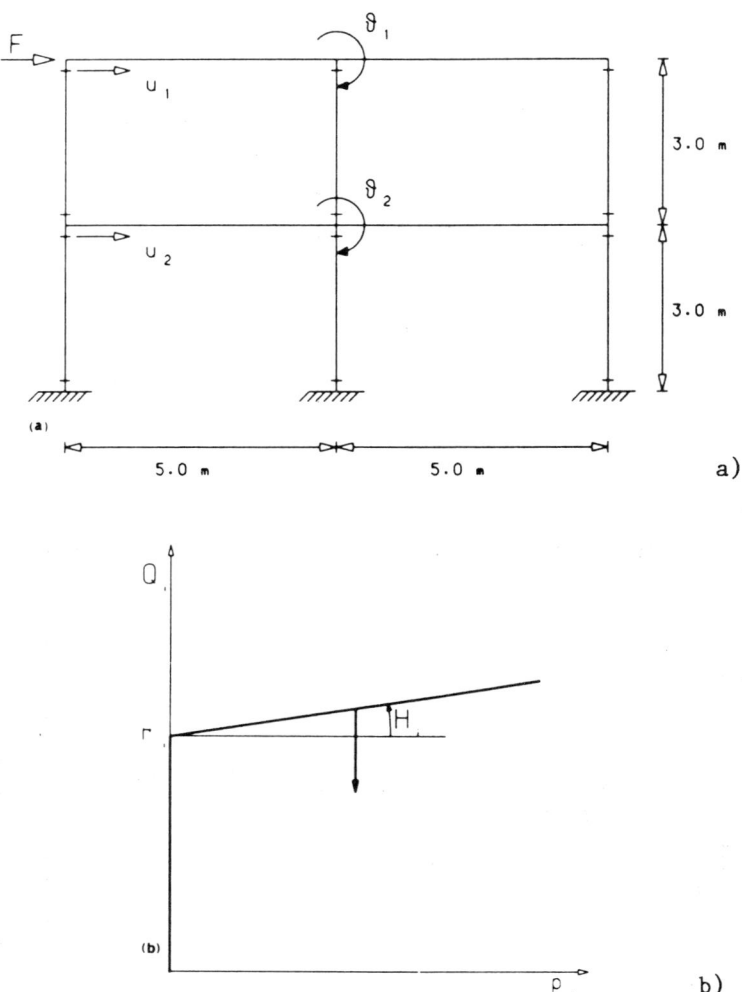

Fig. 8 - Frame used for numerical tests (a) and assumed constitutive law
         for critical sections (b)

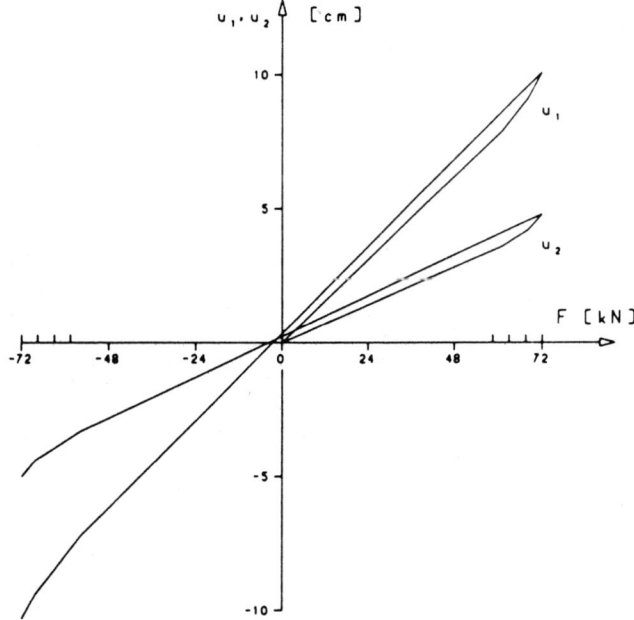

Fig. 9 - Displacements u_1 and u_2 vs. load in simulated experiment

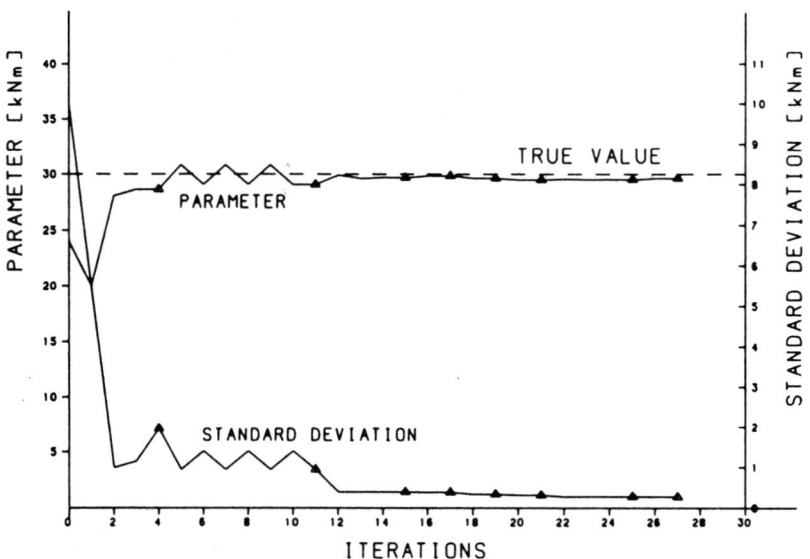

Fig. 10 - Estimation path for parameter P_2 and its standard deviation
        (case 1 of Tab. 4)

factor. The plot of Fig. 10 is concerned with the yield limit for the mid column, R_2. The second column of the Tab. 5 specifies the number of iterations required in order to account for nonlinearities at each load factor. The subsequent columns indicate the values after processing each set of measurements sequentially. The figures in brackets (last three columns of the table) refer to the yield modes which are found to be active during the identification process.

(ii) The assumed variances are two orders of magnitude greater than before and an increased uncertainty is found in the estimated parameters (higher values of their final standard deviations).

(iii) The error variances coincide with the variances of the first case, but the set of 'measured' variables now includes also the rotations θ_1 and θ_2 (i.e., the rotations at the central nodes, as shown in Fig. 8a). More available information produces an improvement of the estimation accuracy (as expected). Tab. 4 shows that the estimates of the flexural strength P_2 of the central columns are more accurate than the others, since the corresponding variances are much smaller. This is due to the fact that the loading process 'activates' a higher number of yielding modes related to P_2 than to P_1 and P_3 (see figures in brackets in Tab. 5) and, hence, 'more information' concerning P_2 than P_1 and P_3 is available and processed in the identification procedure.

Aknowledgements

The advice of S. Bittanti, G. Gioda and G. Maier and the support from C.N.R. are gratefully acknowledged.

References

1. Pister, K.S.," Constitutive modelling and numerical solution of field problems", Nucl. Eng. Design, 1974, 28, 137

2. Eikhoff, P., System identification, John Wiley, Chichester, 1979

3. Ibañez, P., "Identificaion of dynamic parameters of linear and non-linear structural models from experimental data", Nucl. Eng. Design, 1972, 25, 30.

4. Collins, J.D., Hart, G.C., Hasselmann, T.K., Kennedy B. "Statistical identification of structures", AIAA J., 12, 1974, 185-190.

CASE	MEAS.D DISP.S	INITIAL ESTIMATES			STANDARD DEVIATIONS			VARIANCES OF MEAS. ERRORS	FINAL ESTIMATES			FINAL ST.D DEVIATIONS		
		P_1	P_2	P_3	σ_1	σ_2	σ_3		P_1	P_2	P_3	σ_1	σ_2	σ_3
1	$u_1\ u_2$	32	24	32	10	10	10	.0000001	39.4	29.7	39.3	7.1	.84	7.1
2	$u_1\ u_2$	32	24	32	10	10	10	.00001	39.1	29.5	39.1	7.4	2.6	7.4
3	$u_1\ u_2$ $\theta_1\ \theta_2$	32	24	32	10	10	10	.0000001	39.8	29.6	39.1	5.0	.26	5.1

Tab. 4 - Data and results of three tests on yield limit identification by Kalman Filtering (units are and m)

LOAD FACTOR	ITERATIONS	P_1[KN m]	P_2[KN m]	P_3[KN m]	σ_1[KN m]	σ_2[KN m]	σ_3[KN m]
1.3	7	37.2	28.6	37.2	7.2(0)	5.2(0)	7.3(0)
1.4	7	38.6	29.1	38.6	7.1(1)	2.8(2)	7.2(1)
1.5	3	38.6	29.9	38.4	2.1(1)	1.2(2)	2.2(1)
1.6	2	38.8	30.0	38.7	2.1(2)	1.2(4)	2.2(2)
- 1.3	2	39.1	29.6	39.1	2.1(0)	1.0(2)	2.1(0)
- 1.4	2	39.2	29.6	39.1	2.1(0)	0.9(2)	2.1(1)
- 1.5	2	39.3	29.6	39.1	2.1(0)	0.8(2)	2.1(1)
- 1.6	2	39.4	29.7	39.3	2.1(2)	0.8(4)	2.1(2)

Tab. 5 - Estimation process for the first case of Tab. 4.

5. Beliveau, J.G., "Identification of viscous damping in structures from modal information", J. Appl. Mech., 98, 2, 1976, 335-339.

6. Goodwin, G.C., Payne, R.T., Dynamic system identification. Experiment design and data analysis, Academic Press, New York 1977.

7. Hart, G.C., Torkamani, M.A.M., "Structural system identification", in Stochastic problems in mechanics, Eds. Ariaratuam S.T., Leipholz M.M.E., Univ. of Waterloo Canada, 1977, 207-228.

8. Hart, G.C., Yao, J.T.P., "System identification in structural dynamics", J. Eng. mech. Div., Proc. ASCE, 103, 6, 1977, 1089-1104.

9. Natke, H.G., "Die Korrectur des Rechnenmodelles eines Elastomechanischen systems mittels gemessener erzungener Schwingungen", Ing. Arch., 46, 1977, 169.

10. Liu, S.C., Yao, J.T.P., "Structural identification concept", J. Struct. Div., Proc. ASCE, 1978, 104, (12), 1845.

11. Fillod, R., Piranda, J., "Identification of eigensolution by Galerkin techniques", 79-DET-35, Design Engineering Techn. Conference, ASME, 79-DET-75, 1979.

12. Isenberg, J., Collins, J.D., Kavarna, J., "Statistical estimations of geotechnical material model parameters from in situ test data", Proc. ASCE Spec. Conf. on Probabilistic Mechanics and Structural Reliability, Tucson, 1979, 348-352.

13. Wedig, W., "Anwendung einer Spektralfiltertechnik zur Identifikation linearer und nichtlinearer Systeme", Ing. Arcg., 1980, 49, 413.

14. Yun, C.B., Shinozuka M., "Identification of non-linear structural dynamic systems", J. Struct. Mech., 8, 2, 1980, 187-203.

15. Stanton, J.F., McNiven, H.D., "The development of a mathematical model to predict the flexural reponse of reinforced concrete beams to cyclic loads, using system identification", Rep. UCB/EERC-79/02, Univ of California, Berkeley, 1979.

16. Gioda, G., Maier, G., "Direct search solution of an inverse problem in elastoplasticity: Identification of cohesionm friction angle and 'in situ' stress by pressure tunnel tests", Int. J. Num. Meth. Eng., 15, 1980, 1823-1848.

17. Jurina, L., Maier, G., Podolak, K., "On model identification problems in rock mechanics", Proc. Int. Symp. on Geotechnics of Structurally Complex Foundations, Capri, Vol. 1, AGI 1977.

18. Asoako, A., Matsuo, M., "Bayesian approach to inverse problem in consolidation and its application to settlement prediction", Proc. 3rd Int. Conf. on Numerical Methods in Geomechanics, Aachen 1979.

19. Tomizawa, M., "Identification of a one-dimensional model for a soil-layer system during an earthquake", Earthq. Engng. Struct. Dynam., 8, 1980.

20. Gioda, G., Jurina, L., "Numerical identification of soil structure interaction pressure", Int. J. Num. Analyt. Meth. Geomech., 5, 1981, 33-56.

21. Maier, G., Gioda, G., "Optimization methods for parametric identification of geotechnical systems", in Numerical Methods in Geomechanics, Ed. J.B. Martins., Reidel, Dordrecht 1982.

22. Cividini, A., Jurina, L., Gioda, G., "Some aspects of 'characterization' problems in geomechanics", Int. J. Rock Mech. Min. Sci., 18, 1981, 487-503.

23. Maier, G., Nappi, A., Cividini, A., "Statistical identification of yield limits in piecewiselinear structural models", In Proc. Int. Conf. on Computational Methods and Experimental Measurements, ISCME (Washington, D.C. July 1982) Eds. G.C. Keramidas, C.A. Brebbia, Springer, Berlin, 1982, 812-829.

24. Bittanti S., Maier, G., Nappi, A., "Inverse problems in structural elastoplasticity: A Kalman filter approach", in Plasticity Today: Modelling, Methods and Applications, Eds. A. Sawczuck, G. Bianchi, Elsevier A.S.P., Amsterdam, 1984.

25. Cividini, A., Maier, G., Nappi, A., "Parameter estimation of a static geotechnical model using a Bayes' approach", Int. J. Rock Mech. Mining Sci., 20, 1983, 215-226.

26. Maier, G., Giannessi, F., Nappi, A., "Indirect identification of yield limits by mathematical programming", Eng. Struct., 4, 1982, 86-89.

27. Maier, G., "Inverse problem in engineering plasticity: a quadratic programming approach", Atti Acc. Naz. dei Lincei, Cl, Sc. Apr. 1982.

28. Nappi, A., "System identification for yield limits and hardening moduli in discrete elastic-plastic structures by nonlinear programming", Appl. Math. Modelling, 6, 1982, 441-448.

29. Nappi, A., "Identificazione indiretta dei limiti elastici e dei coefficienti di incrudimento in strutture elastoplastiche discrete", Atti IX Convegno AIAS, Trieste, 1981.

30. Maier, G., "Mathematical programming methods in structural analysis", in Variational methods in engineering, Eds. H. Tottenham, C. Brebbia, Southampton Univ., Press 1973.

31. Kunzi, F, Krelle, R., Nonlinear programming, Blaisdell Publ., 1967.

32. Maier, G., "A quadratic programming appoach for certain classes of non linear structural problems", Meccanica, 2, 3, 1968, 121-130

33. Lewis, T.O., Odell, P.L., Estimation in linear models, Prentice-Hall, Englewood Cliff, 1971.

34. Anderson, B.D.O., Moore, J.B., Optimal Filtering, Prentice-Hall, New York, 1979.

35. Kalman, R.E.A., "A new approach to linear filtering and prediction problems", Trans. ASME. J. Basic Engng., 82, 1960, 35-45.

36. Kalman, R.E., Bucy, R.S., "New results in linear filtering and prediction theory", Trans. ASME. J. Basic Engng. 83D 1961, 95-108.

37. Gelb, A., (Ed.), Applied optimal estimation, MIT Press, Cambridge, Mass., 1974.

APPLICATIONS IN CIVIL ENGINEERING:
MODAL PARAMETER IDENTIFICATION OF AN OFFSHORE PLATFORM AND IDENTIFICATION OF THE AERODYNAMIC ADMITTANCE FUNCTIONS OF TALL BUILDINGS

H.G. Natke
Universität Hannover, Hannover, F.R.G.

Introduction

System identification is an important tool in civil engineering, as also shown by J.T.P. Yao in his lecture /1/. However, many problems are still unsolved or not approved. The reasons are manifold: some are due to the large dimensions and their consequences (frequency range, realizable excitation), other to local behaviour (and therefore difficulties in modelling), and others to parameter sensitivity.

Applications are given here with regard to two structures. Eigenfrequencies and damping ratios are estimated

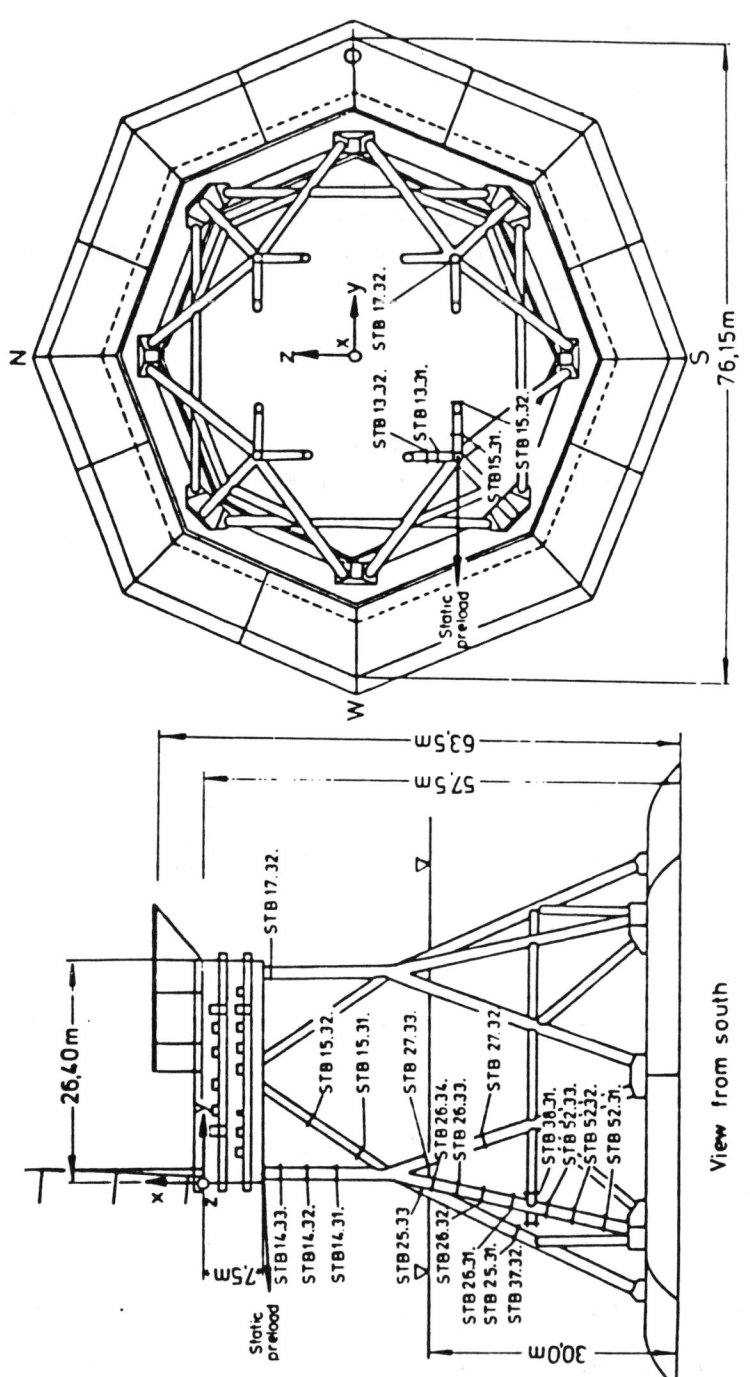

FIG. 1 RESEARCH PLATFORM WITH ITS MAIN DIMENSIONS

View from south

for the offshore platform NORDSEE from a test under bad weather conditions. The identified eigenfrequencies then serve for analytical model improvement /2/. The second application concerns input identification, especially the identification of the aerodynamic admittance function of a tall building. The Munich TV tower was chosen to demonstrate the method /3, 4/.

In addition to these examples from the author's experience there are, of course, many other applications that exist and have been published (see e.g. /5/6/7/). The conclusion drawn from this experience is that - as stated at the beginning - system identification applied in practice leads to useful results if

- a priori knowledge is available (the physics are well understood)
- statistical methods are used
- error estimates (known confidence of results) are performed.

1. Modal Parameter Identification of an Offshore Platform

1.1 <u>System and Model Description</u>. The system is the "Forschungsplattform NORDSEE" (FPN). Fig. 1 shows its main dimensions. It is a steelframed structure consisting of

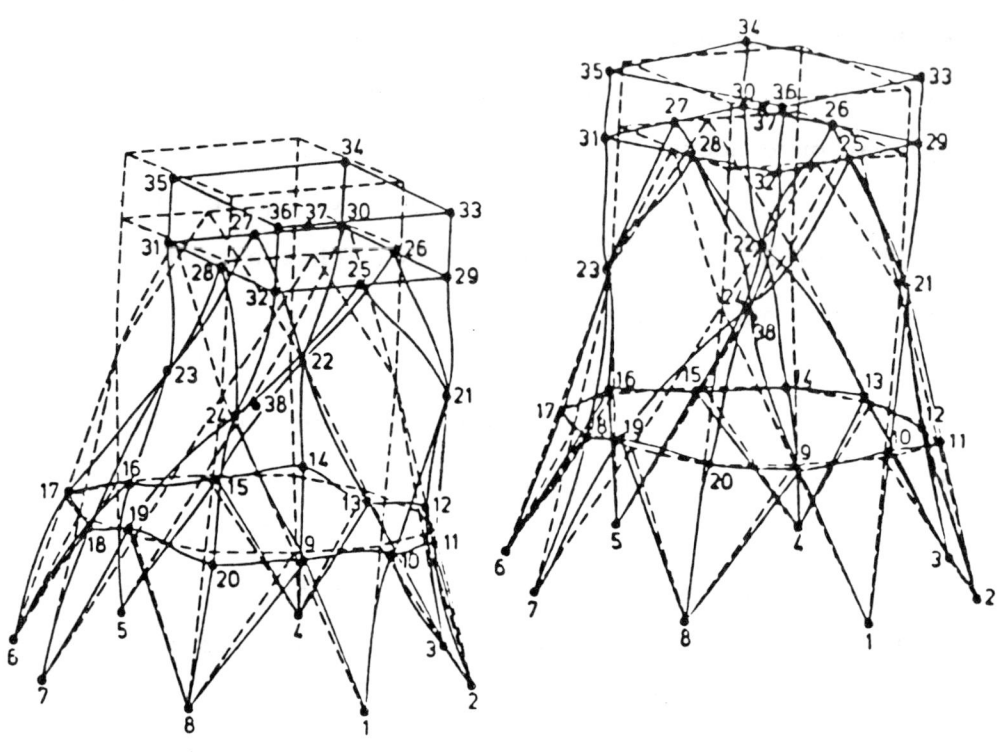

FIG. 2a FUNDAMENTAL BENDING MODE   FIG 2b FUNDAMENTAL TORSION MODE

steel pipes with diameters of 1016 and 1420 mm. It is
gravity founded. The legs are inclined in order to increase
the area of the foundation. The deck body consists of
different levels. The storage space for fuel and lubricants
as well as for fresh and waste water is on the tank deck
underneath the lower deck. A detailed description is contained in Ref. /8/.

The analytical model is built up with beam elements for the framework construction and with massless disk elements for the deck. The deck masses are modelled as lumped masses located at the outer deck boundary in order to form the inertias for torsional motion. The legs are rigidly supported in the rigid foundation, and the foundation soil is also assumed as rigid. This original model consists of 798 degrees-of-freedom. Because only the first lower eigen-frequencies and normal modes are of interest, the model was reduced with respect to the number of degrees-of-freedom by simplifying the deck model and by omitting intermediate points of the beam elements. The result was a 176 degree-of-freedom model with the following eigenfrequencies:

Table 1: Calculated eigenfrequencies (SAP4)

i	f_{oi} (Hz)	Description
1	1.96	first bending mode
2	3.15	first torsion mode
3	3.76	-

The first two modes are shown in Fig. 2

1.2 Identification of Eigenfrequencies and Damping
Ratios. The platform was preloaded excentrically by a
static force applied to the deck as indicated in Fig. 1. The
loading was realized by a tuck and a connecting hawser. The
force was measured by a special tensile pickup, abrupt
unloading was achieved by cutting the hawser. Fig. 3
contains the measured force. The measurement of the dynamic
response ensued (unsuitably) with bounded strain gauges
(Fig. 1). Some of these quantized signals are plotted in
Fig. 4. As can be seen, the quantization is insufficient.
Unfortunately, the test had to be interrupted because of
bad weather and rough sea and could not be repeated.
Therefore only a few signals with large quantization errors
are available for the identification of eigenfrequencies
and damping ratios.

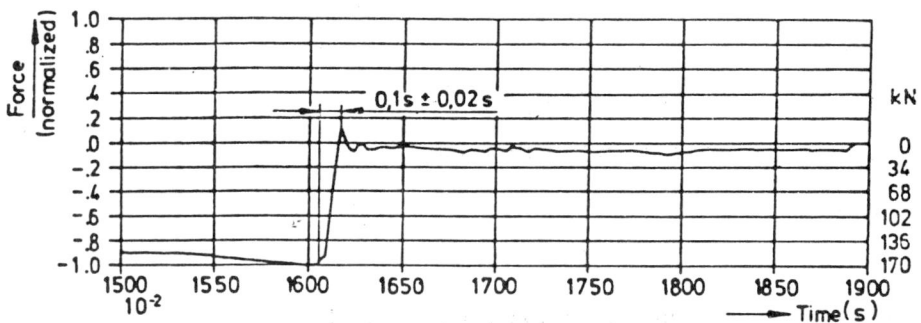

FIG. 3 MEASURED PRE - AND UNLOADING

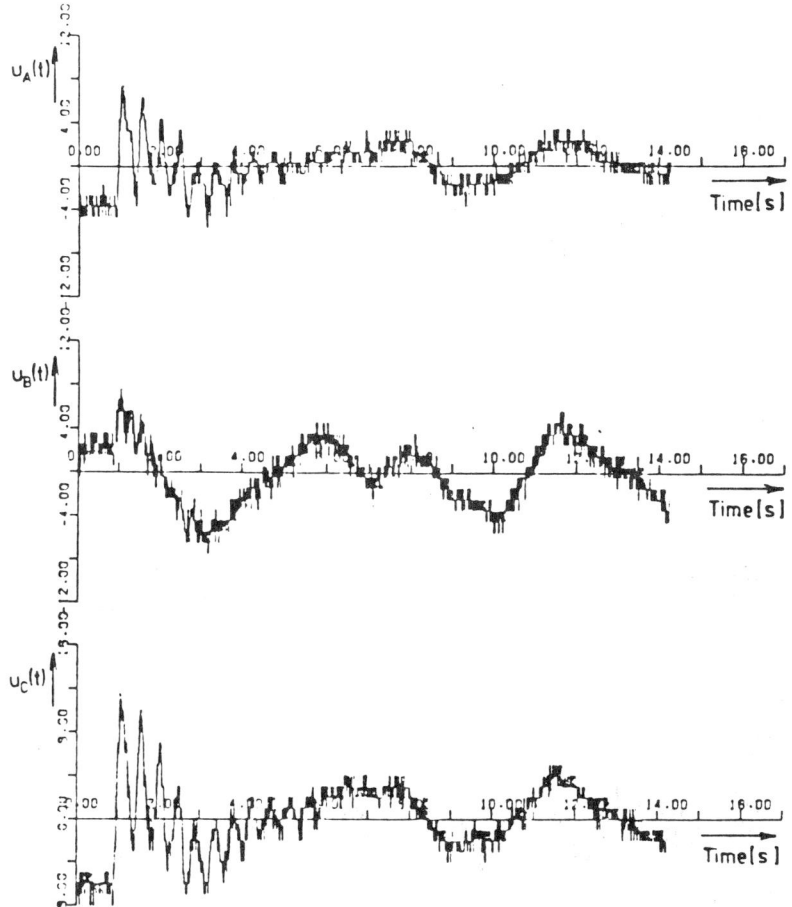

FIG.4 DIGITIZED MEASURED SIGNALS

The signals show the decaying due to the abrupt unloading of about 6 s superimposed by the dynamic response caused by the rough sea. Laplacian transform with an exponential window which reduces the measured signal after 6 s to, for example, 1 % of the maximum amplitude, would result

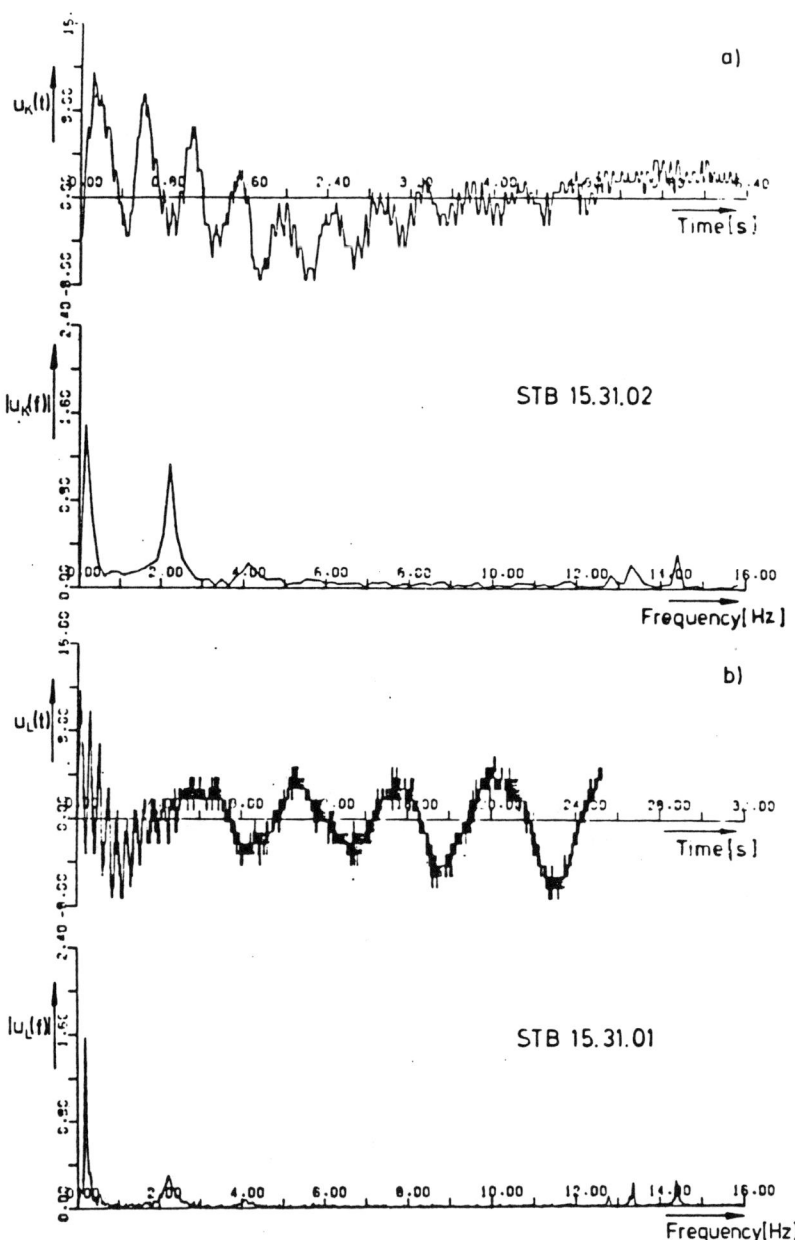

FIG.5 EXAMPLES OF SIGNALS IN THE TIME
AND IN THE FREQUENCY DOMAIN

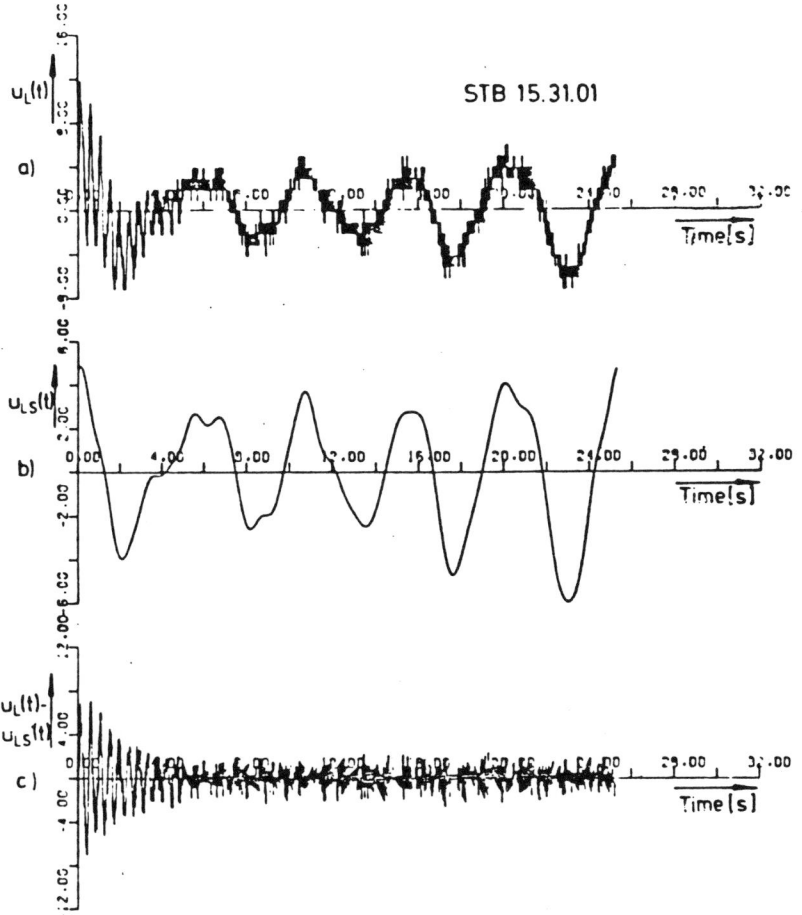

FIG.6 a) MEASURED SIGNAL
b) SIGNAL OF SEA EFFECT ONLY
c) DECAYING SIGNAL WITHOUT SEA EFFECT

result in a large additional damping and therefore lead to a bad resolution. Instead of choosing this procedure, the large effect of the sea was eliminated. From the Fourier transformed signals (Fig. 5) it can be seen that the sea effects cause large amplitudes in a frequency interval up

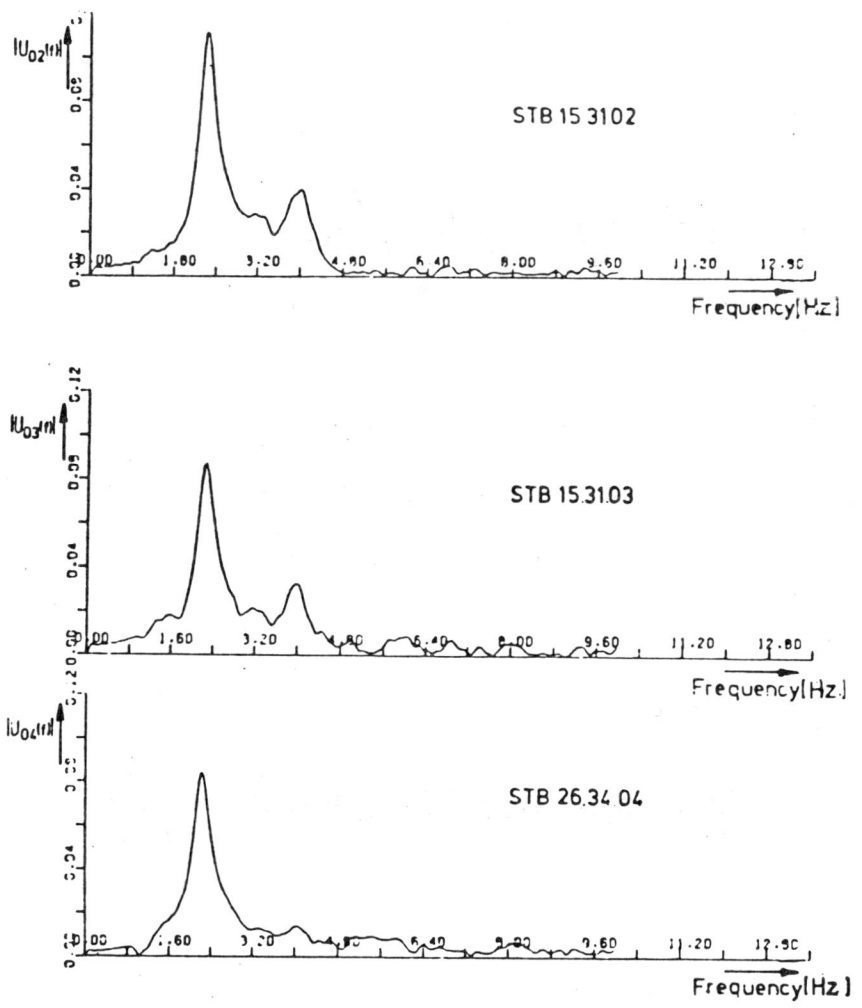

FIG. 7 LAPLACE TRANSFORMED
SIGNALS WITHOUT SEA EFFECT

to 0.2 Hz which do not interfere with the fundamental
eigenfrequency of about 2 Hz. All the measured signals in
the time interval up to 25.3 s were Fourier transformed,
and the results were set equal to zero for frequencies
higher than 1 Hz. Then they were retransformed and sub-

tracted from the original time signals: the results are the decaying time signals without (approximately) sea effects (Fig. 6). These signals were extended by zeroes in order to obtain closer frequency spacings and then Laplace transformed, so that 1 % of the max. amplitude remains at 6 s (Fig. 7). The transformed signals serve as a data basis for the identification of eigenfrequencies and damping ratios using the procedure described in /9/, Chapter 5.3.2.1. The result is contained in Table 2.

Table 2: Identified Values

k	Eigenfrequencies $\overset{\Delta}{f_k} \overset{+}{-} \overset{\Delta}{\tau}_{fk}$ (Hz)	Damping ratios $\overset{\Delta}{\alpha}_k \overset{+}{-} \overset{\Delta}{\tau}_{\alpha n}$ (p.c.)	Description
1	$2.22 \overset{+}{-} 0.01$	$2.8 \overset{+}{-} 0.4$	fundamental bending
2	$3.34 \overset{+}{-} 0.12$	$2.^{*}$	local vibration
3	$4.03 \overset{+}{-} 0.04$	$2.3 \overset{+}{-} 0.7$	fundamental torsion

[*] it was not possible to estimate the standard deviation

Denoting the complex eigenvalues $\lambda_k =: \lambda_k^{re} + j\lambda_k^{im}$ (j: imaginary unit), the eigenfrequency is defined by $f_k =: \lambda_k^{im} / (2\pi)$ and the damping ratio by $\alpha_k = -\lambda_k^{re} / (2\pi f_k)$.

Verification of the results was done by subtracting the Fourier transformed signals of the degrees-of-freedom with reference to the identified values from the "cleaned" original signals and by looking at the remaining degrees-of-freedom which could be identified. Recalculation (synthesized response) with the estimated values and comparison with the measured responses also show sufficient agreement.

This example proves that the developed identification procedures work, despite the unsatifactory quality of the digitized data.

1.3 Sensitivity Investigations. The next step is the parameter improvement of the (prior) undamped mathematical model with 176 degrees-of-freedom using the two identified eigenfrequencies. The largest deviation exists between calculated and identified fundamental torsional eigenfrequencies. Next, the question is to be answered concerning which parameters have to be improved. Which are the most uncertain assumptions, which influence the eigenfrequencies most (significantly)? Here linearized differential error calculations should help. Table 3 shows the varied parameters and the resulting eigenfrequencies. The effective beam lengths are considered, because the connection of several pipes with the given diameters of more than 1,000 mm, which are one-dimensionally modelled, creates beam lengths that are too large due to their node points. As can be seen, the eigenfrequencies are not very sensitive due to the varied quantities, and they do not explain the deviations. Therefore the deviations must have other causes. A further look at the test conditions reveals that the mass state during the test was not well known. It therefore follows that the inertias have to be improved.

Table 3: Sensitivity investigations

Model with 176 dof:

parameters varied	f_1 (Hz)	f_2 (Hz)	f_3 (Hz)
-	1.96	3.15	3.76
100 p.c. displaced water masses included	1.95	3.06	3.15
vertical soil stiffness included	1.89	3.14	3.76
effective beam lengths taken into account	2.02	3.17	4.22

1.4   <u>Inertia Improvement of the Mathematical Model</u>.

The inertia improvement is performed as described in /9/,

Chapter 6.2.1.3. The inertia matrix of the system is parti-

tioned into three submodels:

Submodel 1:   deck masses related to bending motion,

                correction factor a_{M1}

Submodel 2:   deck masses concerning torsion motion,

                correction factor a_{M2}

Submodel 3:   remaining inertia matrix, choosing factor

                $a_{M3} = 1$

The iterative calculation was completed in 2 steps with the

results:

Table 4: Mass parameters

                 Prior Model   Improved Model

Submodel 1: 1x140t: 140t $(3.095 \overset{+}{-} 0.097) \times 140t : (433 \overset{+}{-} 14) t$

Submodel 2: 12x140t: <u>1680t</u> $(0.585 \overset{+}{-} 0.006) \times 12 \times 140t : \underline{(983 \overset{+}{-} 10) t}$

            total   1820t $(1416 \overset{+}{-} 17) t$
                      =====   ==========

In spite of the insufficient test data and knowledge of the

test conditions, an improvement of the model parameters is

obtained so that it describes the measured eigenfrequencies.

2.   Identification of the Aerodynamic Admittance Functions
     of Tall Buildings

As described in /10/, input identification with a
structured model is reduced to parameter estimation.
Structured input identification of this kind is applied in
civil engineering in order to identify the aerodynamic
admittance function of a tall building related to along-
wind loads.

2.1 Using Davenport's Concept /11/. With the structural admittance function (equal to the structured response
function) $F_{(m)}$ the input/output relation of the building in
the frequency domain can be described by

$$S_{pu}(\omega) = S_{pp}(\omega)\, F_{(m)}^{T}(j\omega) \tag{2.1}$$

and

$$S_{uu}(\omega) = F_{(m)}^{*}(j\omega)\, S_{pp}(\omega)\, F_{(m)}^{T}(j\omega) \tag{2.2}$$

and as shown in Fig. 8.

The star indicates conjugate complex. The along-wing loads
are described stochastically by a random mean free velocity
$v'(z,t)$ and a mean velocity $\bar{v}(z)$ dependent on height z in
superposition:

$$v(z,t) = \bar{v}(z) + v'(z,t). \tag{2.3}$$

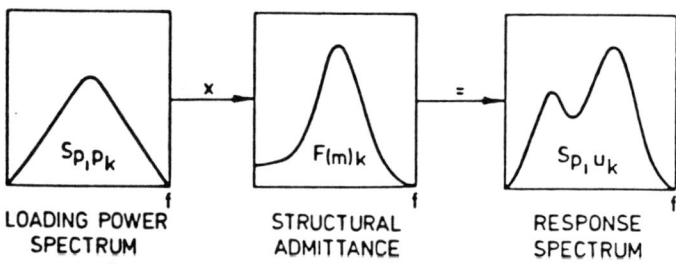

FIG.8   INPUT / OUTPUT RELATION
        OF THE STRUCTURE

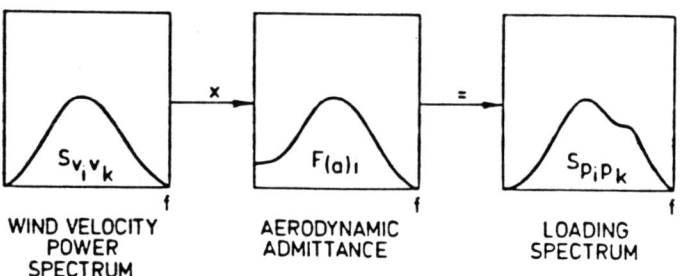

FIG.9   INPUT/ OUTPUT RELATION
        BETWEEN WIND VELOCITY
        AND WIND LOADING

Assuming ergodicity and applying the Wiener-Khintchine transformation to the correlation functions of wind velocity and the generalized force of the building, one obtains the relationship between the spectrum of the wind velocity and the wind loading with the aerodynamic admittance $F_{(a)}(j\omega)$:

$$S_{pv}(\omega) = S_{vv}(\omega)\, F_{(a)}^{T}(j\omega)$$

$$(2.4)$$

and

$$S_{pp}(\omega) = F^*_{(a)}(j\omega) \, S_{vv}(\omega) \, F^T_{(a)}(j\omega) \qquad (2.5)$$

and as shown in Fig. 9.

Combining the aerodynamic with the structural equation in the frequency domain yields the description of the entire problem.

Assumptions /3/ with respect to
- the normal modes to be excited (only 1)
- the coupling of modal coordinates (nothing)
- stochastical wind velocity (linearization)
- the imaginary part of the cross-wind spectrum (negli-bible)
- independence of the power spectrum of wind velocity from the height z
- validity of Hellmann's law (1916) with power α
- product ansatz of velocity coherence function (valid for the difference between the heights) incl. exponent C_{1z}
- using Davenport's approximation for the joint acceptance (as a part of aerodynamic admittance)

we obtain independent procedures for estimating α, C_{1z} and for estimating the pressure coefficient C_D and damping ratio D_1 (of the fundamental degree-of-freedom).

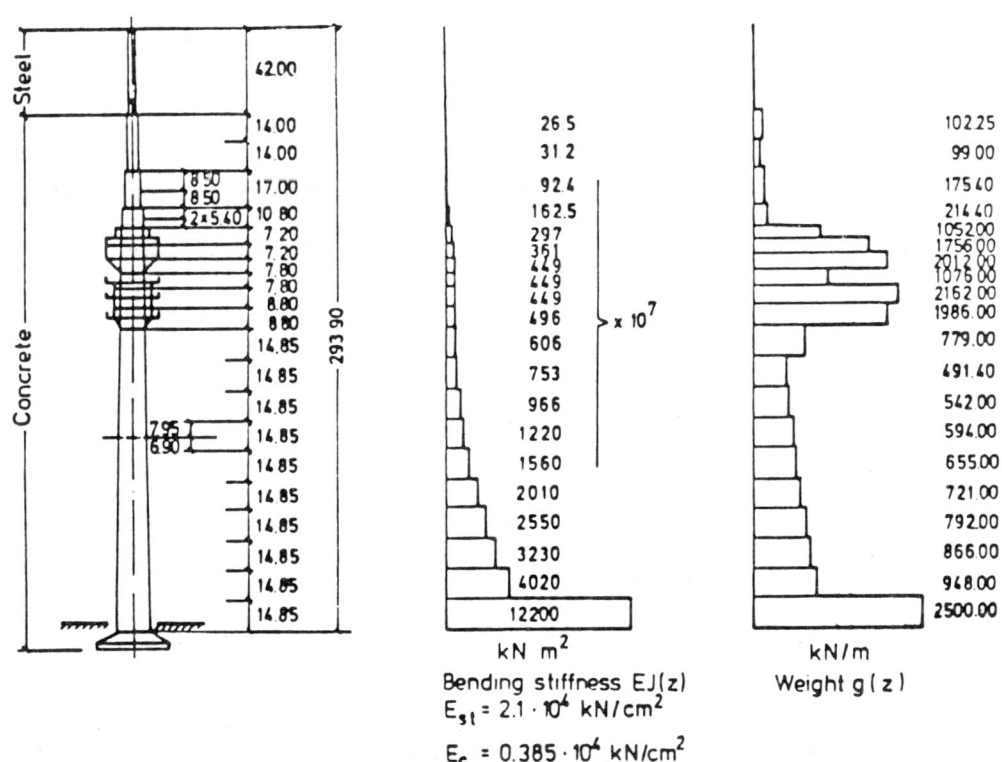

FIG.10 MUNICH TV TOWER WITH
ITS MAIN CHARACTERISTICS

The application to the Munich TV tower is based on
long-term measurements /12/. The main characteristics are
contained in Fig. 10. The estimate of the power α is $\overset{\Delta}{\alpha}$ =
0.496 $\overset{+}{-}$ 0.007 instead of about 0.3, as published elsewhere.
The reason for this may be found in the power law itself

/13/. The mean value $\overset{\Delta}{C}_{1z}$ = 1.706 $\pm$ 0.039 lies within other measured values. The last estimation procedure yields the estimates $\overset{\Delta}{C}_D$ = 0.55 $\pm$ 0.18 and $\overset{\Delta}{D}_1$ = 0.010 $\pm$ 0.004 for given f_{o1}. The estimates of the standard deviations are large, but within the scope of the values given elsewhere.

2.2 <u>Using only General Assumptions /4/</u>. Taking the least squares of the residuals of the measured dynamic responses and modelled responses, which consist of the frequency response matrix multiplied by the modelled force vector

$$P'(j\omega) = \rho_0\, C_D\, \text{diag}\, (\bar{v}_i A_i)\, V(j\omega) \qquad (2.6)$$

with air density ρ_O, the Fourier transformed vector $V(j\omega)$ of wind velocity depending on height coordinate, A_i denotes the i-th partial area affected by along-wind loadings measured at the i-th measuring point, and $\bar{v}_i$ the corresponding mean value of wind velocity. The frequency response matrix is denoted by

$$F_{(m)}(j\omega_r, D_k) = \sum_{k=1}^{n} \frac{\hat{u}_{0k}\, \overset{\wedge T}{u}_{0k}}{m_{gk}\left[(\omega_{0k}^2 - \omega_r^2) + 2j D_k \omega_r \omega_{0k}\right]} \qquad (2.7)$$

with

$\hat{u}_{ok}$ the vector of the k-th normal mode

m_{gk} the k-th generalized mass.

The resulting loss function is

$$J(C_D, D_k) = \frac{1}{2} \sum_{r=1}^{N} trM \left[\hat{S}_{uu}(\omega_r) + c_D^2 F_{(m)}^*(\omega_r, D_k) \hat{A}(\omega_r) F_{(m)}(\omega_r, D_k) \right.$$

$$\left. - c_D \left((\hat{B}(\omega_r) F_{(m)}(\omega_r, D_k))^* + \hat{B}(\omega_r) F_{(m)}(\omega_r, D_k) \right) \right] \qquad (2.8)$$

with the measured quantities

$$\hat{S}_{uu}(\omega) = F\left\{ E\left(u(t) u^T(t+\tau) \right) \right\} = \hat{S}_{uu}^{*T}(\omega)$$

$$\hat{S}_{vv}(\omega) = F\left\{ E\left(v'(t) v'^T(t+\tau) \right) \right\} = \hat{S}_{vv}^{*T}(\omega)$$

$$\hat{S}_{uv}(\omega) = F\left\{ E\left(u(t) v'^T(t+\tau) \right) \right\} = \hat{S}_{vu}^{*T}(\omega) \qquad (2.9)$$

$$\hat{A}(\omega) := \rho_0^2 \, diag(\bar{\hat{v}}_i A_i) \, \hat{S}_{vv}(\omega) \, diag(\bar{\hat{v}}_i A_i) = \hat{A}^{*T}(\omega)$$

$$\hat{B}(\omega) := \rho_0 \hat{S}_{uv}(\omega) \, diag(\bar{\hat{v}}_i A_i) \; .$$

The quadratic functional (2.8) is weighted with the inertia matrix M, making use of the orthogonality relations to improve the condition of the Hessian matrix by minimizing the number of its off-diagonal elements. For $N \geq n$ and $N > 1$ the Hessian matrix is positive definite, i.e. a strong minimum of (2.8) exists for C_D and D_k, $k = 1(1)n$.

As can be noted, only a few assumptions are necessary (linearity and Eq. (2.6) compared with Davenport's concept). However, many more measured data are needed. The practibility of these methods has still to be checked.

References

1. Yao, J.T.P.; Application of System Identification in Civil Engineering, within these Lecture Notes

2. Natke, H.G.; and Schulze, H.; Parameter Adjustment of a Model of an Offshore Platform from Estimated Eigenfrequencies Data; J. Sound and Vibration (1981) 77 (2), 271-285

3. Natke, H.G.; Identification of the Aerodynamic Admittance Function; IFAC Identification and System Parameter Estimation 1985, York, UK, 1985

4. Natke, H.G. and Cottin, N.; On the Input Identification of Tall Buildings without the usual Limiting Assumptions; Proc. of the 3rd Internat. Conf. on Computional Methods and Experimental Measurements 1986, Porto Caras, Greece

5.   Yao, J.T.P.; Safety and Reliability of Existing
     Structures; Pitman Advanced Publ. Program, 1985

6.   Cifuentes, A.O., Nelson, B. (Edts.); System Identi-
     fication of Hysteretic Structures; Calif. Institute of
     Technology, EERL 84-04, Pasadena, Ca., 1984

7.   Hart, G.C. and Nelson, B. (Edts.); Dynamic Response of
     Structures; Proc. of the 3rd Conf. publ. by ASCE New
     York, 1986, Los Angeles, Ca.

8.   Payer, H.G. and Longree, W.D.; Design, Analysis and
     Construction of the Research Platform "Nordsee";
     Offshore Technology Conf. 1976, Houston, No. 2168

9.   Natke, H.G.; Einführung in Theorie und Praxis der
     Zeitreihen- und Modalanalyse, Friedr. Vieweg & Sohn,
     1983

10. Natke, H.G. and Cottin, N.; Introduction to System
     Identification, within these Lecture Notes

11. Davenport, A.G.; The Dependence of Wind Loads on
     Meteorological Parameters; Intern. Res. Seminar: Wind
     Effects on Buildings and Structures; Ottawa, Can.,
     Sep. 67, Proc., Vol. 1, Univ. of Toronto Press

12. Schneider, F.X., Wittmann, F.H. and Panggabean, H.;
    Zusammenstellung der im Verlauf mehrerer Jahre am
    Münchener Fernsehturm durchgeführten Wind- und Schwin-
    gungsmessungen; Beiträge zur Anwendung der Aerodynamik
    im Bauwesen (BAAB), Heft 4, 1975, Techn. Univ. München

13. Reichmann, K.H.; Beurteilung der Sicherheit und
    Zuverlässigkeit turmartiger Bauwerke unter Windein-
    wirkung; Beiträge zur Anwendung der Aeroelastik im
    Bauwesen, Heft 19, 1984 Techn. Univ. München

IDENTIFICATION OF STRUCTURAL DAMAGE
IN CIVIL ENGINEERING

J.T.P. Yao
Purdue University, W. Lafayette, U.S.A.

INTRODUCTION

Civil engineering systems such as bridges, buildings, and dams are critical to survival and well-being of our society. Because of my education and background, the emphasis of these lectures is placed on the damage identification of existing structures.

To make mathematical analyses, it is necessary to simplify and idealize the structural system and its environment. The design of a structure follows an iterative process involving both structural analysis and structural design using generalized mathematical models, which are based on experience and available knowledge in the engineering profession. Using field data and other relevant information, a preliminary design is made and the idealized mathematical model is analyzed for

expected or specified loading conditions. Based on these analytical studies, the design may be revised and re-analyzed in an iterative manner
until all design criteria are satisfied. The completed design is then
constructed accordingly.

For structures which have been constructed and are thus existing, it
is necessary at times to assess their respective damage states on the
basis of available information including measured and recorded experimental data.   In addition, it is desirable to evaluate the reliability of
these structures so that rational decisions can be made concerning necessary repairs, replacements, retirement, and other maintenance or rehabil-
itation procedures as shown in Fig. 1.

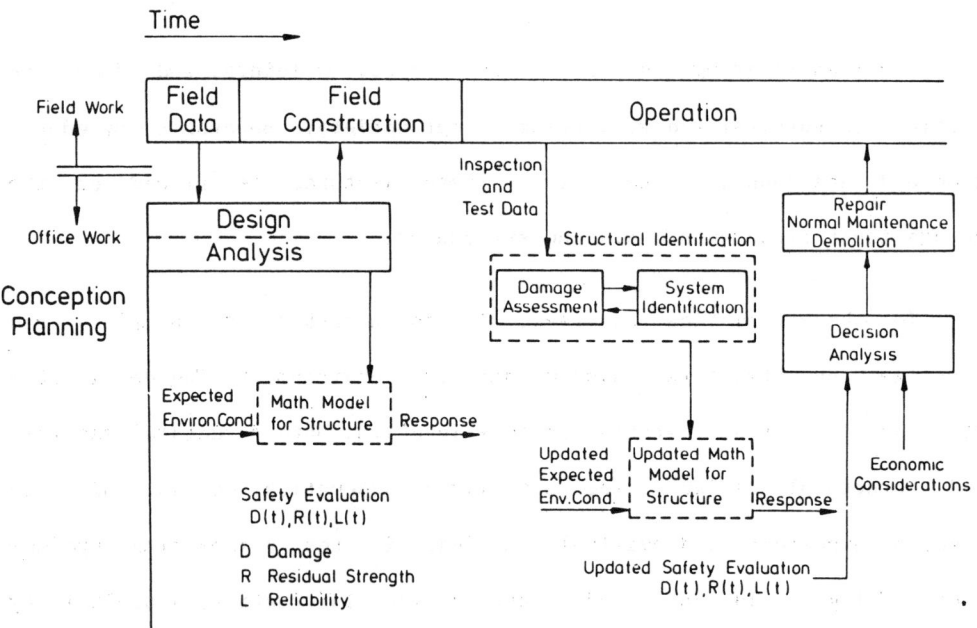

Fig. 1 : Role of Structural Identification (From Yao, 1985)

Because (a) it is difficult to predict future loading conditions and (b) material properties are random in nature, stochastic processes have been used to represent these quantities for the estimation of failure probabilities. However, the as-built structure is usually different from the original mathematical model in the design phase because the real-world structure is an extremely complex system. Even with the use of finite element methods and modern computers, it is usually difficult to consider all the details in the mathematical model of a given structure. Moreover, the damage path and failure behavior of most large structures remain unknown because few experimental studies of full-scale structures have been conducted to-date.

For important structures, nondestructive dynamic tests are conducted for the estimation of dynamic properties of the as-built structure. These test data are then used to obtain "improved" or "more realistic" equations of motion. These equations of motion are applicable within the range of the test amplitude, which is usually small and within the linear behavior of the given structure. Therefore, results of such analysis should not be applied where destructive or damaging loading conditions are considered. Nevertheless, these mathematical representations can be useful for comparison purposes. For example, any change in the measured natural frequencies may be used as an indication of structural damage.

In general, experienced structural engineers can investigate the condition of a particular structure and determine its level of safety. In such investigations, the original design calculations and drawings (if available) are examined and checked. Inspections and testings are conducted, and the resulting data are analyzed. Results of these analyses are then summarized and interpreted by experienced engineers to yield appropriate recommendations. Although it is possible to understand the inspection and testing conducted and the detailed analysis performed in these studies, the decision-making process involved in the determination of (a) specific types of inspection and testing procedures and (b) the summary and interpretation of experimental and analytical results remain privileged information of relatively few experts in the structural engineering profession.

The objective of these lectures is to summarize and discuss the state-of-the-art of several subject areas related to detection of structural damage in civil engineering. The role of system identification and its potential application to structural dynamics and reliability studies are explored. In addition, the possible application of rule-inference methods to damage assessment is introduced and examined.

SYSTEM IDENTIFICATION IN STRUCTURAL DYNAMICS

General Remarks

The mathematical models as used in the design prior to the construction phase do not truly represent the behavior of a given structure. To improve a mathematical model in the simulation of the real structure, response records with or without known forcing functions have been collected and analyzed with the use of system identification techniques during these past two decades. By necessity, these tests are always conducted at small response amplitudes within the serviceability and safety limit states. Consequently, the resulting modified mathematical models are applicable only to the linear or at most slightly nonlinear range of the structural behavior.

At present, it is numerically possible to simulate the structural response to extreme forces such as strong earthquakes or wind storms with the use of digital or hybrid computers, and thus to evaluate the serviceability and safety conditions of the structures. Nevertheless, there exists the paradox that (a) the applicability of "realistic" models of the structure are limited to small-amplitude response range, (b) the catastrophic loading conditions are likely to cause the structures to behave beyond the linear or "near-linear" responses which are usually assumed, and (c) the severe loadings may cause serious damages in the structure and thus change the structural behaviors appreciably. It is important that the extent of damage in structures can be assessed following each major catastrophic event or at regular intervals for the evaluation of effects of aging and deterioration. On the basis of such damage assessment, appropriate decisions can be made as to whether a particular structure can and should be repaired in order to salvage its residual values. The purposes of this lecture are to (a) review and summarize the relevant literature on the methods of system identification in structural dynamics and (b) discuss how the results of system identification studies may be used to obtain a rational procedure for the safety evaluation of existing structures.

System identification is a process for constructing a mathematical model of a physical system when both the input to the system and the

limit states. Consequently, the resulting modified mathematical models are applicable only to the linear or at most slightly nonlinear range of the structural behavior.

At present, it is numerically possible to simulate the structural response to extreme forces such as strong earthquakes or wind storms with the use of digital or hybrid computers, and thus to evaluate the serviceability and safety conditions of the structures. Nevertheless, there exists the paradox that (a) the applicability of "realistic" models of the structure are limited to small-amplitude response range, (b) the catastrophic loading conditions are likely to cause the structures to behave beyond the linear or "near-linear" responses which are usually assumed, and (c) the severe loadings may cause serious damages in the structure and thus change the structural behaviors appreciably. It is important that the extent of damage in structures can be assessed following each major catastrophic event or at regular intervals for the evaluation of effects of aging and deterioration. On the basis of such damage assessment, appropriate decisions can be made as to whether a particular structure can and should be repaired in order to salvage its residual values. The purposes of this lecture are to (a) review and summarize the relevant literature on the methods of system identification in structural dynamics and (b) discuss how the results of system identification studies may be used to obtain a rational procedure for the safety evaluation of existing structures.

System identification is a process for constructing a mathematical model of a physical system when both the input to the system and the corresponding output are known. For most of the civil engineering applications, the input is usually a forcing function and the output is the displacement, velocity, or acceleration response of the structure to these forces. The mathematical model obtained from the identification process should produce a response that in some sense matches closely the measured output, when it is subjected to the same input (see Natke, 1986).

Generally, the system identification technique consists of the following three parts: (a) determination of the form of the model and the system parameters; (b) selection of a criterion function by means of the "goodness of fit" of the model response to the actual response that can be evaluated, when both the model and the actual system are subjected to the same input; (c) selection of an algorithm for modification of the system parameters, so that the discrepancies between the model and the actual system can be minimized. Because of the large size and mass of most real structures, many common techniques for generating a convenient force input and a suitable system output are not practical for the identification of civil engineering structural systems. Only limited source of input, such as vibrations due to earthquakes, strong wind loads, and controlled explosions, are possible to generate sufficiently large excitation for nonlinear structural response. Even for laboratory simulations, the limitations on the types of structure and the types of response to be performed in a laboratory are far greater than an electrical system or a mechanical system, the identification techniques of which

are well developed. In addition, most of the inputs and outputs are random in nature. To extract useful informations from such data posts a new
problem in the subject area of system identification.

System Identification Techniques

System identification techniques have been widely used in many bran-
ches of science and engineering for the estimation of various characteristics of a physical system [e.g.,Eykhoff, 1974; Sage and Melsa, 1971].
Their applications in civil engineering have been studied with increasing
interest during the last two decades. In the available literature, a set
of differential equations (lumped-mass model or simple continuous model
in time domain) or a transfer function (black box model or lumped-mass
model in frequency domain) are often used to represent the structural
behavior. A set of parameters are to be estimated from the measured
response of the real structure to a known disturbance. The application
of system identification techniques to solve structural engineering problems is called structural identification by several investigators [Hart
and Yao,1977; Rodeman,1974; Ting, Chen, and Yao,1978; Toussi, 1982; Natke
and Yao,1986a].

Because of their simplicity, the linear lumped-parameter models are
the most widely used models in structural identification. More complex
models such as the linear continuous-parameter models and nonlinear
models are used only when the lumped-parameter model cannot be used to
provide an adequate representation of the structural behavior. For lumped systems or continuous systems with lumping approximations, the dis-

turbance must also be represented in a discrete form. On the other hand, the disturbances can be either discrete or continuously distributed in a continuous system.

Various least-squares estimation methods (including repeated and generalized least squares), the instrumentation variables method, and the maximum likelihood estimation have been used to identify parameters of linear structural models. The least-squares estimation minimizes the summation of square errors between the predicted response and the measured structural response. In the generalized least-squares method, the criterion function for evaluating the "goodness of fit" is the summation of square generalized errors which is defined to include the additive noise covariance matrix. Repeated application of the least squares method can be used to modify the usual least squares procedure by increasing the order of the mathematical model in an iterative process until the desired accuracy is obtained.

The instrumental variables method is applicable in order to avoid biased estimates as they generally result from least square estimators. The method involves an iterative process in the calculation of revised estimate and instrumental variables matrix function. The maximum likelihood method is widely used for parameter estimation in statistics. It determines the parameter estimate by minimizing criterion function through an iterative procedure. The method appears to have the advantage of providing the best estimation for a wide range of contamination intensity in the external excitation and the structural response [Gersch, 1975].

In contrast with the linear models, relatively little work seems to have been done for nonlinear models [Natke and Yao, 1986b]. It is in part due to the mathematical difficulties in considering the nonlinear terms. Some common techniques in dealing with linear systems, such as the modal expansion and transfer function, are not appropriate in the nonlinear case. Nevertheless, it is possible to apply the modal expansion analysis to obtain approximate solutions for ``slightly´´ nonlinear problems. It is also because the current developments in structural identification have mostly dealt with structural parameters with limited range of application or parameters for highly simplified structural behaviors. For example, in the evaluation of vibratory parameters of structures, the models are often limited to small-amplitude response range and time-invariant structural behaviors. However, the catastrophic loading conditions such as strong earthquakes and severe windstorms are likely to cause the structure to behave beyond the assumed linear range of responses.

Using the theory of invariant imbedding, a best a priori estimate cam be obtained by minimizing an error function [Distefano and Pena-Pardo,1975]. The method is applicable to some general boundary conditions. Dynamic programming filter is a more general method with the invariant imbedding as a special case. Instead of going through the Euler-Lagrange equations to determine the best estimate that minimizes the error function, dynamic programming may be applied directly. In such cases, the decomposition of the error function can lead to a system of partial differential equations. The optimal least squares filter satisfies the governing differential equation which describes the structural model and minimizes the quadratic error function. The error function is defined in terms of observed error vectors (weighting matrices) and the best a priori estimate of the parameters [Distefano,1972; Distefano and Todeschini,1974].

The Kalman filter has been used to obtain optimum sequential linear estimation and an extended filter deals with nonlinear filtering. Its good approximation for high sampling rates has been demonstrated in simulation studies of parameter estimation [Sage and Melsa,1971].

The maximum likelihood method has been applied to both linear and nonlinear systems. It can be used to treat both the measurement noise and the process noise, and may also be used to estimate the covariances of the noises [Rodeman,1974]. It is also suggested that the extended Kalman filter may be introduced in the calculation of the likelihood function. For recent advancement in this direction, see Tomlinson (1986).

An input-output relationship of multiple integral form is assumed to represent the model [Marmarelis and Udwadia,1976]. The kernel functions which represent model parameters can be estimated by a cross-correlation technique. In theory, the relationship can be written in Laplace domain and thus the kernels are identified in terms of the Laplace parameter. Their values in the time domain are then obtained by the usual inversion techniques.

The form of nonlinear models generally varies with the type of excitation and the algorithm employed for numerical calculation. One of the direct extensions of the linear model can be obtained by assuming that

$$m\ddot{x} + g(x,\dot{x}) = F \qquad (1)$$

Where x is the structural displacement response, F is the excitation matrix (usually the external forces) and nonlinear function g may be taken as an odd algebraic function of x and $\dot{x}$. For example, in the case of a single degree of freedom system,

$$g(\dot{x},x) = a_1\dot{x} + a_2\dot{x}^3 + a_3x + a_4x^3 \qquad (2)$$

The integral form of the formulation of the excitation-response relationship has also been used when transfer function is used for the linear model. In an integral formulation, instead of using three constant-parameter matrices, i.e., m, c and k, the model characteristics are lumped in a kernel function $h(\tau)$ in the following form:

$$x(t) = \int_0^\infty h(\tau) \, F(t - \tau) \, d\tau \qquad (3)$$

It is formally easy to extend the integral formulation to include the nonlinear kernels. For example, a second-order model has the form [Marmarelis and Udwadia,1976; Udwadia and Marmarelis,1976],

$$x(t) = \int_0^\infty h_1(\tau) \, F(t - \tau) d\tau$$
$$+ \int_0^\infty \int_0^\infty h_2(\tau_1,\tau_2) \, F(t - \tau_1) \, F(t - \tau_2) d\tau_1 d\tau_2 \qquad (4)$$

However, the computation efforts involved for the second or higher order models are much greater.

Hart and Yao (1977) presented a review of the identification theories and applications in structural dynamics as of 1976. They included identification problems which require a prior structural model with or without a quantification of experimental and modeling errors. The review also contained a brief description of the algorithms and sample data. Ibanez (1979) presented a comprehensive review of various techniques for the improvement of mathematical models in structural dynamics.

Liu and Yao (1978) discussed the concept of structural identification in the context of system identification and unique characteristics in its structural engineering applications. Basically, structural engineers are interested in identifying the damage and reliability functions in addition to the equation of motion. From another viewpoint, the updated equation of motion using test data and system identification can be a tool for the estimation of damage and reliability of existing structures [Natke and Yao, 1986b].

When a structure is inspected for the purpose of making damage assessment, a sequence of tests (or measurements using natural loading) may be conducted and the resulting data can be analyzed accordingly. Quantities which can be measured and recorded in testing structures include the load, the deformation (or strain) and the acceleration. From such data, mechanical properties such as stiffness and strength and dynamic characteristics such as natural frequency and damping can be estimated. in addition, visible damage such as cracks and local buckling in the plastic range can be detected by experienced observers. As a practical example, binoculars have been used by persons looking for color change in window panes in a tall building which indicate the presence of flaws causing the eventual breakage of these window glasses.

For metal structures which are subjected to repeated load applications, dye-check, ultrasonic or x-ray devices may be used to find and measure small and hidden fatigue cracks which indicate structural damage. The effect of detecting such fatigue cracks during a periodic inspection on the structural reliability of aircraft structures was studied by Yang and Trapp (1974). More recent development along this line has been reviewed by Yao et al. (1986).

Many full-scale on-site load tests of building structures have been performed in the United States of America during these past several decades [Hudson,1977]. To-date, most field-load-tests are static in nature and limited to studies of flexural response. Fitzsimons and Lon-ginow (1975) emphasized the fact that a static test cannot be used to reveal such weaknesses of a given structure as those due to corrosion, repeated load, creep, and brittleness. Nevertheless, load tests can be used to improve the reliability estimate [Shinozuka and Yang,1969]. Moreover, valuable information such as the stiffness of the structure can be obtained for the improvement of the mathematical representation of the structure for further dynamic analysis.

When a structure undergoes various degrees of damage, certain characteristics have been found to change. In testing a reinforced concrete shear wall under reversed loading conditions, free vibration tests were performed to estimate the fundamental natural frequency and damping ratio. Results of these tests as given by Wang, Bertero, and Popov

(1975) indicate that (a) the frequency decreased monotonically with damage while the damping ratio increased initially and then decreased, and (b) the repaired specimen was not restored to the original condition as indicated by free-vibration tests data. Similar results were reported by Hudson (1977), Hilgardo and Clough (1974), and Aristizabal-Ochoa and Sozen (1976).

Comprehensive experimental results of dynamic full-scale tests were obtained for a multi-story building structure and a 3-span highway bridge. Galambos and Mayes (1978) tested a rectangular 11-story reinforced concrete tower structure, which was designed in 1953, built in 1958, and tested to failure in 1976. The large-amplitude (and damaging) motions were induced with the sinusoidal horizontal movements of a 60-kip lead-mass which was placed on hardened steel balls on the eleventh floor. This lead-mass can be displaced up to $\pm$ 20 inches and the frequency capacity was 5Hz with the use of a servo-controlled hydraulic actuator, one end of which is fastened to the building frame. The maximum horizontal force range was $\pm$ 30,000 pounds. Test results indicated that the natural frequency decreased with increasing damage in general. Similarly, Baldwin et al. (1978) concluded from their testing of a three-span continuous composite bridge that changes in the bridge stiffness and vibration signatures can be used as an indication of structural damage under repeated loads. Further analyses of such full-scale test data should be useful in understanding the structural behavior as well as in making damage assessment of existing structures.

System identification tests are always conducted at extremely low-level vibrations such that they can be performed as many times as it is needed without causing any apparent damage to the existing structure. In most cases, only earthquake response records are available for the purpose of damage analysis.

In general, the complete record of an earthquake can be separated into the following three portions with different characteristics: (a) strongly-excited portion with higher modes contribution at the beginning of an earthquake, (b) much larger amplitude portion with nonlinear behavior, and (c) very low level vibration portion at the end of an earthquake. In system identification problems, parameters identified from the first portion cannot be very accurate although a contribution of higher modes has been considered in the analysis because of higher irregularity of earthquake input and response data. Because the third portion is equivalent to very low level ambient vibration, natural frequencies identified from the last portion is always higher than and cannot be compared with that from the second portion. However, the period of the first portion and the relatively low amplitude of the last portion cannot be determined because they depend on the duration and intensity of earthquake and structural characters. Consequently, one approach used by Chen (1980) is to deal with the identification of structural characteristics only in the second portion by dividing this portion into several segments in order to study and compare the changes among those characteristics. Method I is applied to find parameters ω_n (natural frequency) and ξ (damping coefficient) as functions of time from two linear equations of motion at time t and time t + Δtb by using measured earthquake and response data, $\ddot{x}_0(t)$ and $\ddot{y}(t)$, respectively. Parameters at any time t can be found as follows:

$$\omega_n^2(t) = \frac{(\ddot{x}_0(t)+\ddot{y}(t))\dot{y}(t+\Delta t)-(\ddot{x}_0(t+\Delta t)+\ddot{y}(t+\Delta t))\dot{y}(t)}{y(t)\dot{y}(t+\Delta t)-y(t+\Delta t)\dot{y}(t)} \tag{5}$$

and

$$\xi(t) = \frac{\ddot{x}_0(t)+\ddot{y}(t)+\omega_n^2(t)y(t)}{2\omega_n(t)\dot{y}(t)} \tag{6}$$

When the denominators of Eqs. 5 and 6 approach zero due to (a) inadequate choice of Δt, (b) nonlinear behavior, (c) higher mode contribution, or (d) measurement noise, it is difficult to estimate the values of $\omega_n(t)$ and $\xi(t)$ using this method. By using the least-square-error-fit, natural frequency, ω_n, and damping ratio, ξ, can be estimated by minimizing the following integral-squared difference, E, between he excitation, x_{0i}, input to a structure and th excitation, $s_{0i}^{(1)}$, calculated from its linear model:

$$E = \sum_{i=n_j}^{n_k} (\ddot{x}_{0i}-\ddot{x}_{0i}^{(1)})^2$$

$$= \sum_{i=n_j}^{n_k} (\ddot{x}_{0i}+\ddot{y}_i+2\xi\omega_n\dot{y}_i+\omega_n^2 y_i)^2 \tag{7}$$

Chen (1980) obtained the following estimates:

$$\omega_n^2 = \frac{(\Sigma\ddot{y}_i\dot{y}_i+\Sigma\ddot{x}_{oi}\dot{y}_i)\Sigma\dot{y}_i y_i-(\Sigma\ddot{y}_i y_i+\Sigma\ddot{x}_{oi}y_i)\Sigma\dot{y}_i^2}{\Sigma\dot{y}_i^2\Sigma y_i^2-(\Sigma\dot{y}_i y_i)^2} \tag{8}$$

and

$$\xi = - \frac{\Sigma \ddot{y}_i y_i + \Sigma \ddot{x}_{oi} y_i + \omega_n^2 \Sigma y_i^2}{2\omega_n \Sigma \dot{y}_i y_i} \tag{9}$$

where $\ddot{x}_{oi}$, $\ddot{y}_i$, $\dot{y}_i$, and y_i are measured response data during an earth-quake, and intervals n_j, n_k are segments of the records to be used for identification of ω_n and ξ. In addition, values ω_{n1} and ξ_{n2}, and ξ_2, ..., ω_{nn} and ξ_n are identified from segments (n_1, n_2), (n_2, n_3), ..., (n_n, n_{n+1}), respectively. Note that Equations 8 and 9 are reduced to Equations 5 and 6, respectively when only two point of record are con-sidered.

The main advantage of using such simple methods and linear models is the ease in checking the results and in preventing unnecessary errors associated with some complicated calculations. In a possible future application to structural control [Yao,1980], such simple methods can be useful. Moreover, the real nonlinear characteristics for any given full-scale structure are not readily known in most cases.

These methods are applied to analyze response records of two buil-dings collected during the 1971 San Fernando Valley earthquake [Foutch et al.,1975]. The Union Bank Building is a 42-story steel-frame structure in downtown Los Angeles. Prior to the 1971 San Fernando Earthquake, strong-motion accelerographs with synchronized timing were installed in the sub-basement, on the 19th floor and on the 39th floor. However, the instruments on the 39th floor failed to function. The S38 $^\circ$ W components of the digitized relative acceleration, velocity and displacement at the 19th floor were used as the response data in the analysis. As shown in

Figure 2 the results of Chen's methods agreed with those of modal minim-

ization method by Beck (1978). The natural frequency is shown to

decrease segment by segment except for the last value obtained from

Method II. The loss of stiffness as indicated by this change in natural

frequency seems to be the results of cracking and other types of damage

in nonstructural elements during the occurrence of large-amplitude earth-

quake response.

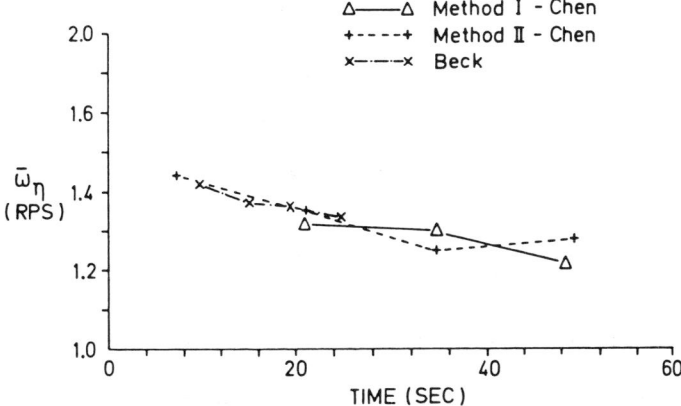

Fig. 2   Comparison of the Natural Frequency Identified from
Different Methods for Union Bank Building (From Chen, 1980)

Building 180 is a 9-story steel-frame structure on the grounds of

the Jet Propulsion Laboratory, Pasadena, California. The S82°E com-

ponents of the ground acceleration, relative acceleration, velocity and

displacement at the roof were used as the excitation and response data.

Figure 3 shown results from Method I and Method II by Chen (1980), and

Modal Minimization method by Beck (1978). The amplitude of the accelera-

tion response of this building during the earthquake was twice that of

the Union Bank but damage was limited to minor nonstructural cracking.

However, very little changes in the natural frequency result in this
analysis due to relatively minor damage involved.

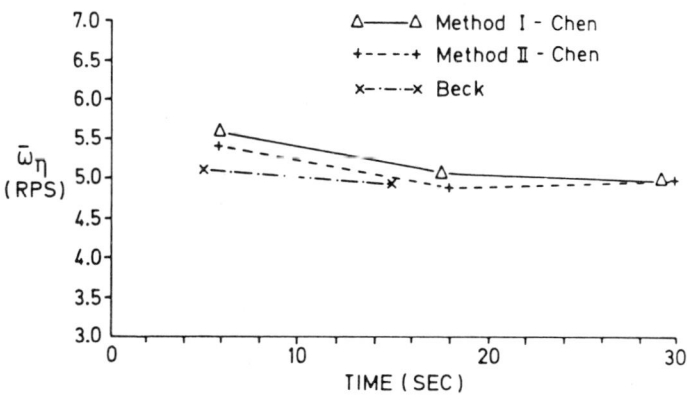

Fig.3 Comparison of the Natural Frequency Identified from
Different Methods for JPL Building 180 (From Chen, 1980)

Application of ``Non-parametric´´ Methods

Consider a single-degree-of-freedom system as shown in Fig. 4(a).
Its spring and damping forces are assumed to be functions of displacement
and velocity response, respectively, as shown in Fig. 4(b). Suppose that
the displacement and velocity response can be obtained from recorded data
as shown in Fig. 4(c). The local maximum and minimum values of the
displacement occur whenever the velocity response is zero. In between
these local maximum and minimum values, polynomial functions such as Eq.
2 can be used to describe the relationships between (a) the damping force
f_p and velocity $\dot{x}$ and (b) the spring force f_k and displacement x [Masri
et al.,1979; 1980].

Toussi (1982) developed such a nonparametric method for a multi-
story building frame, where the relative acceleration y(t) and the

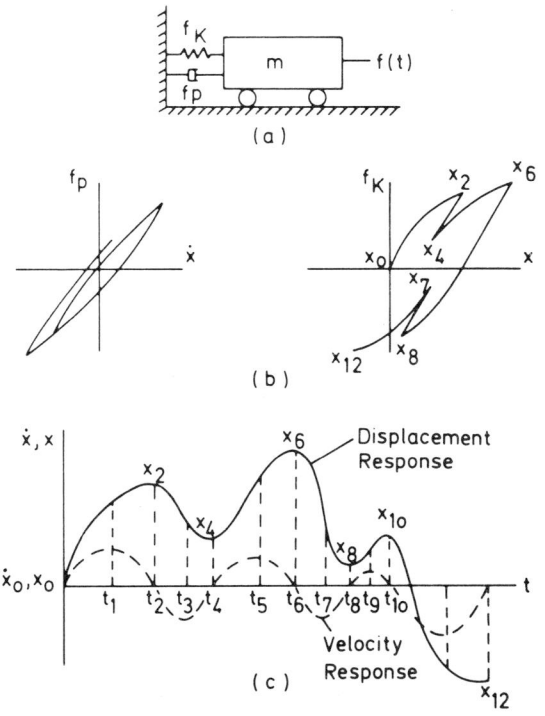

Fig. 4 System's Model (From Toussi, 1982)

applied force f(t) are available as recorded data. The resisting force

is assumed to depend upon the relative displacement and velocity, i.e.,

$$f_R(t) = f_R(y(t), \dot{y}(t)) \tag{10}$$

The resisting force, $f_R(t)$, is often separated into the following two

parts: one part depends primarily on $\dot{y}$ and another part depends mostly

on y. Here, they are referred to as spring and damping forces, which can

be nonlinear functions of $\dot{y}$ and y, respectively, i.e.,

$$f_R(t) = f_p(\dot{y}) + f_K(y) \tag{11}$$

where the damping force, $f_p(\dot{y})$, and the spring force, $f_K(y)$, are defined

as follows:

$$f_K(y) = \sum_{j=0}^{p} a_j y^j (t) \tag{12}$$

$$f_p(y) = \sum_{j=0}^{q} b_j \dot{y}^j (t) \tag{13}$$

and

$$p + q \leq n_i - 1 \tag{14}$$

In which, n_i indicates the number of points (samples) within interval i.

Substitution of Equations 12 and 13 into Equation 11 yields

$$f_R(\dot{y}, y, t) = c + \sum_{j=1}^{p} a_j y^j (t) + \sum_{j=1}^{q} b_j \dot{y}^j (t) \tag{15}$$

where

$$c = a_0 + b_0 \tag{16}$$

Repeating Equation 15 for the response components of each one of these n_i

points results in n_i simultaneous equations as follows:

$$F_R = YA \tag{17}$$

where F_R, Y and A represent the resisting force vector, response matrix

and parameter vector of the ith interval respectively. The elements of

vector F_R are calculated by setting the equilibrium of forces which are

acting on the structural system equal to zero.

$$f_R(t) = m\ddot{y}(t) + f(t) \tag{18}$$

Because $\dot{y}$ and y can be calculated by integrating the recorded accelera-

tion, $\ddot{y}$, the response matrix, Y, is also known. Therefore, the factor A

can be found by pre-multiplying the inverse of matrix Y on both sides of

Equation 17. Thus

$$A = Y^{-1} F_R \qquad (19)$$

after the parameter vector is calculated, the components of the resisting

force can be obtained. It is important to study the measured and recorded

noise and the effect of imperfect mathematical representation of struc-

tural behavior. The presence of noise suggests the use of mathematical

statistics. Because statistical calculations require a sufficient number

of samples, the number of parameters is reduced in Toussi's study instead

of increasing the number of samples (1982).

For the lumped-mass model for a building with masses m_j,

$j=1,2,\ldots,N$, it is assumed that these masses are connected by nonlinear

dashpots C_j, and nonlinear springs, K_j, as shown. The recorded motions

consist of the acceleration of the base, $\ddot{y}_g(t)$, and absolute accelera-

tion of the floors, $\ddot{x}_j(t)$, $j=1,2,3,\ldots N$. The absolute velocity and dis-

placement of each floor is obtained by integrating the corresponding

absolute acceleration. The forces created in the springs and dashpots

are assumed to depend upon the relative displacement and velocity between

the neighboring masses, respectively, i.e.,

$$[f_k(t)]_j = f_k[y_j(t)], \quad j=1,2,\ldots,N \qquad (20)$$

$$[f_p(t)]_j = f_p[\dot{y}_j(t)], \quad j=1,2,\ldots,N \qquad (21)$$

where

$$y_j(t) = x_j(t) - x_{j-1}(t), \quad j=1,2,\ldots N \qquad (22)$$

$$\dot{y}_j(t) = \dot{x}_j(t) - \dot{x}_{j-1}(t), \quad j=1,2\ldots N \qquad (23)$$

Now the equilibrium of forces applied to each floor is formed as follows:

$$m_N \ddot{x}_N(t) + f_p[\dot{y}_N(t)] + f_K[y_N(t)] = 0$$

$$m_{N-1}\ddot{x}_{N-1}(t) + f_p[\dot{y}_{N-1}(t)] + f_K[y_{N-1}(t)] = f_p[\dot{y}_N(t)] + f_K[y_N(t)]$$

$$\vdots$$

$$m_j \ddot{x}_j(t) + f_p[\dot{y}_j(t)] + f_K[y_j(t)] = f_p[\dot{y}_{j+1}(t)] + f_K[y_{j+1}(t)] \tag{24}$$

$$\vdots$$

$$m_2\ddot{x}_2(t) + f_p[\dot{y}_2(t)] + f_K[y_2(t)] = f_p[\dot{y}_3(t)] + f_K[y_3(t)]$$

$$m_1\ddot{x}_1(t) + f_p[\dot{y}_1(t)] + f_K[y_1(t)] = f_p[\dot{y}_2(t)] + f_K[y_2(t)]$$

Summing the top j equations at a time with $j=1,2,\ldots,N$, yields N equations as follows:

$$f_p[\dot{y}_j(t)] + f_K[y_j(t)] = \sum_{K=j}^{N} m_K \ddot{x}_K(t), \quad j=1,2,\ldots,N \tag{25}$$

In Eq. 25, because the quantities on the right-hand side are known, the summation of forces can be estimated at any given time.

The response data of two test structures (MF1 and H2) were used by Toussi (1982) to evaluate the effectiveness and applicability of the proposed method. The MF1 test structure is a one tenth-scale, ten-story, three-bay reinforced concrete structure which was tested by Healey and Sozen (1978) at the University of Illinois. The test procedure included a series of strong base motions simulating a scaled-version of the north-south component of the 1940 El-Centro earthquake. The input

acceleration level was magnified for each one of the three test runs.
The H2 test structure was also a one tenth-scale ten-story reinforced
concrete structure which was tested by Cecen (1979) and subjected to
seven simulated excitations. from the 1940 El-Centro earthquake.

The removal of the noise trend hidden in the measured acceleration
response was accomplished through fitting a polynomial of degree 5 to
each one of the velocity data obtained by integrating the corresponding
acceleration time-histories. Toussi (1982) then applied the hysteresis
identification method to estimate the inter-story load–deflection relationships of the frame. The spring force is restricted to a polynomial of
degree three while the viscous damping is chosen to have a linear form.

The estimated behavior of the seventh floor of H2 test frame for the
seven test runs is shown in Fig. 5. For this frame, the general soften-
ing is concluded and as the intensity of the earthquake excitation
increases, more nonlinearity in the structural behavior appears. Another

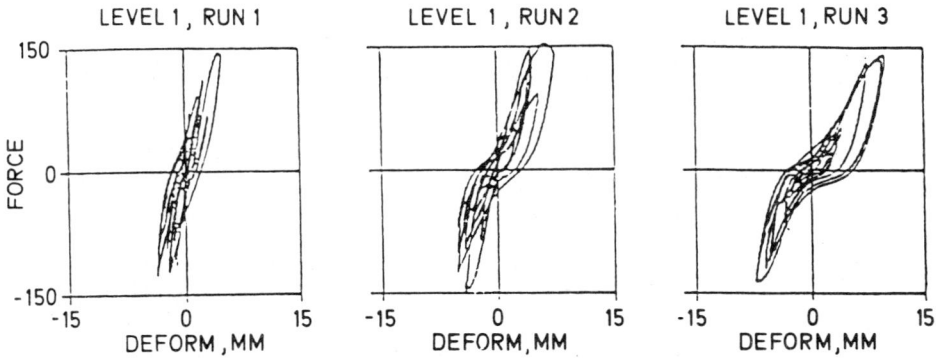

Figure 5 : Estimated force - deformation response , Level 1 ,
Model MF 1 , Runs 1,2 and 3

(From Stephens , 1985 ; using data of Healy and Sozen,1978)

interesting feature is the "soft-to-stiff" type of behavior that the structure experiences under different levels of load. Finally, the identified load-deflection curves become rather wide area-wise which is the indication of the dissipation of energy.

In 1983, Stephens et al. (1983) reviewed available information concerning (a) real-world structures in actual earthquakes, (b) full-size structural systems in controlled tests, and (c) small-size to full-size models of structural systems in laboratory experiments. As examples, results on several experiments of reinforced concrete structural models are summarized in Tables 1 and 2.

Table 1 : University of Illinois, 1/12 Size Reinforced Concrete Model Tests

Experimenters	Test Objectives	Model Configuration [a]	Displacement History [b]
Moehle and Sozen (1980)	investigate the earthquake response of structures having abrupt interruptions in story stiffness	4 models, 9 stories, each model with a central wall of a different height	El Centro NS, 1940, design peak accelerations of 0.4, 0.8, and 1.2 g
Healey and Sozen (1978)	study the nonlinear dynamic response of a reinforced concrete structure subjected to earthquake motions	1 model, 10 stories	El Centro NS, 1940, design peak accelerations of 0.4, 0.95, and 1.2 g
Aristizabal-Ochoa and Sozen (1976)	investigate the response of structures resisting earthquake forces through "cantilever" rather than "frame" action	4 models, 10 stories, 3 models designed to to yield initially in beams, 1 model with stiffened beams	El Centro NS, 1940, Taft N 21 E, 1952 peak accelerations of 0.41 to 1.96 g
Cecen (1979) and Sozen	study both the elastic and inelastic response of reinforced concrete structures subjected to earthquakes	2 models, 10 stories	El Centro NS, 1940, each structure subjected to a different sequence of events
Abrams (1980) and Sozen	investigate the interaction of the frames and walls of reinforced concrete structures subjected to strong earthquake ground motions	4 models, 10 stories, the models contained slender structural walls of different strengths	each structure subjected to three simulated earthquake events of increasing intensity

[a] The basic model configuration consisted of two 1/12 size reinforced concrete frames, 9 or 10 stories high and 3 bays wide. The frames were mounted parallel on the shake table.
[b] In each earthquake event, the models were shaken uniaxially, parallel to the long axis of the frames. The time scale of the events was compressed by a factor of 2.5. Each model was subjected to a series of events of increasing intensity.

(From Stephens and Yao, 1983)

Table 2 : University of California, Berkeley ; 0.7 Size Reinforced Concrete Frame Tests[a]

Experimenters	Test Objectives	Model Configuration[b]	Displacement History
Hidalgo and Clough (1974)	study the inelastic behavior of reinforced concrete structures under seismic loads	one model (RC 1), damaged during testing, repaired and retested	Taft N69W, 1952, El Centro NS, 1940 uniaxial shaking along the primary axis of the frames, each event of increasing intensity
Clough and Gidwani	investigate the inelastic response of reinforced concrete structures under seismic loads	one model (RC 2) identical in design to RC 1	Taft, 1952, uniaxial shaking along the primary axis of the frames, first event of damaging intensity
Oliva (1980)	study the inelastic biaxial column buckling in reinforced concrete frames subjected to seismic loads	one model (RC 5), identical in design to RC 2; damaged during testing, repaired and retested	Taft, 1952, model skewed 25 deg with respect to the directon of table motion

[a] Documentation was found in the (available) literature only for tests RC 1, RC 2 and RC 5.

[b] The basic model used for all the tests was a 0.7 size concrete frame, two stories high and one bay wide.

(From Stephens and Yao, 1983)

Evaluation of Seismic Damage

To make use of the estimated load–deformation relations from structural response records, it is desirable to find a suitable failure theorem. On the basis of a low–cycle failure criterion [Yao and Munse, 1962], Stephens (1985) developed the following damage function:

$$D = \sum_{i=1}^{n} [(\frac{\Delta\delta_{pt}}{\Delta\delta_{pf}})^a]_i \qquad (26)$$

where the fatigue damage exponent, $a = 1-(b*rl)$; b = deformation ratio coefficient; rl = relative deformation ratio defined as the ratio of the negative change in plastic deformation in cycle i, $\Delta\delta_{pc}$, to the positive change in plastic deformation in one–cycle test to failure conducted at the relative deformation ratio of cycle i. The value of the cumulative damage, D, is postulated to range from 0 to 1.0, with a value of 0 corresponding to no damage (safe), a value of 1.0 to failure (critically damaged).

To develop such a damage function for structural systems, it was
assumed that deformation under reversed cyclic load in one direction pro-
duced "indirect damaging response" (e.g. "less-than-ultimate" compression
strains) at some locations and "direct damaging response" (e.g. tension
strains) at other locations in the structure. The response condition at
each location reversed when the direction of deformation reversed.
Effects of "indirect damaging response" were considered through the
cycle-shape-dependent parameters $r1$ and $\Delta\delta_{pf}$. Damage was accumulated
independently in each deformation direction for both forms of the expressions.   Total damage was conservatively estimated as the larger value of
the damage indices as calculated in both directions.   Several acceleration records of structural response to earthquake ground motions were
processed and analyzed to estimate the force-deformation response of the
structure as shown in Figures 6 through 8.   Building structures studied

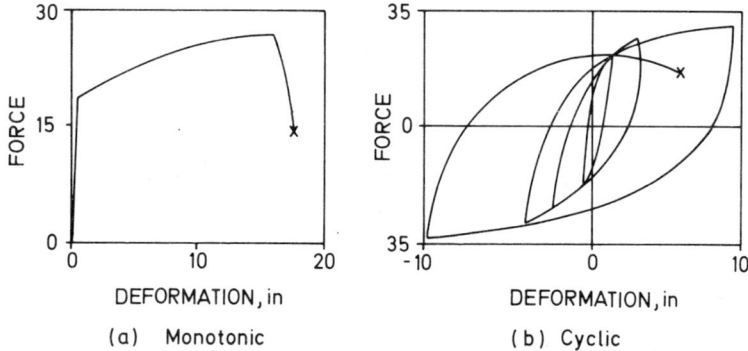

Figure 6 : Measured force - deformation response from
beam - column component tests
(From Burns and Siess, 1966)

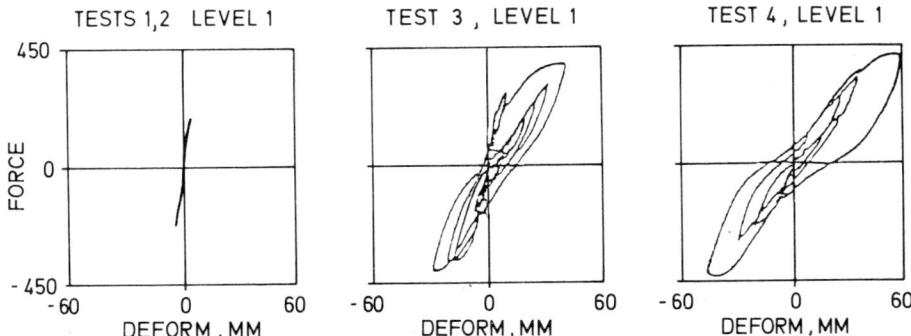

Figure 7: Force - deformation response, U.S.- Japan test
structure, ground floor
(From Stephens, 1985)

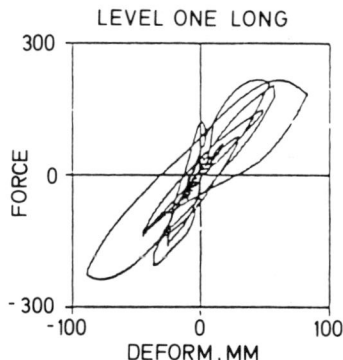

Figure 8 : Estimated force - deformation response, Imperial
County Services Building , ground floor
(From Stephens , 1985)

included laboratory test models (e.g., Healey and Sozen 1978; Okamoto et

al. 1982) and the Imperial County Services Building (McJunkin and Rags-

dale, 1980). Information from the force—deformation response is then

substituted into damage functions to obtain quantitative measures of the

damage conditions of the structure. Damage indices using damage indices

of Stephens (1985) and Park et al. (1984) are listed in Tables 3 through

6.

Table 3 : Damage indices at failure of beam-column
components (From Stephens , 1985)

Specimen	New Index		Park and Ang Index	Drift Ratio
	Direct 1	Direct 2		
J-8	1.21	1.08	2.84	0.16
J-6	1.03	0.19	1.04	0.13
J-5	0.69	0.10	0.79	0.13
J-5[a]	0.93	0 14	1.00	0.13

[a] recalculated

Table 4 : Damage indices and damage classifications
Model MF 1, Level 1 (From Stephens, 1985)

Test Event	Damage Expression	Assumed monotonic drift to failure			Toussi and Yao (1982/83) Damage condition[b]	
		5%	10%[a]	15%	Algorithm	Opinion
Run 1	New Index Direct 1 Direct 2	0.46 0.48	0.19 0.21	0.12 0.03	S	L
	Park and Ang S = 0.25 0.50 1.00	0.57 0.80 1.26	0.28 0.40 0.63	0.19 0.27 0.42		
Run 2	New Index Direct 1 Direct 2	1.23 1.21	0.46 0.45	0.27 0.26	D •	D
	Park and Ang S = 0.25 0.50 1.00	1.41 2.12 3.54	0.71 1.06 1.77	0.47 0.70 1.18		
Run 3	New Index Direct 1 Direct 2	2.88 2.71	1.04 0.91	0.59 0.49	C	C
	Park and Ang β = 0.25 0.50 1.00	2.21 3.59 6.55	1.10 1.79 3.17	0.74 1.20 2.12		

[a] best estimate
[b] S - safe, L - lightly damage, D - damaged, C - critically damaged

Table 5 : Damage indices following the fourth test on the
U.S. - Japan structure (From Stephens, 1985)

Damage Expression	Assumed monotonic drift to failure		
	4%	8%[a]	12%
New Index Direct 1 Direct 2	0.32 0.11	0.15 0.04	0.10 0.02
Park and Ang β = 0.25 0.50 1.00	0.86 1.24 1.99	0.43 0.63 1.01	0.29 0.40 0.79
Slope Ratio	0.26		
Drift Ratio	0.019		

[a] best estimate

Table 6 : Damage indices , Imperial County Services Building ,
ground floor , longitudinal direction
(From Stephens , 1985)

Damage Expression	Assumed monotonic drift to failure		
	5%	10%[a]	15%
New Index			
Direct 1	0.26	0.09	0.05
Direct 2	0.26	0.09	0.05
Park and Ang			
β = 0.25	0.85	0.42	0.28
0.50	1.21	0.59	0.40
1.00	1.94	0.94	0.63
Slope Ratio	0.50		
Drift Ratio	0.025		

[a] best estimate

The damage indices obtained for these structures were also correlated with independently formulated descriptive damage assessments of the
type safe, lightly damaged, damaged, and critically damaged (Toussi and
Yao, 1982, 1983). Based on these correlations, the damage functions pro-
duced reasonable and potentially useful measures of the damage conditions
of the structures. However, the available data were considered to be
insufficient to reliably determine the specific correlation between damage index and damage state (Stephens 1985).

EXPERT SYSTEMS FOR DAMAGE ASSESSMENT

General Remarks

It is usually difficult to consider all the details in the mathematical model of a given structure. Moreover, the failure behavior of most
large structures subjected to various loading conditions remain unknown

because few experimental studies of full-scale structures are available for comprehensive studies to-date [e.g., see Okamoto et al. (1982)].

Many structural engineers specialize in the condition evaluation of a particular structure and the determination of its level of safety. In such investigations, the available design calculations and drawings are examined and checked. Inspections and tests are conducted, and the resulting data analyzed. The results of these analyses are then summarized and interpreted to yield appropriate recommendations. The decision-making process involved in the determination of (a) specific types of inspection and test procedures to be used and (b) the summary and interpretation of analytical and experimental results require extensive and specialized judgment and experience of professional engineers.

In this section, a unified approach to safety evaluation of existing structures is presented and discussed. The emphasis is placed on the application of expert systems and fuzzy sets, which are used for the summary and interpretation of results of inspection, testing, and analysis.

Expert Systems

Experienced structural engineers are capable of making accurate measurements and precise calculations. However, the summary and interpretation of such results in order to obtain meaningful conclusions and practical decisions remain an art in most cases. In the current practice of structural evaluation, it is desirable to obtain available building documentation, visual examination, structural analysis, field and/or laboratory testing, and continual evaluation and interpretation of the collected information by human experts through their experience, intuition, and judgment [Moses and Yao, 1984; Yao, Bresler and Hanson, 1984].

SPERIL (Structural PERIL) is an expert system for the assessment of structural damage. To-date, two preliminary versions of this system have been formulated for the purposes of demonstration and practical implementation.

SPERIL-I is a production system for the diagnosis of seismic damage of existing structures [Ishizuka et al., 1983]. This system consists of a knowledge base and an inference machine. In the knowledge base, useful information for the damage assessment is stored as production rules which are obtained mainly from the visual inspection of structures and the analyses of acceleration records during the earthquake. The most distinct feature of this system is that separate evidential observations are integrated using Dempster and Shafer's combination rule and fuzzy sets. Possible damage classifications include (a) no damage, (b) slight damage, (c) moderate damage, (d) severe damage, (e) destructive damage, and (f) no appropriate answer.

In SPERIL-II, metarules are adopted for the selection of the rule group and a suitable inference method [Ogawa et al. 1985]. The inference machine has several reasoning methods such as forward reasoning and backward reasoning. The knowledge base consists of rules for damage assessment and metarules. In the memory, information stored include inspection data, testing results, drawings and documents, and historical records.

To compare the results of more recently, an expert systems is being developed for evaluation of structural durability [Furuta et al. 1986]. Other applications of expert systems in structural engineering are reviewed by Furuta et al. (1985).

Application of Fuzzy Sets

The theory of fuzzy sets began with Zadeh (1965,1973,1983). An elementary introduction was given by Brown and Yao (1983). Ishizuka et al. (1983), Ogawa et al. (1985), and Zadeh (1983) among others have applied fuzzy logic in various expert systems.

Many interrelating factors affect the fatigue behavior of a welded structure [Bowman and Yao, 1983]. While it is possible to assign numerical values to some of these variables, it is difficult to obtain a complete mathematical model of the fatigue behavior of existing structures [Bowman et al. 1985]. Consequently, the method of linguistic assessment is suitable for solving such complex problems.

There are the following four basic steps in a linguistic damage assessment procedure: (a) Assessment of each variable in linguistic terms; (b) Translations of linguistic terms into fuzzy sets through the use of a dictionary; (c) Inference of the fuzzy damage state; (d) Translation of fuzzy damaged state into linguistic terms [Watada, 1983; Watada et al. 1984]. Hinkle et al. (1986) demonstrated the usefulness of using such a procedure with actual test data.

CONCLUDING REMARKS

Because (a) the failure behavior of full-size structures is highly nonlinear and dependent on loading history, and (b) there are insufficient data on loading condition for reliable prediction of future loads, much work remains to be done in estimating the damage and reliability of structural systems. In particular, existing and deteriorating structures are extremely complex systems, the damage state of which is difficult to evaluate. In spite of recent advances in finite-element analyses and computer technology, it is still difficult to mathematically model the behavior of structural systems. Moreover, there exist various uncertainties in many theorems of cumulative damage. The application of expert systems is suggested along with system identification techniques in which many sources of data, calculation, and other information concerning the structure may be considered. Note that the practical implementation of such expert systems remains to be a challenging task.

ACKNOWLEDGMENT

I wish to thank my senior colleague Professor H. G. Natke for his constructive suggestions and special considerations. Mrs. Vicki Gascho has been most helpful in the typing and preparation of these lecture notes. Moreover, I wish to acknowledge the collaboration of all my co-workers and the National Science Foundation and a NATO Research Grant No. 625/84 for my knowledge in this subject area.

REFERENCES

Abrams, D.P., and Sozen, M.A., (1980) "Experimental Study of Frame-Wall Interaction in Reinforced Concrete Structures Subjected to Strong Earthquake Motions," Structural Research Series, No. 460, Department of Civil

Engineering, University of Illinois, Urbana, IL.

Aristizabal-Ochoa, J.D., and Sozen, M.A., (1976) Behavior of a Ten-Story Reinforced Concrete Walls Subjected to Earthquake Motions, SSRR No. 431, Department of Civil Engineering, University of Illinois, Urbana, IL.

Baldwin, J.W., Jr., Salane, H.J., and Duffield, R.C., (1978) Fatigue Test of a Three-Span Composite Highway Bridge, Study 73-1, Department of Civil Engineering, University of Missouri-Columbia.

Beck, J.L. (1978) "Determining Models of Structures from Earthquake Records," Report No. EERL 78-01, California Institute of Technology, Pasadena, CA.

Burns, N.H., and Siess, C.P., (1966a) "Plastic Hinging in Reinforced Concrete," Journal of the Structural Division, ASCE, 92, No. ST5, pp. 45-64.

Burns, N.H., and Siess, C.P. (1966b), "Repeated and Reversed Loading in Reinforced Concrete," Journal of the Structural Division, ASCE, 92, No. ST5, pp. 65-78.

Bowman, M.D., Nordmark, G.E., and Yao, J.T.P., (1985), "Fuzzy-Logic Approach in Structural Fatigue," presented at NAFIPS '85, Atlanta, GA, 24-25 October 1985.

Bowman, M.D., and Yao, J.T.P., (1983), "Fatigue Damage Assessment of Welded Structures," Proceedings, W.H. Munse Symposium on Behavior of Metal Structures, Research to Practice, Edited by W.J. Hall and M.P. Gaus, ASCE National Convention, Philadelphia, PA, May 17, 1983, pp. 45-69.

Brown, C.B., and Yao, J.T.P., (1983), "Fuzzy Sets in Structural Engineering," to appear in Journal of the Structural Engineering Division, ASCE.

Cecen, H., (1979) Response of Ten-Story Reinforced Concrete Frames to Simulated Earthquakes, Ph.D. Thesis, School of Civil Engineering, University of Illinois, Urbana, IL.

Chen, S.J.H., (1980) System Identification and Damage Assessment of Existing Structures, Ph.D. Thesis, School of Civil Engineering, Purdue University, W. Lafayette, IN.

Distefano, N., and Pena-Pardo, B., (1975) "System Identification of Frames under Seismic Loads," ASCE National Structural Engineering Meeting, New Orleans.

Distefano, N., (1972) "Some Numerical Aspects in the Identification of a Class Nonlinear Viscoelastic Materials," ZAMM 52, p. 389.

Distefano, N., and Todeschini, R., (1974) "Modeling Identification and Prediction of a Class of Nonlinear Viscoelastic Materials," _Int. J. Solids Struct._ 9, I, pp. 805-818; II, pp. 1431-1438.

Dubois, D., and Prade, H., _Fuzzy Sets and Systems: Theory and Applications_, Academic Press, 1980.

Eykhoff, P., (1974), _System Identification-Parameter and State Estimation_, John Wiley & Sons.

FitzSimons, N., and Longinow, A., (1975) "Guidance for Load Tests of Buildings," _Journal of the Structural Division_, ASCE, v. 101, n. ST7, pp. 1367-1380.

Foutch, D.A., Housner, G.W., and Jennings, P.C., (1975) _Dynamic Responses of Six Multistory Buildings During the San Fernando Earthquake_, Report No. EERL 75-02, California Institute of Technology, Pasadena, CA 1975.

Furuta, H., Fu, K.S., and Yao, J.T.P., (1985), "Structural Engineering Applications of Expert Systems," Technical Report No. CE-STR-85-11, School of Civil Engineering, Purdue University, W. Lafayette, IN.

Furuta, H., Shiraishi, N., and Yao, J.T.P., (1986), "An Expert System for Evaluation of Structural Durability," presented at the 5tgh International Symposium on Offshore Mechanics and Arctic Engineering, Tokyo, JAPAN.

Galambos, T.V., and Mayes, R.I., (1978), "Dynamic Test of a R/C Building," Department of Civil Engineering and Applied Science, Washington University, St. Louis, Missouri.

Gersch, W., (1975) "Parameter Identification: Stochastic Process Techniques," _Shock and Vibration Digest_, 1975.

Hinkle, A.J., and Yao, J.T.P., (1986), "Linguistic Assessment of Fatigue Damage of Welded Structures," presented at NAFIPS `86, New Orleans, 1-4 June 1986.

Hart, G.C., and Yao, J.T.P., (1977) "System Identification in Structural Dynamics," _Journal of the Engineering Mechanics Division_, ASCE, v. 103, n. EM6, pp. 1089-1104.

Healey, T.J., and Sozen, M.T., (1978), "Experimental Study of the Dynamic Response of a Ten-Story Reinforced Concrete Frame with a Tall First Story," Civil Engineering Studies, Structural Research Series No. 450, University of Illinois.

Hidalgo, P., and Clough, R.W., (1974), _Earthquake Simulator Study of a Reinforced Concrete Frame_, Report No. EERC 74-13, Earthquake Engineering Research Center, University of California, Berkeley, California.

Hudson, D.E., (1977), "Dynamic Tests of Full-Scale Structures," Journal of the Engineering Mechanics Division, ASCE, v. 103, n. EM6, pp. 1141-1157.

Ibanez, P., et al., (1979), Review of Analytical and Experimental Techniques for Improving Structural Dynamic Models, Bulletin 249, Welding Research Council, New York, June 1979, 44 pages.

Ishizuka, M., Fu, K.S., and Yao, J.T.P., (1983a), "Rule-Based Damage Assessment System for Existing Structures," Solid Mechanics Archives, Vol. 8, Martinus Nijhoff Publishers, The Netherlands, pp. 99-118.

Ishizuka, M., Fu, K.S., and Yao, J.T.P., (1983b), "Computer-Based System for the Assessment of Structural Damage," Proceedings of Informatics in Structural Engineering, IABSE, Zurich, Switzerland, pp. 89-98.

Liu, S.C., and Yao, J.T.P., (1978), "Structural Identification Concept," Journal of the Structural Division, ASCE, v. 104, n. ST12, December, pp. 1845-1858.

Marmarelis, P-Z., and Udwadia, F.E., (1976), "The Identification of Building Structural Systems - II. The Nonlinear Case," Bulletin of the Seismological Society of America, v. 66, n. 1, pp. 153-171.

McJunkin, R.D., and Ragsdale, J.T., (1980), "Compilation of Strong-Motion Records and Preliminary Data from the Imperial Valley Earthquake of 15 October 1979," Preliminary Report No. 26, Office of Strong-Motion Studies, California Division of Mines and Geology, Sacramento, California, 1980.

Marsi, S.F., and Anderson, J.C., (1980), "Identification/Modeling Studies of Non-Linear Multidegree Systems," Vol. 3, Analytical and Experimental Studies of Non-Linear System Modeling, - A Progress Report A_T (49-24-0262), U.S. Nuclear Regulatory Commission, April, 1980.

Masri, S.F., and Caughey, T.K., (1979), "A Nonparametric Identification Technique for Non-Linear Dynamic Problems," Journal of Applied Mechanics, ASME, Vol. 46, pp. 433-447.

Moses, F., and Yao, J.T.P., (1984) "Safety Evaluation of Buildings and Bridges," The Role of Design, Inspection, and Redundancy in Marine Structural Reliability, Edited by D. Faulkner, M. Shinozuka, R.R. Fiebrandt, and I.C. Franck, Committee on Marine Structures, NRC, Washington, D.C.,, pp. 549-572.

Natke, H.G., and Cottin, N., (1986), "Introduction: Fundamentals and Survey", CISM - Course on Application of System Identification in Engineering, Udine, Italy, 7-11, July 1986.

Natke, H.G., and Yao, J.T.P., (1986a), "Research Topics in Structural Identification," _Dynamic Response of Structures_, American Society of Civil Engineers, March 31 - April 2, 1986, pp. 542-550.

Natke, H.G., and Yao, J.T.P., (1986b), "System Identification Approach in Structural Damage Evaluation," to be presented at the Structural Engineering Congress, American Society of Civil Engineers, New Orleans, 15-18 September 1986.

Ogawa, H., Fu, K.S., and Yao, J.T.P., (1985), "A Knowledge-Based Approach for Safety Assessment of Existing Structures," _Proceedings_, 4th International Conference on Structural Safety and Reliability, IASSAR, Kobe, Japan, Vol. IV, 27-29 May 1985, pp II597-600.

Okamoto, S., et al., (1982), "A Progress Report on the Full-Scale Seismic Experiment of a Seven Story Reinforced Concrete Building — Part of the U.S. Japan Cooperative Program," Building Research Institute, Ministry of Construction, Japan.

Park, Y.J., (1984), "Seismic Damage Analysis and Damage-Limiting Design of R.C. Buildings," _Report No._ UILU-ENG-84-2007, SRS 516, University of Illinois, Urbana, IL, 1984.

Rodeman, R., (1974), _Estimation of Structural Dynamic Model Parameters_, Ph.D. Thesis, School of Civil Engineering, Purdue University, W. Lafayette, IN.

Sage, A.P., and Melsa, J.L., (1971), _System Identification_, Academic Press.

Shinozuka, M., and Yang, J.N., (1969), "Optimum Structural Design Based on Reliability and Proof-Land Test," _Annals of Assurance Science_, Proceedings of the Reliability and Maintainability Conference, v. 8, pp. 375-391.

Stephens, J.E., and Yao, J.T.P., (1983), "Survey of Available Structural Response Data for Damage Assessment," Technical Report No. CE-STR-83-23, School of Civil Engineering, Purdue University, W. Lafayette, IN.

Stephens, J.E. (1985), _Structural Damage Assessment Using Response Measurements_, Thesis, Presented to Purdue University, at West Lafayette, IN, in partial fulfillment of the requirements for the degree of Doctor of Philosophy.

Tomlinson, G., (1986), Lecture at CISM - Course on Application of System Identification in Engineering, Udine, Italy, 7-11 July 1986.

Ting, E.C., Chen, S.J. Hong, and Yao, J.T.P., (1978), "System Identification Damage Assessment and Reliability Evaluation of Structures," Technical Report No. CE-STR-78-1, School of Civil Engineering, Purdue University, 62 pages.

Toussi, S., (1982), System Identification Methods for the Evaluation of Structural Damage, Ph.D. Thesis, School of Civil Engineering, Purdue University, W. Lafayette, IN.

Toussi, S., and Yao, J.T.P., "Assessment of Structural Damage Using the Theory of Evidence," Structural Safety, Vol. 1, (1982/1983), pp. 107-121.

Udawadia, F.E., and Marmarelis, P.Z., (1976), "The Identification of Building Structural Systems - I. The Linear Case," Bulleting of the Seismological Society of America, v. 66, n. 1, February 1976, pp. 125-151.

Wang, T.Y., Bertero, V.V., and Popov, E.P., (1975), Hysteretic Behavior of Reinforced Concrete Framed Walls, Report No. EERC 75-23, Earthquake Engineering Research Center, University of California, Berkeley, CA.

Watada, J., (1983) Theory of Fuzzy Multivariate Analysis and Its Applications, Ph.D. Thesis, University of Osaka.

Watada, J., Fu, K.S., and Yao, J.T.P., (1984), "Damage Assessment Using Fuzzy Multivariant Analysis," Technical Report No. CE-STR-84-4, School of Civil Engineering, Purdue University, West Lafayette, IN.

Yang, J.N., and Trapp, W.J., (1974), "Reliability Analysis of Aircraft Structures Under Random Loading and Periodic Inspection," AIAA Journal, v. 12, n. 12, pp. 1623-1630.

Yao, J.T.P., (1980), "Identification and Control of Structural Damage," Solid Mechanics Archives, v. 5, Issue 3, Sijthoff & Noordhoff International Publishers, Netherlands, pp. 325-345.

Yao, J.T.P., (1985), Safety and Reliability of Existing Structures, Pitman Advanced Publishing Program, Boston, London, Melbourne, 1985.

Yao, J.T.P., Bresler, B., and Hanson, J.M., (1984) "Condition Evaluation and Interpretation for Existing Concrete Buildings," presented at the Symposium on Evaluation of Existing Concrete Buildings, ACI Convention, Phoenix, AZ, 8-9 March 1984.

Yao, J.T.P., Kozin, F., Wen, Y.K., Yang, J.N., Schueller, G.I., and Ditlevesen, O., (1986) "Stochastic Fatigue, Fracture, and Damage Analysis," presented at RILEM Symposium on Stochastic Methods in Materials and Structural Engineering, UCLA, April 3-4, 1986; to appear in Structural Safety.

Yao, J.T.P., and Munse, W.H., (1962), "Low-Cycle Fatigue Behavior of Mild Steel," Special Technical Publication No. 338, American Society for Testing and Materials, pp. 5-24.

Zadeh, L.A., "Fuzzy Sets," (1965), Information and Control, v. 8, pp. 338-353.

Zadeh, L.A., (1973), "Outline of a New Approach to Analysis of Complex Systems and Decision Processes," IEEE Transactions on Systems, Man and Cybernetics, Vol. SMC-3, No. 1, pp. 28-44.

Zadeh, L.A., (1983), "The Role of Fuzzy Logic in the Management of Uncertainty in Expert Systems," Fuzzy Sets and Systems, Vol. 11, pp. 199-227.

APPLICATIONS IN AEROSPACE AND AIRPLANE ENGINEERING: ESTIMATION OF MODAL QUANTITIES AND MODEL IMPROVEMENT

H.G. Natke
Universität Hannover, Hannover, F.R.G.

Here the author reports on older experience of ground[1] and flight[2] vibration tests of airplanes which, however, in its statements is still valid. Today better relative pickups are available than 10 years ago, and more effective mini- and μ-processors, including software, can be used for data processing. The power amplifiers of the electromagnetic exciters are much smaller and need no water-cooling any

----

[1] Test on the ground simulating the free-free boundary conditions of the plane in order to determine its eigen-quantities

[2] Test during flight of an airplane to estimate eigenfrequencies and damping ratios depending on Mach number and speed

more. This is through the use of transistors instead of
tubes etc. The excitation possibilities are also greater
than they were 20 years ago. But when one looks at the
ground vibration test, especially in France and the Federal
Republic of Germany, one finds that the "official" insti-
tutions such as ONERA and DFVLR still mainly use the phase
resonance technique with harmonic excitation /1/2/ as
described in /3/4/. The flight vibration tests, including
their quick-look application, make use of newer data processing methods, such as FFT and phase separation techniques
/5/. The modal identification of satellites in free-free
boundary simulation is done similarly as for airplanes.
However, in satellite testing in the earth-gravity
environment it is more difficult to fix exciters at those
points which are necessary for the appropriate force excitation. In the clamped-free boundary condition (such as a
cantilever) in which the attached part is generally excited,
the special excitation has to be taken into account.

1.   Experimental Modal Analysis

      Nothing is to be repeated here concerning the phase
resonance method (see /3/4/5/). The suspension can be done
(satellites, small rockets etc.) by steel cables (or wires)
which can work in the vertical direction as soft springs,

and in the horizontal direction with the test object as a
rigid mass as a pendulum (through which their lengths are
determined): Fig. 1. Rubber cables are inexpensive for light
test objects (material models), but their disadvantage is
the creep of the cables (Fig. 2). Instead of cables, bellow
elements can be used (air-springs), which may be displace-
ment-controlled when testing is performed in different mass
configurations (Fig. 3). Fig. 4 shows the application of
airsprings to a cargo-plane and Fig. 5 to a (vertical
take-off) fighter. The preparation of the payload of the
Europa I launcher for ground vibration testing is shown in
Fig 6. In Fig. 7 a quite different support can be seen for a
helicopter, which can be suspended by servo-controlled
pneumatic springs or can stand on grassy soil (looking for
ground resonance effects). Sometimes, with less expenditure,
the plane can stand on softened deflated tires in order to
simulate the free-free condition. The suspension should be
fixed at structure points with fewer displacements and in
such a way that the suspended structure is stable. The
free-free condition is obtained when the ratio of suspension
eigenfrequency to the fundamental frequency of the elastic
degrees-of-freedom is about 1/3 or 1/4 (see /4/).

Fig. 8 shows an electromagnetic exciter and Fig. 9 a
servo-hydraulic device, both being used for sinusoidal
excitation; slow frequency sweep and stepped sine was

Fig. 1 Suspension by steel cables /6/

Fig. 2   Suspension by rubber cables

Fig. 3a Suspension by steel cables
combined with air springs

Fig. 3b Suspension by steel cables
combined with air springs

Fig. 4a Air springs used for a cargo plane

Fig. 4b Air springs used for a cargo plane

Fig. 4c Air springs used for a cargo plane

Fig. 5   Simulation of the free-free condition for a fighter

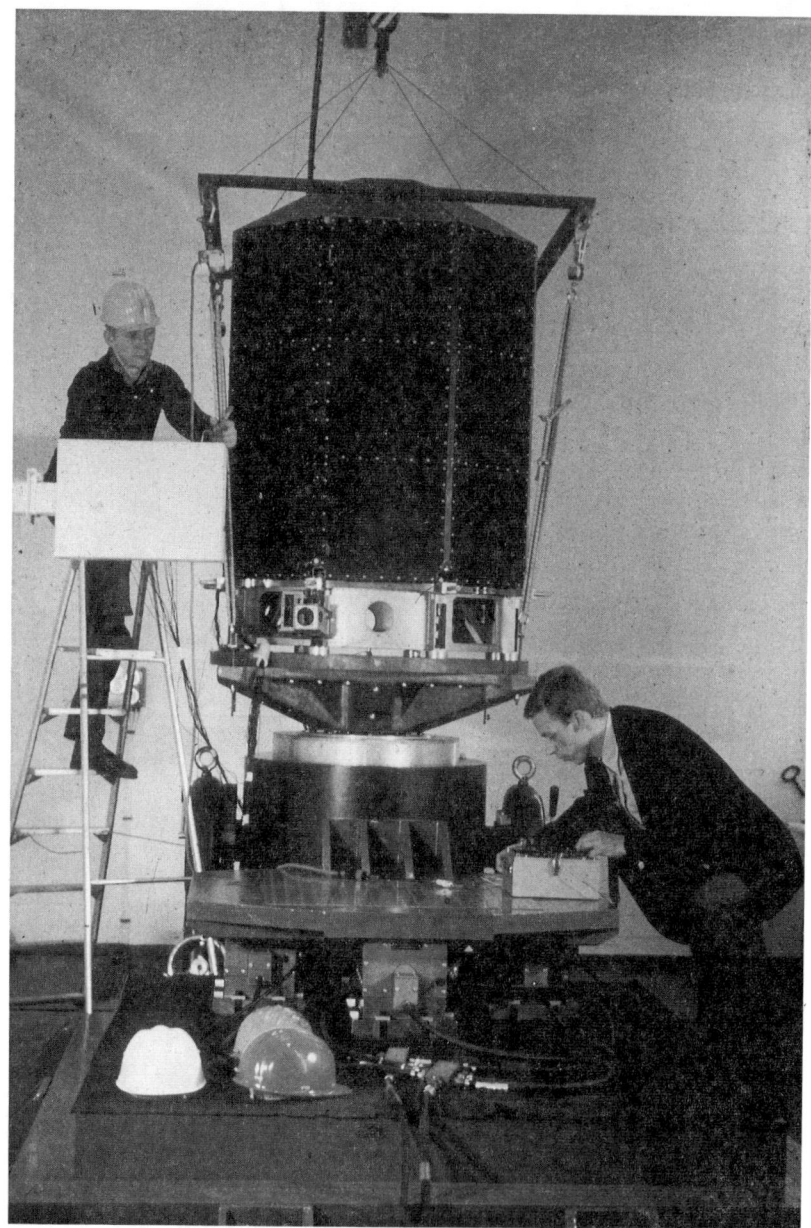

Fig. 6   Vibrator test with the
payload of a launcher

Fig. 7   Simulation of boundary conditions for a helicopter

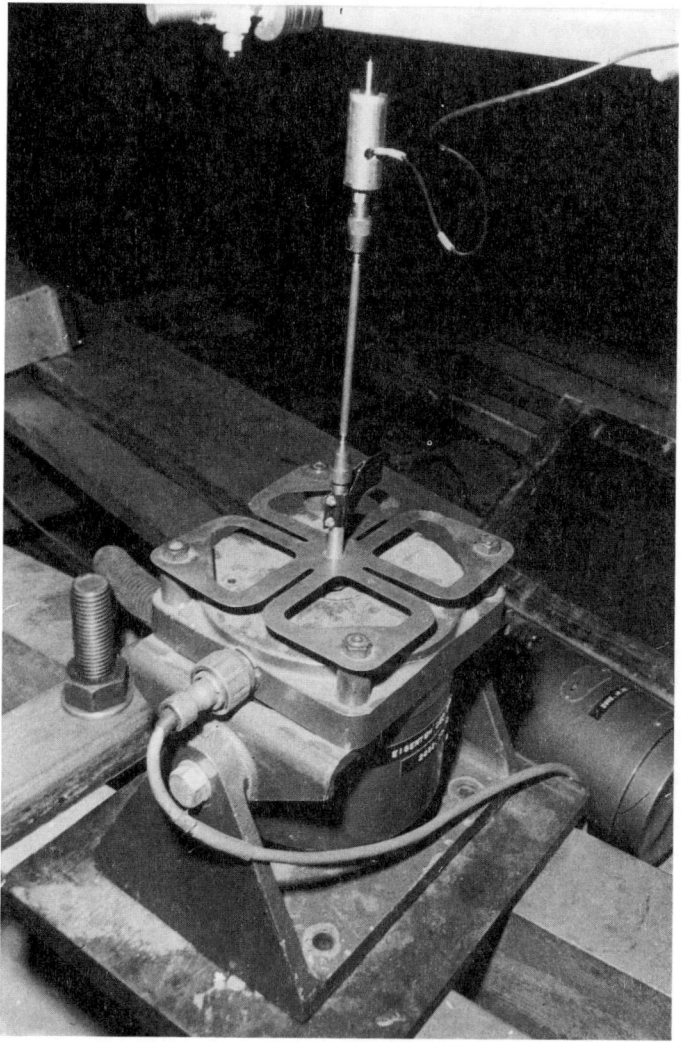

Fig. 8   Electro-magnetic exciter

Fig. 9 Servo-hydraulic device

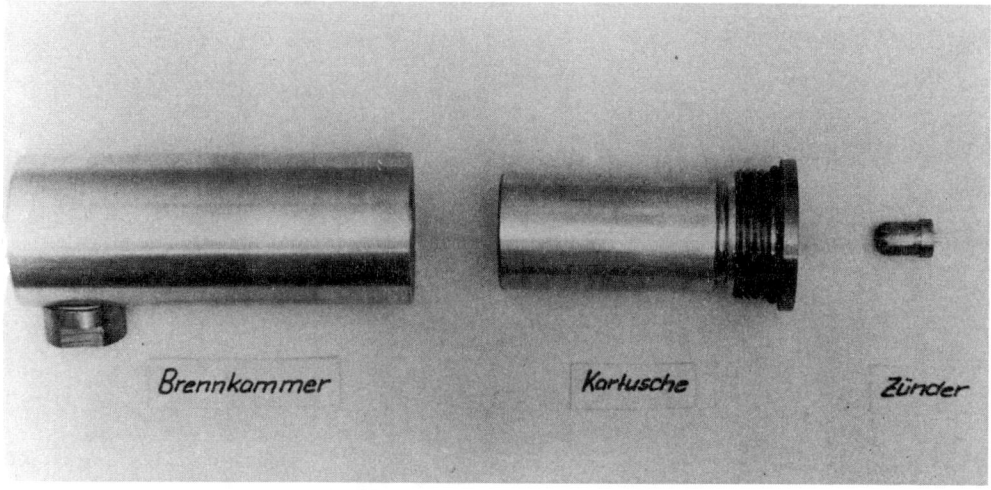

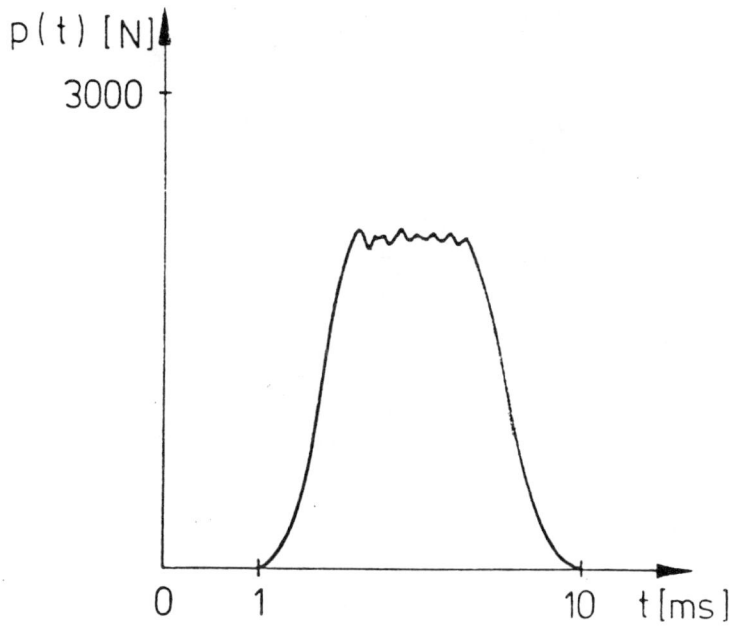

Fig. 10 Cartridge used in flight vibration
        testing (top) with its burn diagram

Fig. 11 Vanes for flight vibration testing

applied /6/. Fig. 10 contains a cartridge used in flight
vibration testing with its burn diagram. Fig. 11 gives an
impression of vanes for harmonic excitation during flight.
In addition to impulse excitation by stick jerking (Fig. 12)
and inertia based harmonic electromagnetic excitation,
randomly controlled exciters are used. The synchronous
working of the exciters is important for in flight multi-
point excitation (problem when using cartridges!), and it is
also important that they cause no plastic deformation or
similar damage.

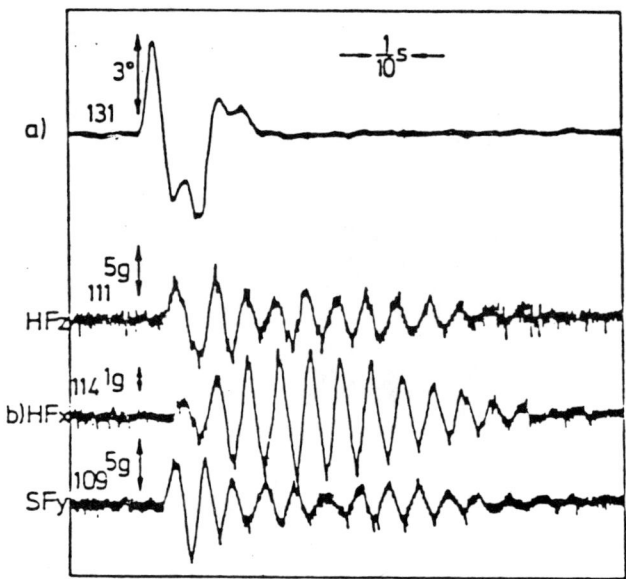

Fig. 12 : Dynamic response due to
stick jerking
a) of aileron
b) of tail unit

For measurement and pickups see /3/, Chapter 2. Tele-
processing is not a technical problem. Data handling in
general (see indicator function for detecting eigenfrequencies etc.) is also commonly used.

The application of phase separation techniques (also
with harmonical excitation) is treated cautiously by French
and German ground vibration test teams /1/2/. However, the
author gained some experience when he and his co-workers
developed these methods. Table 1 contains the measured
eigenfrequencies and damping ratios for the wing model of
Fig. 2 for a given frequency range of the classical phase
resonance test (K2) and of the phase separation technique

Table 1 Estimates of two ground vibration tests

Tafel 1: Erregungsfrequenzen N_j [Hz] und Teilergergebnisse aus dem
Standschwingungsversuch nach dem klassischen und
versuchsmäßig-rechnerischen Verfahren   Flügelmodell   VFW 614

Frei-heits-grad	K2		R1.2.5.m	R1.1.4.30		R1.2.5.30	
	Eigenfrequ. N_{E_i} [Hz]	eff. Dämpf. g_i [1]	Erregungs-fr. N_j [Hz]	Eigenfrequ. N_{E_i} [Hz]	eff. Dämpf. g_i [1]	Eigenfrequ. N_{E_i} [Hz]	eff. Dämpf. g_i [1]
1	2	3	4	5	6	7	8
1	12,09	0,014	12,14	11,79±0,18	0,021±0,012	12,07±0,01	0,004±0,001
2	18,00	0,009	18,01	17,99±0,01	0,014±0,005	17,99±0,01	0,007±0,002
3	(18,2)	—	18,25	—	—	18,21±0,01	0,050±0,009
4	19,51	0,034	19,62	19,36±0,05	0,055±0,011	19,35±0,04	0,054±0,009
5	21,10	0,013	21,11	21,05±0,02	0,014±0,007	21,07±0,01	0,011±0,002

Table 2 Comparison of measured and calculated dynamic responses

Flügelmodell VFW 614

Tafel 2: Vergleich der gemessenen und errechneten Imaginärteilantworten für N= 18,1 Hz

$R_1 = 0{,}7335$ kp, $P_s = 0$ kp, $s = 2,3,\ldots\ldots,30$

| k | $(Y_k^{im}/981)$ cm s^{-2} | | | | Prozentuale Fehler bezogen auf die gemessenen Antworten von | | |
| | Meßwert | R1.2.5.30 | R1.1.4.30 | K2 | R1.2.5.30 | R1.1.4.30 | K2 |
1	2	3	4	5	6	7	8
1	0,269	0,270	0,115	-0,038	-0,37	57,24	114,12
2	-1,112	-1,084	-0,923	-0,423	2,51	16,99	61,96
3	-0,366	-0,340	0,059	-0,003	7,10	83,87	99,18
4	0,209	0,224	0,726	0,426	-7,17	-247,36	-103,82
5	0,231	0,242	0,761	0,422	-4,76	-229,43	-82,86
6	0,145	0,155	0,510	0,296	-6,89	-251,75	-104,13
7	0,151	0,165	0,528	0,289	-9,27	-249,66	-91,39
8	0,021	0,024	0,117	0,073	-14,28	-457,14	-24,76
9	-0,040	-0,040	-0,041	0,018	0	-2,50	-145,00
10*	-0,008	0,002	-0,002	-0,003	-125,00	75,00	62,50
11	-0,047	-0,034	-0,049	0,023	27,65	-4,08	-149,93
12	-0,034	-0,030	-0,070	0,046	11,76	-105,88	-235,29
13	-0,062	-0,046	-0,064	0,033	25,80	-3,22	-153,22
14	-0,051	-0,045	-0,078	0,040	11,76	52,94	-178,43
15	-1,126	-1,073	-0,908	-0,422	4,70	19,36	62,52
16	-0,669	-0,643	-0,541	-0,255	3,88	19,13	61,88
17	0,084	0,081	0,072	-0,032	3,57	14,28	139,28
18	0,102	0,110	0,085	-0,415	-7,84	16,66	506,86
19**	-0,091	0,005	-0,031	0,085	-	-	-
20	0,154	0,141	-0,137	0,082	8,44	188,96	46,75
21	-0,182	-0,160	-0,066	0,024	12,08	63,73	113,18
22	0,010	0,014	0,003	0,036	-40,00	70,00	-260,00
23	0,034	0,037	0,065	-0,040	-8,82	-91,17	217,64
24	0,132	0,131	0,104	-0,048	0,75	21,21	136,36
25	0,302	0,298	0,190	-0,083	1,32	37,08	127,48
26	0,410	0,407	0,241	-0,108	0,73	41,21	126,34
27	-0,532	-0,485	0,082	-0,022	8,83	115,41	95,86
28	0,533	0,531	0,229	-0,074	0,37	57,03	113,88
29	-0,006	-0,007	-0,002	0,0001	-16,67	66,66	-101,42
30	-0,019	-0,013	-0,004	0,002	31,57	78,94	-110,52

* Meßwert liegt außerhalb der Meßgenauigkeit.
**Dieser Meßplatz fiel während der Messung aus.

using an effective number of degrees-of-freedom n_{eff} (order of the model used, see /3/) equal to 4 (R1.1.4.30) and equal to 5 (R1.2.5.30) /7/. The classical phase resonance test suggested an eigenfrequency of 18.2 Hz, but it was not clear and it yielded no damping ratio. A comparison of the estimated standard deviations of the eigenfrequencies and damping rations obtained by the phase separation method indicates the presence of 5 degrees-of-freedom in the frequency range considered. Table 2 shows that it is possible to make a comparison of the percentage errors of the recalculated dynamic responses with the corresponding identified values related to the measured response at a given frequency not used in phase separation identification (final prediction error). Column 8 shows unacceptably large errors for the results of classical phase resonance testing, and the same holds true for the results of the phase separation method with n_{eff} = 4 (column 7). However, the results of column 6 show much smaller errors than the others; the large error of measuring point 10, for example, is due to the small measured value (column 2).

Table 3 shows that the phase separation method /4/ works well, not only in the case of closely spaced eigenfrequencies, but also in the case of well separated eigenfrequencies. An extensive survey of ground vibration testing is contained in /8/.

Table 3 Estimates of the phase separation technique
        (Natke-method) applied in ground vibration test

Tafel 3: Erregungsfrequenzen N_j [Hz] und Ergebnisse
        nach dem versuchsmäßig-rechnerischen Ver-
        fahren   Hubschrauber H3-E2

Frei- heits- grad	Erregungsfr. N_j [Hz]	Eigenfrequ. N_{Ej} [Hz]	eff. Dämpfung g_j [1]
1	2	3	4
1	11,98	11,48 ± 0,05	0,085 ± 0,008
2	14,66	14,56 ± 0,02	0,055 ± 0,006
3	16,92	16,89 ± 0,01	0,038 ± 0,003

In b) Fig. 12 shows the time signals of the vertical
and horizontal stabilizer due to the manual aileron input a)
of an airplane during flight. Fig. 13 shows one of the
Laplacian transforms corresponding to Fig. 12b). Not all the
signals in the time and frequency domain clearly show $n_{eff} =$
2. The estimation of eigenfrequencies and damping ratios for
amplitudes of the Laplacian transforms at different chosen
frequencies leads to results dependent on these frequencies
/9/. The results are plotted in Fig. 13 for n_{eff} equal to 1
and 2. The damping ratios (β_{v1}) show a strong dependence on
the chosen frequencies for the wrong number of n_{eff}.

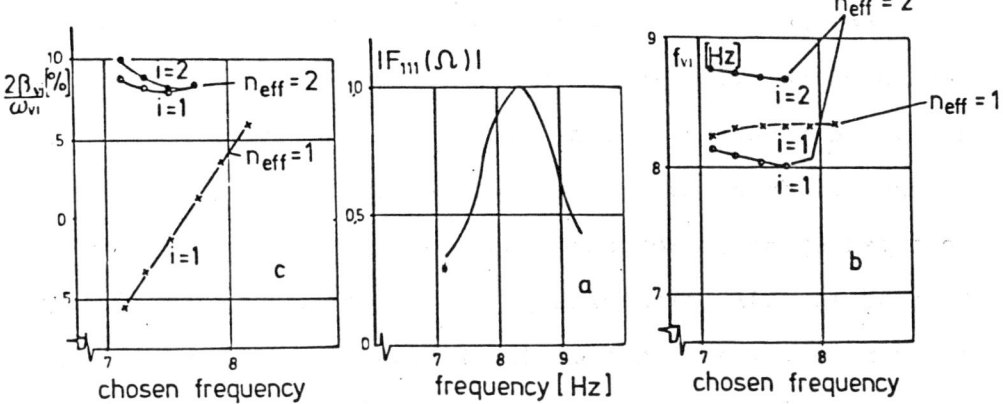

Fig. 13 : Phase separation technique applied in a
          flight vibration test :
          criterion for determining n_{eff}

These examples conclude the section on experimental
modal analysis. The reader can find further applications in
/8/ and the given refs. The method published in /10/ by Link
and Vollan[1] should also be mentioned here.

----

[1] See in this context: Natke, H.G. and Cottin, N.: Some
     remarks on the Application of Phase Separation Technique;
     Z. Flugwiss. Weltraumforsch. 2 (1978) 3, 199-200

2.   Model Adjustment

The applications of parameter improvement methods are
published in /11/. Publications /12/13/ should be emphasized
here. The experience gained in /14/ (adjustment of elements
of the stiffness matrix with measured eigenfrequencies,
undamped system, weighted least squares (WLS)) is similar
to that in /13/ (flexibility submatrices and measured eigenfrequencies, least squares (LS)). The airplane model /13/
was improved by measured eigenfrequencies up to 20 in number
while taking submatrices of flexibility influence coeffi-

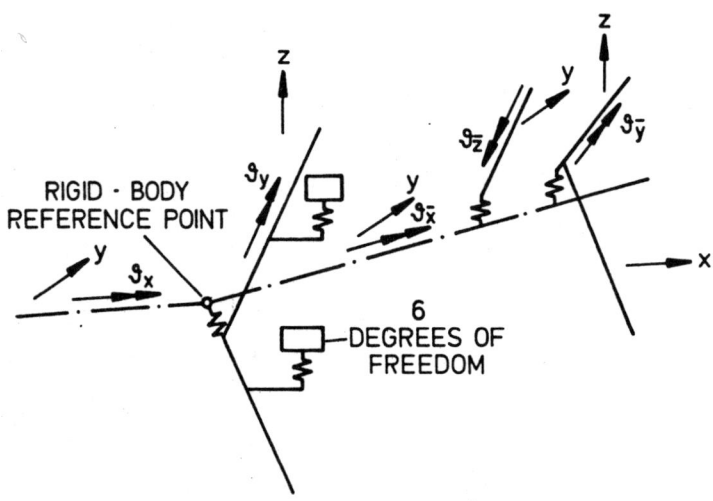

Fig. 14 : Wing - fuselage model

cients up to a maximum of I = 9. An example of the improvement is shown in Figs. 14 - 16. Fig. 14 shows the modelling in the form of beams and a substitution of the undetermined connected wing-fuselage system by a statically determined model. Fig. 15 provides the convergence of the nonlinear procedure and Fig. 16 shows some of the results. As can be seen, not all the eigenfrequencies converge to the measured eigenfrequencies. But it must be stated that the information content only of eigenfrequencies used for improvement is limited, and that it is less than that of eigenfrequencies, damping ratios and eigenvectors or suitable input/output measurements. Improvement simulations using input and output residuals combined with the WLS demonstrate the practicability of these methods /15/16/ (for a limited number of degrees-of-freedom and a limited number of parameters to be estimated). The same statement holds true for the equation error of the matrix eigenvalue problem combined with an orthonormality relation /16/.

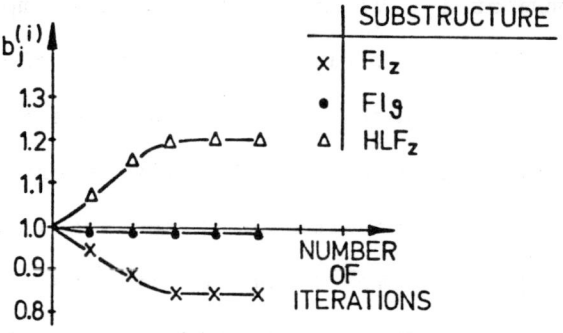

Fig. 15 : Convergence of correction factors
on iteration steps

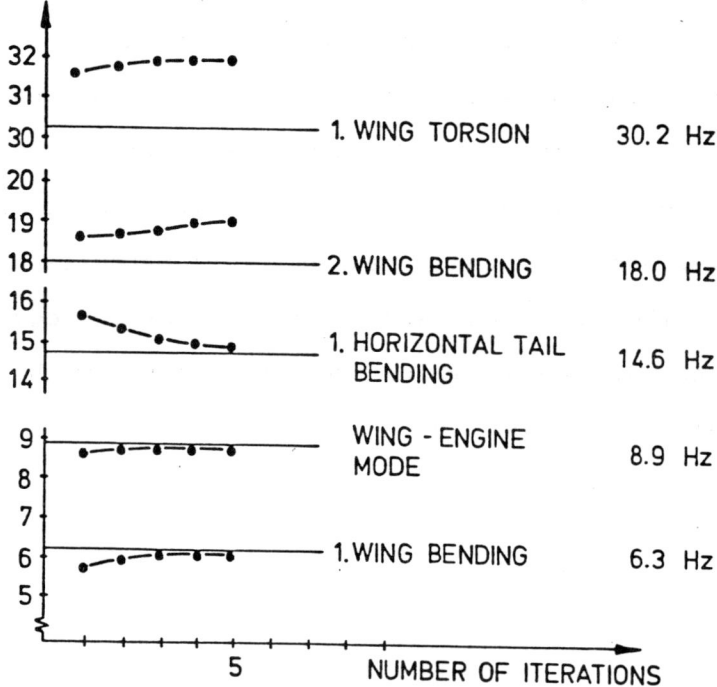

Fig. 16 : Improvement of eigenfrequencies depending on
iteration steps
(horizontal line indicates measured values)

References

1.   Niebdal, N.: Advances in Ground Vibration Testing Using
     a Combination of Phase Resonance and Phase Separation
     Methods; 2nd Internat. Symposium on Aeroelasticity and
     Structural Dynamics, Aachen, 1985, DGLR-Bericht 85-02,
     523-528

2.   Dat, R. and Lubrina, P.: The Methods Implemented at
     ONERA to Improve Airplane Ground Vibration Tests; Proc.
     4th Internat. Modal Analysis Conf., Los Angeles, CA,
     1986, Vol. II, 844-849

3.   Natke, H.G. and Cottin, N.: Introduction to this Course

4.   Natke, H.G.: Einführung in Theorie und Praxis der Zeit-
     reihen- und Modalanalyse; Friedr. Vieweg & Sohn, 1983

5.   -   International Symposium on Aeroelasticity, Nürn-
     berg, 1981, DGLR-Bericht 82-01

6.   Natke, H.G.: Schwingungsversuche mit der dritten Stufe
     der "EUROPA" I; LRT 16 (1970) Nr. 3, 5 7-64

7.   Natke, H.G.: Anwendung eines versuchsmäßig - rechneri-
      schen Verfahrens zur Ermittlung der Eigenschwingungs-
      größen eines elastomechanischen Systems bei einer
      Erregerkonfiguration; Z. Flugwiss. 18 (1970), Heft 8,
      290-303

8.   Gimmestad, D.W. (Editor): An Improved Ground Vibration
      Test Method, Vol. I: Research Report; AFWAL-TR-80-3056,
      Sept. 1980. Boeing Military Airplane Comp. Seattle,
      Wash. and Flight Dynamics Lab. Air Force Wright Aero-
      nautical Lab, Air Force Command, Wright-Patterson Air
      Force Systems Base, Ohio

9.   Strutz, K.-D., Cottin, N. und Eckhardt, K.: Anwendungen
      und Erfahrungen mit einem digitalen Auswerteverfahren
      zur Bestimmung der dynamischen Kenngrößen eines line-
      aren elastomechanischen Systems aus Impulsantworten;
      Z. Flugwiss. 24 (1976), Heft 4, 209-219

10.   Link, M. and Vollan, A.: Identification of Structural
      System Parameters from Dynamic Response Data;
      Z. Flugwiss. Weltraumforsch. 2 (1978), Heft 3, 165-174

11.   Natke, H.G. and Cottin, N.: Updating Mathematical Models
      on the Basis of Vibration and Modal Test Results -
      A Review of Experience; 2nd Internat. Symposium on

Aeroelasticity and Structural Dynamics, Aachen 1985,

DGLR-Bericht 85-02, 625-631

12. Collins, J.D., Hart, G.C., Hasselman, T.K. and Kenne-
    dy, B.: Statistical Identification of Structures; AIAA
    Journal, Vol. 12, No. 2, 1974, 185-190

13. Zimmermann, H., Collmann, D. and Natke, H.G.: Erfahrun-
    gen zur Korrektur des Rechenmodells mit gemessenen
    Eigenfrequenzen am Beispiel des Verkehrsflugzeuges
    VFW 614; Z. Flugwiss. Weltraumforsch. 1 (1977), Heft 4,
    278-285

14. Demchak, L. and Harcrow, H.: Analysis of Structural
    Dynamic Data from Skylab; Martin-Marietta-Corp., NASA
    CR-2727, Vol. 1, 1976

15. Cottin, N., Felgenhauer, H.-P. and Natke, H.G.: On
    the Parameter Identification of Elastomechanical
    Systems Using Input and Output Residuals; Ingenieur-
    Archiv 54, 1984, 378-387

16. Cottin, N. and Natke, H.G.: On the Parameter Identifi-
    cation of Elastomechanical Systems Using Weighted Input
    and Modal Residuals; Ingenieur-Archiv 56, 1986, 106-113

AIRCRAFT SYSTEM IDENTIFICATION —
DETERMINATION OF FLIGHT MECHANICS PARAMETERS

E. Plaetschke, S. Weiss
Institut für Flugmechanik, Braunschweig-Flughafen, F.R.G.

INTRODUCTION

Identification of aircraft stability and control derivatives from flight test data is of growing importance in the design, testing and certification of modern aircraft. The greater need for these derivatives has the following reasons:

- They are used to improve mathematical models for ground and in-flight aircraft simulators.
- They serve as a basis for the design of flight control systems.
- They define a given aircraft and can be used for verification of specified flying/handling qualities.
- They are used for correlation with analytical and wind tunnel data.
- They help to reduce prototype testing time and costs.

The most commonly used procedure for aircraft system identification is shown in figure 1 (cf. Hamel[1]). This paper covers some aspects indicated in the figure, such as parameter estimation methods, optimal input

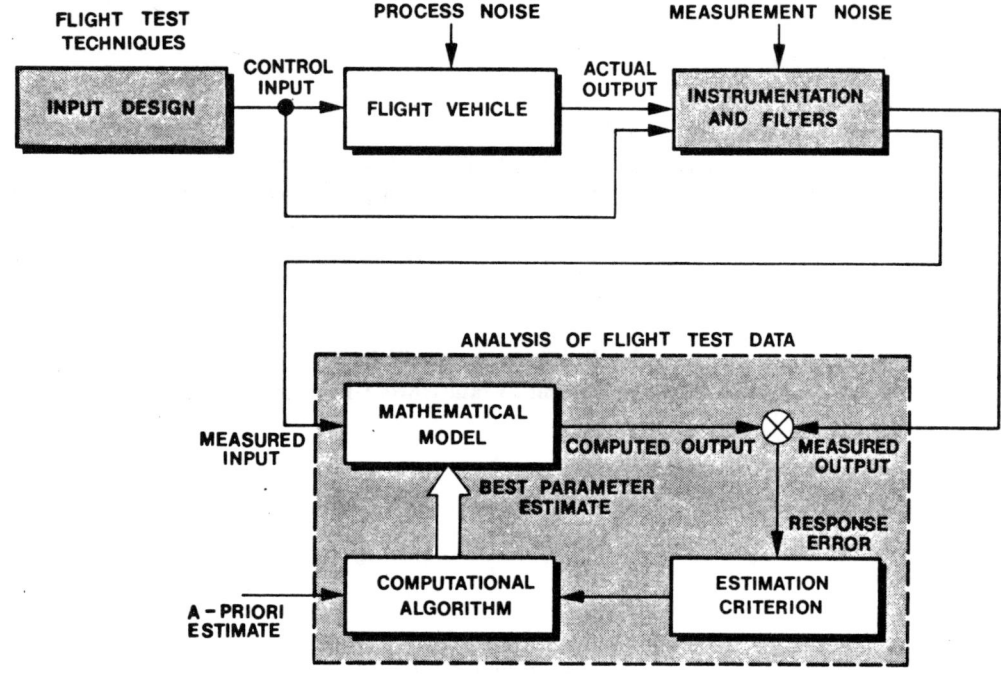

Figure 1   Identification procedure

design and error sources in aircraft system identification. Because of
space limitation we will not deal with instrumentation and filters,
though the accuracy of the instrumentation system highly affects the
quality of the identification results. Flight mechanics nomenclature is
given in the appendix.

PARAMETER ESTIMATION METHODS

Maximum likelihood estimation concept

The aircraft parameter estimation problem can be defined quite simply in general terms (cf. figure 1). The flight vehicle under investigation is assumed to be modeled by a set of dynamic equations containing

unknown parameters. To determine the values of these parameters, the system is excited by a suitable input, and the input and actual system response are measured. The measured response of the flight vehicle is compared with the estimated response of the mathematical model and the difference between these responses is called the response error. For each possible estimate of the unknown parameters, a probability that the model response attains values near the observed response values, can be defined. The estimates which are referred to as the maximum likelihood estimates, are chosen such that this probability is maximized.

Maximum likelihood estimation has many desirable statistical characteristics; it yields asymptotically unbiased, consistent, and efficient estimates. The maximum likelihood estimator also provides a measure of the reliability of each estimate. This measure, analogous to the standard deviation, is called the Cramer-Rao bound or uncertainty level (cf. Iliff[2]).

Parameter estimation in the time domain

The dynamic system, the parameters of which are to be estimated, is described by the following model:

$$\dot{x}(t) = f\left[x(t), u(t), \beta\right] \; ; \; x(t_o) = x_o \tag{1}$$

$$y(t) = g\left[x(t), u(t), \beta\right] . \tag{2}$$

x, u and y are the vectors of the state, control and observation variables, β denotes the vector of the unknown parameters. The observation variables are measured at N discrete time instants

$$z(k) = y(k) + v(k) \quad , \quad k = 1, \ldots, N . \tag{3}$$

In this equation z denotes the model output and v the measurement error which is assumed to be stationary Gaussian white noise with zero mean and covariance matrix R.

In general the initial values x_o of the state variables are unknown and have to be estimated together with the model parameters β. In addition to this, one has to account for systematical errors Δu and Δz in the measured input and output variables. All these unknowns together form the parameter vector

$$\theta^T = (\beta^T, x_o^T, \Delta u^T, \Delta z^T) \ . \tag{4}$$

Usually not all components of θ can be identified separately and the vector has to be reduced to an identifiable subset of unknowns.

The unknown parameters are obtained by maximizing the likelihood functional which is defined as the conditional probability density of the measurements z for given θ and R. An equivalent problem is the minimization of the negative logarithm of the likelihood functional. For the mathematical model (1) - (3) this leads to minimizing

$$L = \frac{1}{2} \sum_{k=1}^{N} \left[z(k) - y(k)\right]^T R^{-1} \left[z(k) - y(k)\right] + \frac{N}{2} \ln|R| \tag{5}$$

with respect to θ.

The minimization of the cost function (5) is a nonlinear optimization problem which can be solved by direct minimum search methods such as the algorithms of Powell, Rosenbrock or Fletcher (cf. Jategaonkar and Plaetschke[3]). Such methods, as they are available in software libraries, are very robust but do usually need a lot of computation time.

An alternative algorithm is the quasi-linearization or modified Newton-Raphson method. This method leads to a system of linear equations for the parameter increments $\Delta\theta$

$$\left[\sum_k \left(\frac{\partial y}{\partial \theta}\right)^T R^{-1} \frac{\partial y}{\partial \theta} \right] \Delta\theta = \sum_k \left(\frac{\partial y}{\partial \theta}\right)^T R^{-1} (z - y) \ . \tag{6}$$

The solution of this system leads to new estimates of the unknown parameters. These are used to update the mathematical model of the flight vehicle, providing a new estimated response and, therefore, a new response
error. The updating of the mathematical model continues until a convergence criterion is satisfied.

Solving (6) requires the calculation of the sensitivity matrix

$$\left(\frac{\partial y}{\partial \theta}\right)_{ij} = \frac{\partial y_i}{\partial \theta_j} \ . \tag{7}$$

Integration of the sensitivity equations

$$\left(\frac{\partial x}{\partial \theta}\right)^{\cdot} = \frac{\partial f}{\partial x} \frac{\partial x}{\partial \theta} + \frac{\partial f}{\partial \theta} \tag{8}$$

$$\frac{\partial y}{\partial \theta} = \frac{\partial g}{\partial x} \frac{\partial x}{\partial \theta} + \frac{\partial g}{\partial \theta} \tag{9}$$

which are derived from the model equations (1), (2) by differentiation
w.r.t. θ, gives a solution for (7) but these sensitivity equations have
to be derived explicitly for each specific nonlinear model. Therefore an
approximation of the sensitivity coefficients by finite differences

$$\frac{\partial y}{\partial \theta} \approx \frac{y(\theta + \delta\theta) - y(\theta)}{\delta\theta} \tag{10}$$

is preferred to an explicit solution of (8), (9).

The quasi-linearization method also yields direct information about
the accuracy of the estimated parameters. The matrix occurring in the
system of linear equations (6) is equal to the information matrix

$$J = \sum_k \left(\frac{\partial y}{\partial \theta}\right)^T R^{-1} \frac{\partial y}{\partial \theta} \ . \tag{11}$$

As the maximum likelihood estimate is asymptotically unbiased and efficient the inverse of J is a good estimate for the estimation error covariance matrix

$$P \approx J^{-1} .$$ (12)

(6) is solved in each iteration and therefore the Cramer-Rao bound for
the estimation error is available in each iteration step.

Linearization of the model equations (1), (2) in the neighbourhood
of a reference point and shifting of the state variables by their initial
values leads to the following model (cf. Plaetschke and Mackie[4]):

$$\dot{x} = A x + B u + b_x \quad ; \quad x(t_o) = 0$$ (13)

$$y = C x + D u + b_y .$$ (14)

x, u and y are now deviations of the state, control and observation variables from the reference state. The unknown coefficients β occur in the
matrices A, B, C and D in linear form. These coefficients are called the
stability and control derivatives. The unknown initial conditions and
systematical errors are estimated as lumped parameters b_x and b_y, all
components of which are identifiable if the system is observable.

The cost function that has to be minimized is the same as for non-
linear models (cf. (5)), therefore quasi-linearization again leads to the
iterative solution of (6). For linear systems the sensitivity equations
(9), (10) can be derived by simple matrix operations

$$\left(\frac{\partial x}{\partial \theta}\right)^{\cdot} = A \frac{\partial x}{\partial \theta} + \frac{\partial A}{\partial \theta} x + \frac{\partial B}{\partial \theta} u + \frac{\partial b_x}{\partial \theta}$$ (15)

$$\frac{\partial y}{\partial \theta} = C \frac{\partial x}{\partial \theta} + \frac{\partial C}{\partial \theta} x + \frac{\partial D}{\partial \theta} u + \frac{\partial b_y}{\partial \theta} .$$ (16)

These equations are valid for all linear systems and hence no approximation of the sensitivity coefficients by finite differences is necessary.

Often it is necessary to use several maneuvers with different information to determine one set of derivatives. The bias vectors (b_x and b_y)
or the initial values and systematical errors (x_o, Δu and Δz) will then
differ for each maneuver. Another extension is the inclusion of time delays for several signals as additional parameters.

Parameter estimation in the frequency domain

There are two main possibilities for model building in the frequency
domain. In the state equation representation the model

$$j\omega\, x(\omega) = A\, x(\omega) + B\, u(\omega) \qquad (17)$$

$$y(\omega) = C\, x(\omega) + D\, u(\omega) \qquad (18)$$

is derived from the linear model in the time domain by discrete Fourier
transformation. Solving these equations yields

$$x(\omega) = (j\omega\, I - A)^{-1}\, B\, u(\omega) \qquad (19)$$

$$y(\omega) = \left[C\, (j\omega\, I - A)^{-1}\, B + D\right] u(\omega) = F(j\omega)\, u(\omega) \qquad (20)$$

where I denotes the identity matrix and F the transfer function matrix.
For this model the parameter vector consists of the unknown elements
occurring in the matrices A, B, C and D. No bias parameters need to be
estimated. The cost function that has to be minimized in this case is

$$L = \sum_{\omega_k} (y - F\, u)^{*}\, S^{-1}\, (y - F\, u) + M \ln|S|\; . \qquad (21)$$

ω_k, k = 1, ..., M are discrete frequencies and S denotes the output error covariance matrix. As can be seen from (21) the optimization criterion for this model formulation is the fit of the frequency responses of flight vehicle and mathematical model (cf. Marchand and Fu[5]).

Another possibility is the model formulation by transfer functions

$$F_i(j\omega) = \frac{a_{0,i} + a_{1,i}(j\omega) + a_{2,i}(j\omega)^2 + \ldots}{b_0 + b_1(j\omega) + b_2(j\omega)^2 + \ldots} . \tag{22}$$

The parameter vector for this model consists of the unknown coefficients of the numerator and denominator polynomials occurring in the transfer functions (22). The cost function for this case is

$$L = \sum_i \sum_k \frac{|F_i(j\omega_k) - F_{im}(j\omega_k)|^2}{|F_i(j\omega_k)|^2} \tag{23}$$

with M discrete frequencies ω_k and F_{im} denoting the measured transfer functions (i.e. transfer functions computed from the measured data).

OPTIMAL INPUT DESIGN

Within the procedure of aircraft parameter identification the design of input signals is the first step. Thereby the limitations of the following steps - as there are: selection of the instrumentation system, flight testing under constraints and disturbances, choice of an appropriate identification algorithm - have to be taken into account. For the design of test signals the following aspects are of importance:

● The inputs have to excite the aircraft modes appropriately, such that parameter variations cause variations of the measured time response (Parameter Sensitivity). In this case the parameters can be identified accurately.

- Amplitude, bandwidth and slope of the input signals have to be bounded such that first, the signals can be realized by the specific actuators (Realizability), and second, the modeling assumptions of linearized decoupled equations of motion are not violated (Linearity).

- The noise characteristics and bandwidth of the sensors as well as of the disturbance process (gusts) have to be considered.

- The inputs should be simple enough to be implemented easily by a pilot (unless a fly-by-wire system is available) and the resulting response of the aircraft should not endanger the pilot (Pilot Acceptability).

- For optimal input signal design good a-priori models of the aircraft motion have to be available as to minimize the estimation error variance of the derivatives (A-Priori Model).

Input design methods

The design of input signals can be performed in the frequency domain and in the time domain considering system criteria and estimation error criteria (cf. Plaetschke and Schulz[6]). Starting with identifiability investigations in the frequency domain, a first method yields multistep input signals which fulfill specific spectral requirements (cf. Koehler[7]). In figure 2 such a multistep signal and its power spectral density are presented. Because of its characteristic shape it is called "3211"-signal (3 time units positive, 2 negative, 1 positive, and 1 negative). By the choice of the time unit Δt the spectrum can be shifted to match the required frequency region.

A second way of input design is based on the optimization of different measures of the estimation error covariance matrix P of the parameters. The volume V of the estimation error ellipsoid is a convenient measure of the concentration of the probability density about its mean.

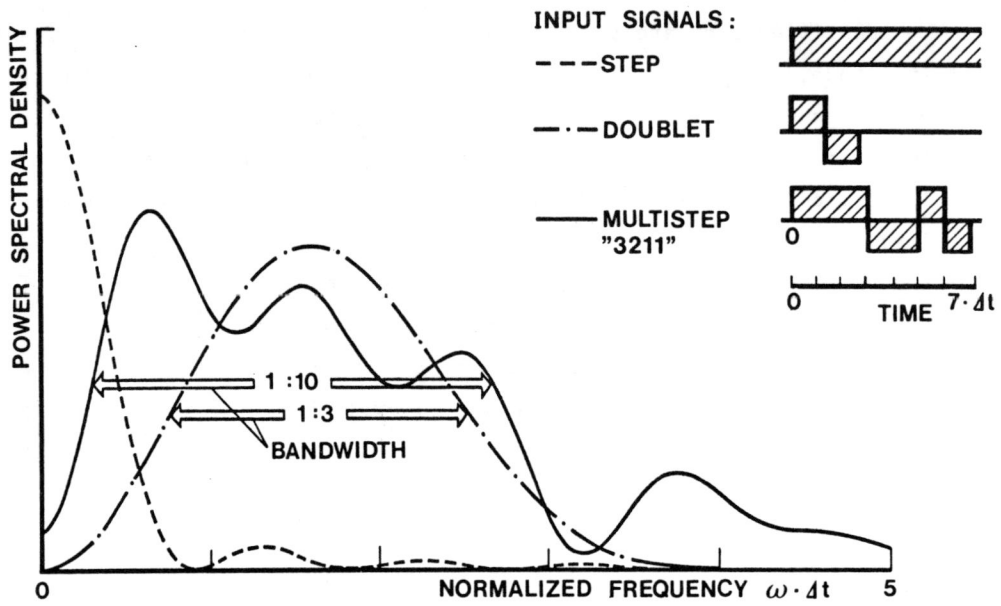

Figure 2   Frequency domain comparison of various input signals

Since $V \sim \sqrt{\det(P)}$, the minimization of $\det(P)$ gives a relevant optimization criterion:

$$\min_{u(t)} \{\det(P)\} \rightarrow u_{opt}(t) .$$ (24)

For efficient estimates the estimation error covariance matrix P is equal to the inverse of the Fisher information matrix J. Therefore criterion (24) is equivalent to the maximization of the determinant of J:

$$\max_{u(t)} \{\det(J)\} \rightarrow u_{opt}(t) .$$ (25)

Based on this criterion, Mehra[8] developed a method for designing optimal inputs for linear systems. The method yields optimal frequencies and amplitudes for the inputs.

Other criteria for optimal input design were used by Schulz[9]

$$\max_{u(t)} \{tr(J)\} \to u_{opt}(t) \tag{26}$$

and by the Delft University of Technology (DUT, cf. Mulder[10])

$$\min_{u(t)} \{tr(P)\} \to u_{opt}(t) \; . \tag{27}$$

Application of optimal input signals in flight test

Five optimal input signals – doublet, "3211"-input, Mehra-input, Schulz-input and DUT-input – were applied in a joint Dutch/German air-craft parameter identification flight test program (cf. Plaetschke, Mulder and Breeman[11]). One of the goals of that program was to study the effect of the various input signals on the identification results. The test aircraft – a De Havilland DHC-2 "Beaver" – was equipped with an electro-hydraulic control system. Thus each flight test maneuver could be exactly reproduced any number of times, allowing statistical analysis of the resulting data.

Figure 3 shows the time histories of five optimal elevator inputs used in flight test. In order to avoid measurement errors due to the low pass flight test instrumentation filters, each control input signal re-ceived the same filtering before being applied. Moreover, without pre-filtering the step-type input signals (doublet and "3211") could not have been realized by the electro-hydraulic actuators.

The evaluation of aircraft parameters from flight test data was car-ried out independently by DFVLR and DUT using different identification methods and models. At DFVLR a maximum likelihood identification algo-rithm for linear systems was applied. The DUT method of parameter identi-

fication is based on a decomposition into separate aircraft state estimation followed by aerodynamic model identification. The solution of the
state estimation problem results in accurate aircraft state trajectories.
Aerodynamic parameter estimation then requires only the solution of six
separate linear regression problems.

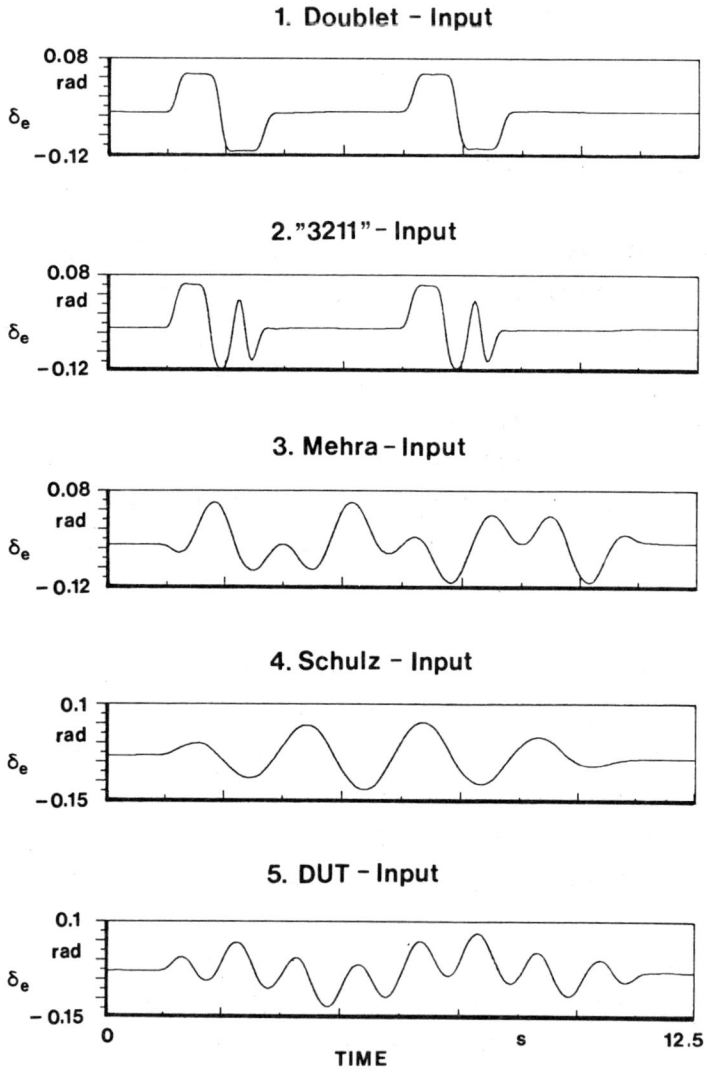

Figure 3   Time histories of five optimal elevator inputs

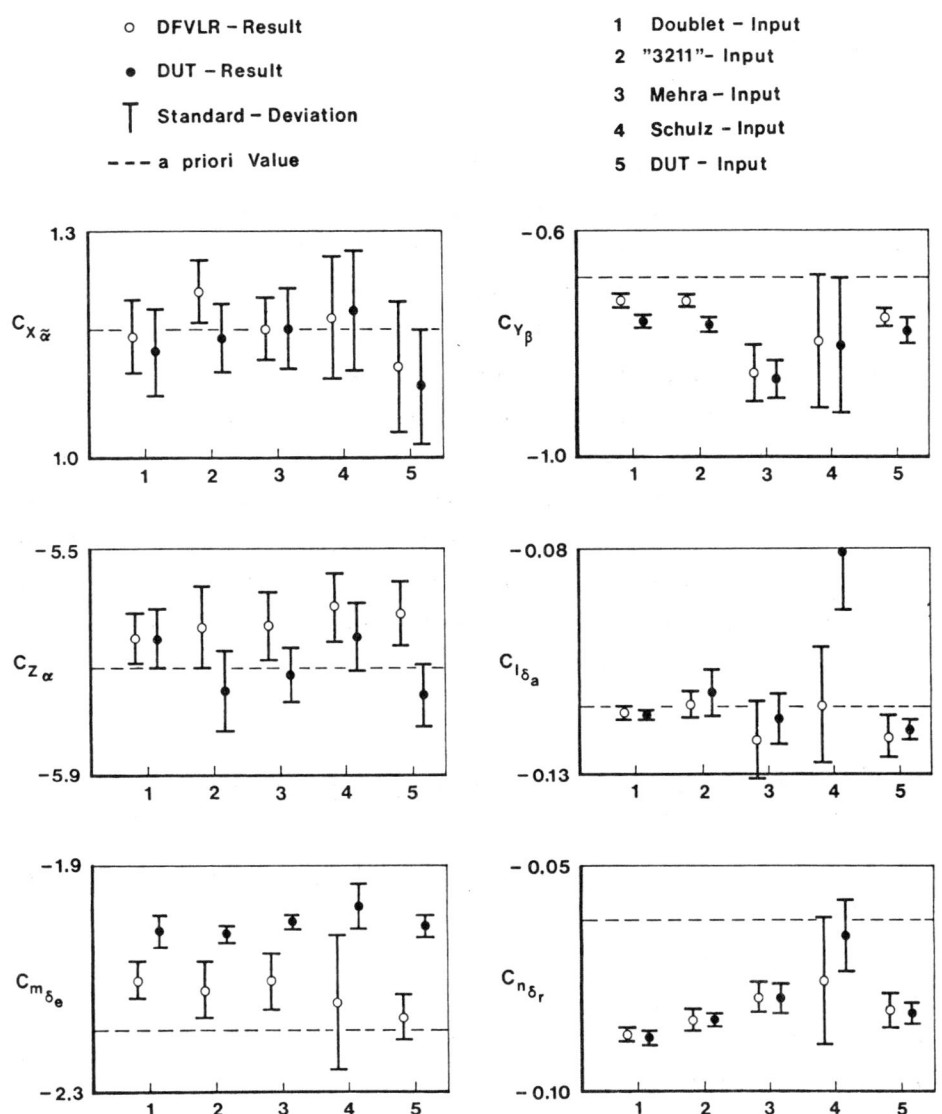

Figure 4   Identification results (Beaver)

Identification results for one nominal flight test condition defined by true airspeed of 45 m/s at 6,000 ft pressure altitude are presented in figure 4. It shows six of the most important longitudinal and lateral-

directional derivatives for the two data analysis methods and the five
input signals. As each input signal was applied nine resp. ten times,
sample means and standard deviations were calculated and presented in the
figure. For the longitudinal derivatives the results differ mainly due to
the data analysis method and the aerodynamic model used, while for the
lateral-directional derivatives the results differ mainly due to the input signal type. In the longitudinal case the differences are attributed
to different mathematical model structures. In the lateral-directional
case however, the mathematical models were nearly identical. Consequently, sample means and standard deviations correspond closely. The correlation between parameter estimates and input signal type is attributed to
the input signal amplitudes being too large for the validity of the linear models used. Finally, as can be seen from figure 4 some of the identification results deviate considerably from their a-priori values. In this
respect, all input signals, which were designed on the basis of those a-
priori values, are only suboptimal.

ERROR SOURCES IN AIRCRAFT SYSTEM IDENTIFICATION

Erroneous identification results are mainly derived from incorrect
data and/or incorrect modeling. Possible data errors are due to
- incorrect sensor calibration,
- sensor zero shifts,
- time shifts,
- incorrect sensor location,
- disregarding sensor and filter dynamics.

Modeling errors may result from the use of
- linear models to describe nonlinear phenomena,
- decoupled equations of motion when cross-coupling effects can
  not be neglected,
- models disregarding process noise to analyse flight test data
  obtained in turbulent air,
- low order models to desribe high order systems,

- incorrect stability augmentation system models in closed loop system identification.

In this section some examples will be given which show how the errors mentioned above affect the identification results and what can be done to account for them. For more examples we refer to Iliff.[2]

Data compatibility checking

Incorrect sensor calibration, zero shifts and time shifts can be detected and eliminated by data compatibility checking (sometimes referred to as flight path reconstruction). For this purpose the nonlinear kinematic equations of the aircraft motion are used. They define the correlation between the various data signals. When high accuracy measurements are made of specific aerodynamic forces A_x, A_y, A_z and rotation rates p, q, r more reliable time histories of the noisy air data (true airspeed V, angle of attack α, angle of sideslip β, dynamic pressure $\bar{q}$ etc.) can be estimated. Unknown calibration factors, zero shifts and time shifts are estimated simultaneously.

For the purpose of illustrations, data compatibility checking is carried out for the variables in the longitudinal motion of the DFVLR research aircraft HFB 320 "Hansa" (cf. Jategaonkar and Plaetschke[3]). The kinematic relations including calibration factors and zero shifts read:

State equations

$$\dot{u} = A_{xm} - \Delta A_x - (q_m - \Delta q)w - g \sin\theta \quad , \quad u(t_o) = u_o$$

$$\dot{w} = A_{zm} - \Delta A_z + (q_m - \Delta q)u + g \cos\theta \quad , \quad w(t_o) = w_o \qquad (28)$$

$$\dot{\theta} = q_m - \Delta q \quad\quad\quad\quad\quad\quad\quad\quad , \quad \theta(t_o) = \theta_o \quad .$$

Observation equations

$$V_m = \sqrt{u_s^2 + w_s^2} + \Delta V$$

$$\alpha_m = \arctan(w_s/u_s)$$

$$\Theta_m = k_\Theta \, \Theta \tag{29}$$

$$\bar{q}_m = k_{\bar{q}} \frac{\rho}{2} (u^2 + w^2) + \Delta \bar{q}$$

with

$$u_s = u + (q_m - \Delta q)z_s$$

$$w_s = w - (q_m - \Delta q)x_s \tag{30}$$

where the subscript m indicates measured variables. Airspeed V and angle of attack α were measured by a flight log at the nose boom with offset distances x_s = 10.992 m and z_s = 0.556 m from the center of gravity.

A comparison of the measured and reconstructed time histories showed that the measurements of airspeed V, angle of attack α and dynamic pressure $\bar{q}$ had slight time delays. In a further identification, the unknown time delays were included in the model as additional parameters (cf. Plaetschke[12]). Identification results obtained with the extended model are shown in figure 5. As can be seen, the measurements are now fitted well by the reconstructed time histories.

Filter dynamics

The effect of filter dynamics on the identification results is shown in an example given by Marchand.[13] In a joint German/French research pro-

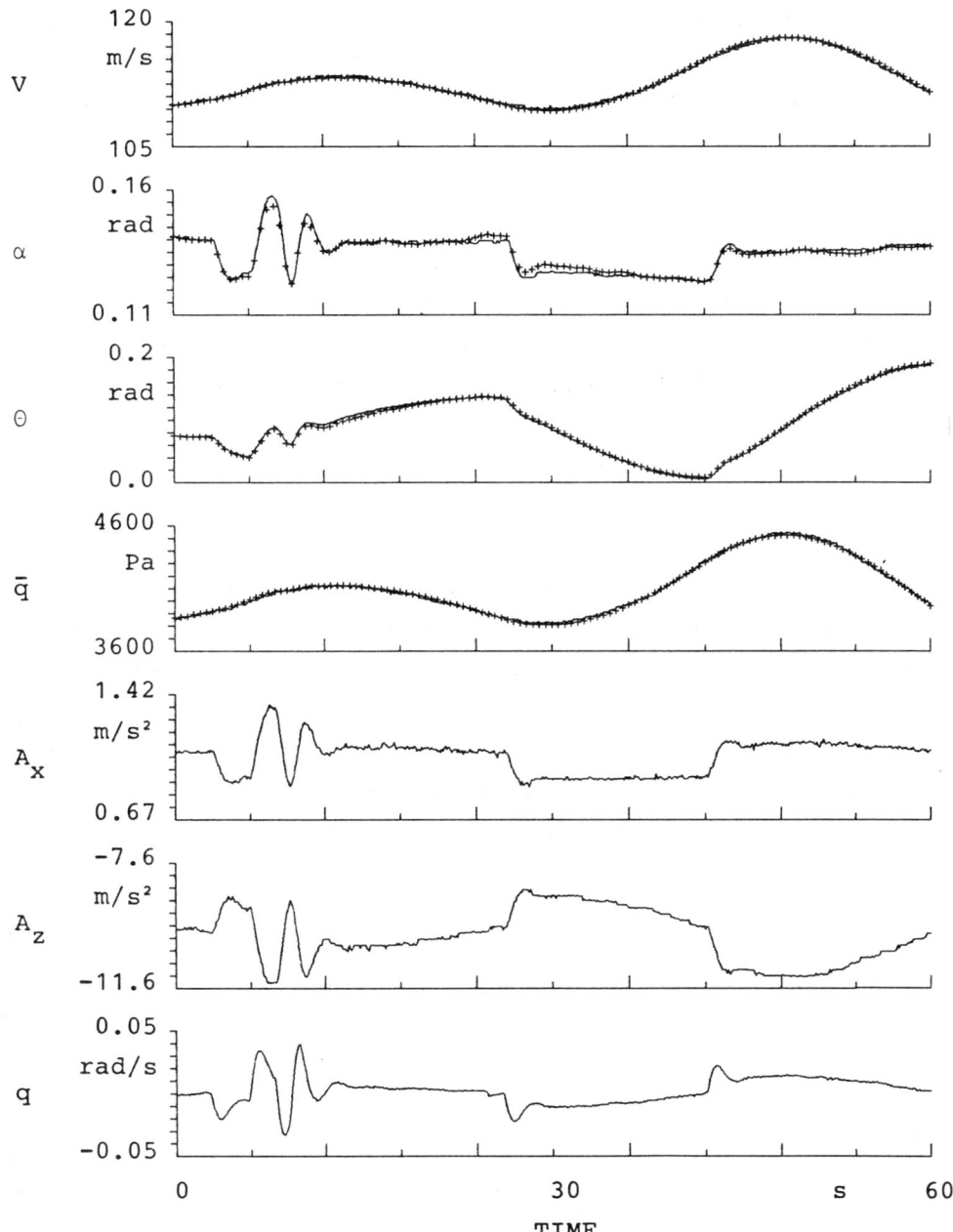

Figure 5 Data compatibility checking with estimation of
time delays in V, α and $\bar{q}$ (HFB 320)
— measured   +++ reconstructed

gram the aerodynamic derivatives of a Do-28-TNT model were to be identified from free flight tests. As can be seen from figure 6, first analysis of the flight test data resulted in poor curve fitting for the pitch acceleration. This phenomenon could be explained by the fact that the pitch rate gyro output was filtered. As no information on the filter dynamics was available, a first order filter with unknown time constant was modeled and included in the identification. Figure 6 shows that by this an excellent fit between measured and identified pitch acceleration could be achieved. The time constant was identified as T = 40 ms. This value was confirmed by a later investigation of the measurement system.

Nonlinear modeling

For determination of the aerodynamic derivatives of a Do-28-TNT model, a series of experiments was conducted in the facility for dynamic simulation in wind tunnels. Analysis carried out here is related to the experiments in which the model was fixed in the vertical axis but having pitching motion freedom. The model was excited by a doublet evelator input. The following linear system representation was initially considered (cf. Jategaonkar and Plaetschke[3]):

State equations:

$$\dot{q} = M_q \, q + M_\theta \, \theta + M_{\delta_e} \, \delta_e$$
$$\dot{\theta} = q \quad . \tag{31}$$

Observation equations:

$$\dot{q}_m = M_q \, q + M_\theta \, \theta + M_{\delta_e} \, \delta_e$$
$$q_m = q \tag{32}$$
$$\theta_m = \theta \quad .$$

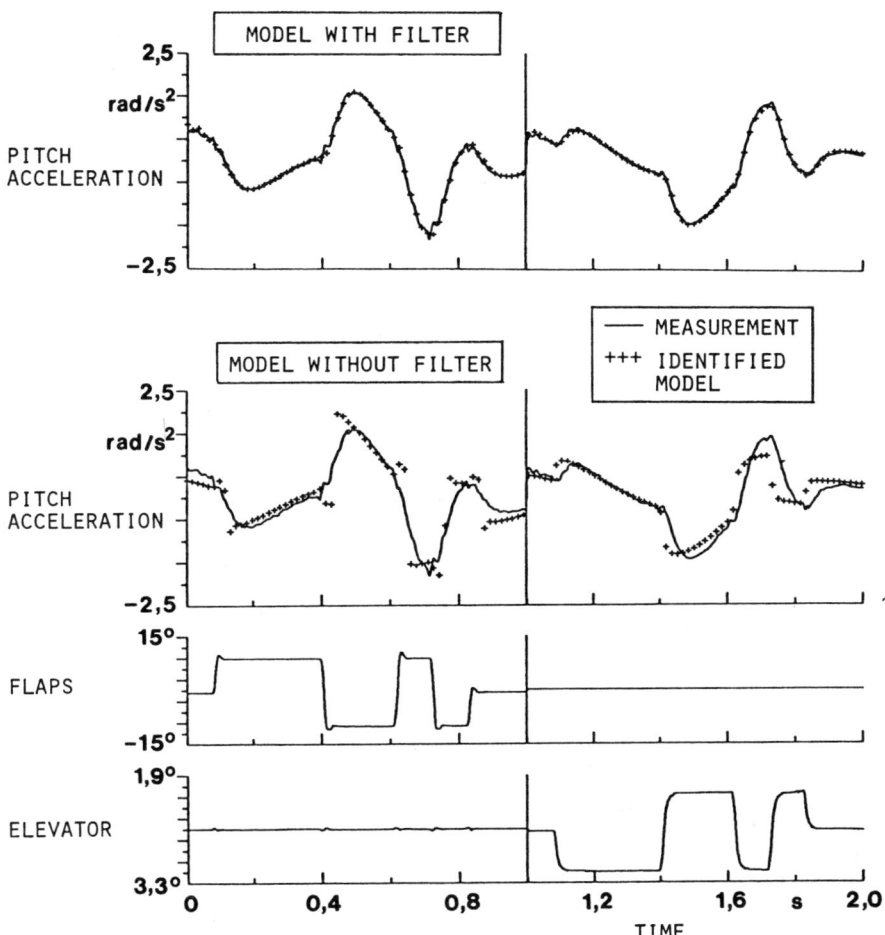

Figure 6   Effect of filter dynamics (Do–28–TNT free-flight model)

Estimation results based on the above linear model are shown in the upper part of figure 7. The agreement between the measured and the estimated responses is poor.

By inspection of the measured time histories, it was detected that the pitch angle response did not change linearly with the elevator input, instead deviated after a definite angle θ. This phenomenon is attributed to the fact that one of the feeder cables was touching the body of the

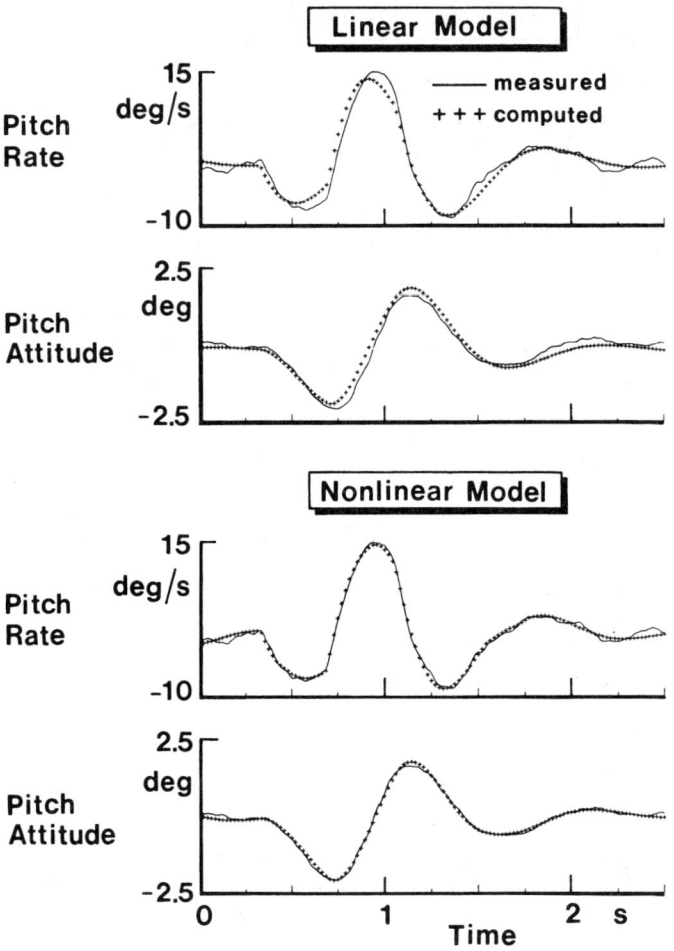

Figure 7   Effect of nonlinear modeling

model for small angles of pitch. That nonlinear effect has been modeled by considering two independent values for the derivative M_Θ as follows:

$$M_\Theta = \begin{cases} M_{\Theta 1} & \text{for} \quad \Theta > \Theta_g \\ M_{\Theta 2} & \text{for} \quad \Theta \leqq \Theta_g \end{cases} \tag{33}$$

This results in a system which is linear in the state and control variables but discontinuous in one of the parameters. In addition to the derivatives, the kink point θ_g needs to be estimated. Identification of this nonlinear system yielded the results shown in the lower part of figure 7. The agreement between the predicted and the measured responses is now good.

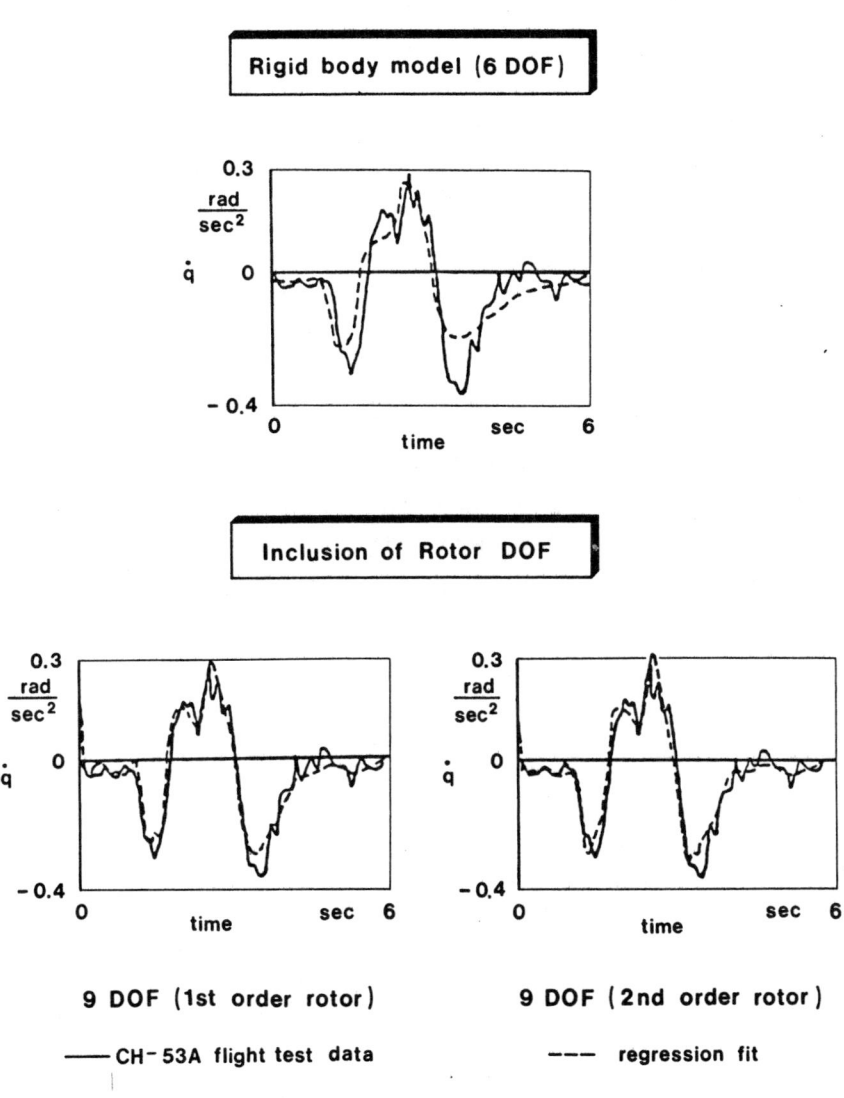

Figure 8   Rotorcraft modeling

High order systems

In general, one should try to separate the different modes of motion
of the dynamic system under investigation. This leads to several low order subsystems with a small number of parameters which can be identified
independently. For example, the equations of motion of a conventional
aircraft operating at normal flight conditions can be decoupled into the
longitudinal and the lateral-directional motion. For the helicopter however, separation into these modes is in general not possible because of
the rotor coupling. The full six-degree-of-freedom equations of motion
have to be used to model the fuselage dynamics. In addition, the rotor
dynamics has to be included. It can be approximated by the motion of the
tip path plane which has three DOF. Thus a nine DOF model with a large
number of parameters results.

Figure 8 shows the necessity of explicit rotor modeling (cf. Hall,
Gupta and Hansen[14]). Here, flight test data of a Sikorsky CH-53A helicopter are compared with the pitch acceleration response of three different
models. When only the fuselage motion is modeled the fit is poor (upper
part of the figure). An inclusion of the rotor dynamics leads to a significant increase in model accuracy (lower part of the figure). Thereby it
is sufficient to model each rotor DOF as first order differential equation. Modeling as second order differential equation leads to no further
improvement.

Closed loop system identification

Most modern aircraft are flown with engaged stability augmentation
systems (SAS). As for security reasons it is often not allowed to remove
the SAS, identification experiments have to be performed on aircraft operating in closed loop. In order to obtain a correct model of the augmen-
ted aircraft also the SAS must be identified.

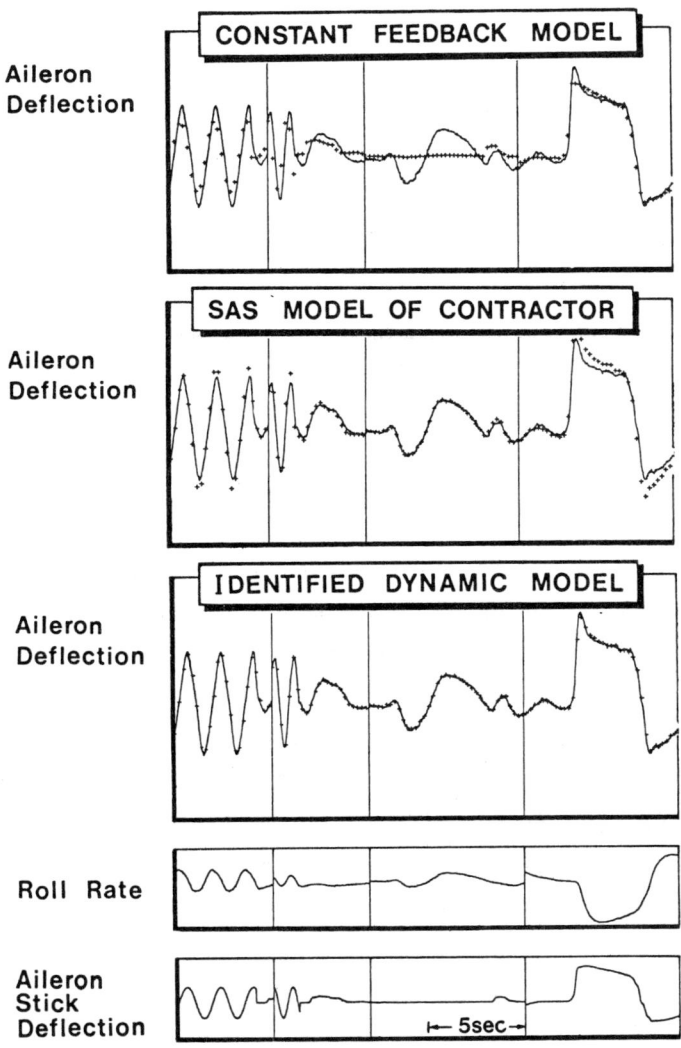

Figure 9   Identification of roll stability augmentation system
           —— Flight test data   +++ Model output

An example of stability augmentation system identification is given
by Koehler and Wilhelm.[15] In figure 9 flight-measured and computed data
of a roll augmentation system with two model structures are compared. In
addition, the response of the SAS model provided by the contractor is

shown. Inputs to the roll augmentation system are roll rate and aileron stick deflection. The identification was carried out applying multiple maneuver evaluation (four time segments). It can be seen that the constant feedback model results in a good fit in the first and fourth time segment, whereas the fit in the two middle time segments is unacceptable. Hence, the dynamics of the roll SAS can not be neglected in this case. The third diagram shows flight test data and identified model response when augmentation system dynamics is taken into account. Here the curve fit is excellent in all time segments.

APPENDIX - NOMENCLATURE

Flight mechanics notation is shown in figure 10 (cf. Etkin[16]). Here x, y, z denote the axes of the body-fixed reference frame, X, Y, Z longitudinal, lateral and vertical aerodynamic force, u, v, w longitudinal, lateral and vertical velocity, L, M, N rolling, pitching and yawing moment, p, q, r roll, pitch and yaw rate, δ_a, δ_e, δ_r aileron, elevator and rudder deflection. The Euler angles, i.e. roll, pitch and yaw angle are denoted by Φ, Θ, Ψ (not indicated in the figure). The flight velocity vector $\underline{V}$ can be given either by the components (u, v, w) or by its magnitude V and the two flow angles, i.e. angle of attack α and sideslip angle β (see figure 11).

The aircraft equations of motion result from the application of Newton's laws of motion. For our purposes, the aircraft is considered to be a rigid body and symmetric w.r.t. the x,z-plane. The resulting force and moment equations are completed by the kinematic relations between the angular rates and the Euler angles. This leads to a system of differential equations with state variables u, v, w, p, q, r, Φ, Θ, Ψ and control variables δ_a, δ_e, δ_r, ... The aerodynamic forces and moments are expanded into Taylor series w.r.t. the state and control variables. The coefficients of the Taylor expansions are called the "aerodynamic derivatives".

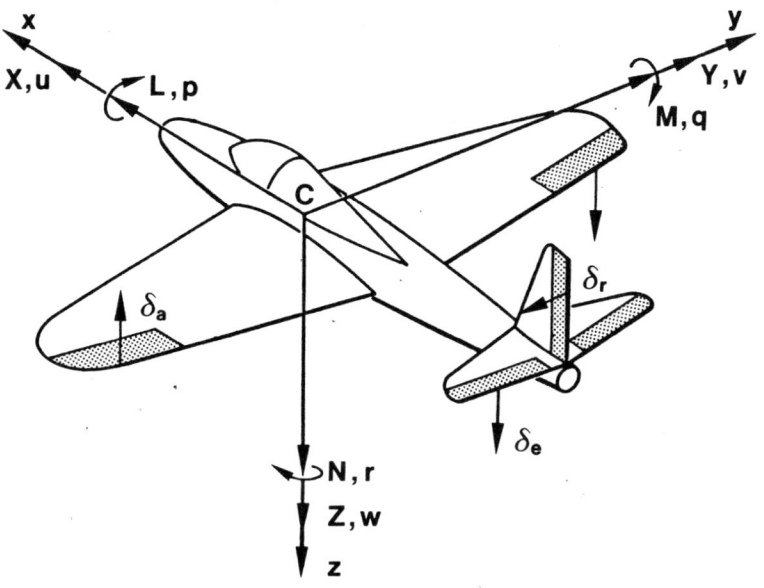

Figure 10   Flight mechanics notation

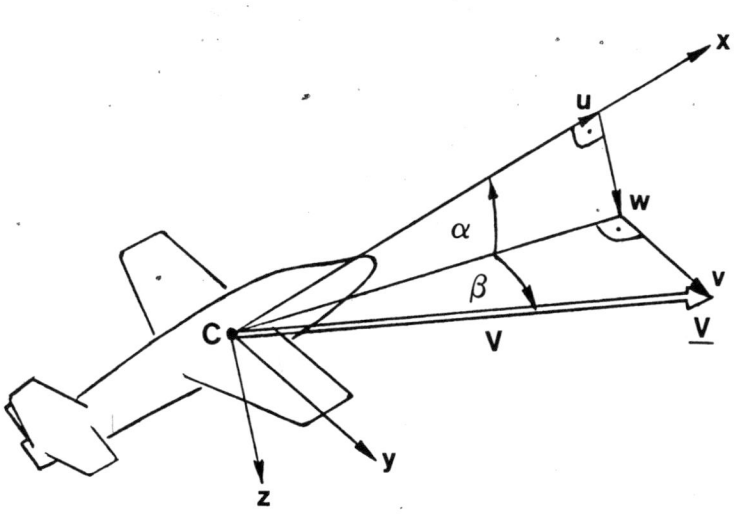

Figure 11   Flow angles

REFERENCES

1. Hamel, P.G., Aircraft parameter identification methods and their applications - Survey and future aspects, in *AGARD Lecture Series No. 104 Parameter Identification*, AGARD, London, 1979, 1-1.

2. Iliff, K.W., Aircraft identification experience, in *AGARD Lecture Series No. 104 Parameter Identification*, AGARD, London, 1979, 6-1.

3. Jategaonkar, R.V. and Plaetschke, E., Maximum likelihood estimation of parameters in nonlinear flight mechanics systems, in *IFAC Identification and System Parameter Estimation 1985*, Barker, H.A. and Young, P.C., Eds., Pergamon Press, Oxford and New York, 1985, 663.

4. Plaetschke, E. and Mackie, D.B., *Maximum-Likelihood-Schätzung von Parametern linearer Systeme aus Flugversuchsdaten - Ein FORTRAN-Programm*, DFVLR-Mitt. 84-10, 1984.

5. Marchand, M. and Fu, K.H., Frequency domain parameter estimation of aeronautical systems without and with time delay, in *IFAC Identification and System Parameter Estimation 1985*, Barker, H.A. and Young, P.C., Eds., Pergamon Press, Oxford and New York, 1985, 669.

6. Plaetschke, E. and Schulz, G., Practical input signal design, in *AGARD Lecture Series No. 104 Parameter Identification*, AGARD, London, 1979, 3-1.

7. Koehler, R. and Wilhelm, K., *Auslegung von Eingangssignalen für die Kennwertermittlung*, Report IB 154-77/40, DFVLR Institut für Flugmechanik, Braunschweig, F.R. Germany, 1977.

8. Mehra, R.K., *Frequency-Domain Synthesis of Optimal Inputs for Linear System Parameter Estimation*, Report TR 645, Division of Engineering and Applied Physics, Harvard University, Cambridge, Mass., 1973.

9. Schulz, G., Entwurf optimaler Eingangssignale für die Systemidenti-
   fizierung unter Berücksichtigung von Meß- und Systemrauschen,
   Regelungstechnik, 25, 324, 1977.

10. Mulder, J.A., *Design and Evaluation of Dynamic Flight Test Manoeu-
    vres*, Delft University of Technology, Delft, The Netherlands, to be
    published.

11. Plaetschke, E., Mulder, J.A. and Breeman, J.H., Flight test results
    of five input signals for aircraft parameter identification, in *IFAC
    Identification and System Parameter Estimation 1982*, Bekey, G.A. and
    Saridis, G.N., Eds., Pergamon Press, Oxford and New York, 1983,
    1149.

12. Plaetschke, E., *Ein FORTRAN-Programm zur Maximum-Likelihood-Parame-
    terschätzung in nichtlinearen retardierten Systemen der Flugmechanik
    - Benutzeranleitung*, DFVLR-Mitt. 86-08, 1986.

13. Marchand, M., *Bestimmung der Derivative eines Do-28-TNT-Modells aus
    Freiflugversuchen*, DFVLR-FB 82-17, 1982.

14. Hall, W.E., Gupta, N.K. and Hansen, R.S., Rotorcraft system
    identification techniques for handling qualities and stability and
    control evaluation, in *34th Annual Forum of the American Helicopter
    Society*, Washington, D.C., 1978, paper 78-30.

15. Koehler, R. and Wilhelm, K., Closed loop aspects of aircraft identi-
    fication, in *AGARD Lecture Series No. 104 Parameter Identification*,
    AGARD, London, 1979, 10-1.

16. Etkin, B., *Dynamics of Atmospheric Flight*, John Wiley & Sons, New
    York, 1972, chap. 4.

APPLICATION OF SYSTEM IDENTIFICATION
IN NAVAL ENGINEERING

R. Maltese et al.
CETENA, Genova, Italy

1 - Introduction

Nowadays, the vibration problem is one to be faced at an early
stage of ship design to avoid uncomfortable service condition or
heavy troubleshooting interventions. Good vibratory ship performance, related to a maximum allowable level, is usually requested
according to specifications by the ship owners, which refer to the
ISO code under publication in its final form. Moreover, the delivery trials include standard vibration tests to check the vibratory
levels. This has caused the shipyards, over the last ten years, to
seek improved methods of predicting ship vibration in order to
protect themselves, as far as possible, against the risk problems
occuring once the ship has been launched. Today, well-defined
methodologies are available and can predict with increasing accuracy both the magnitude of the exciting forces and the level of the
structural response at all stages of design. However, although
well-established from the theoretical and methodological points of
view, these methods still require continuous improvements in order
to cope with the increasing demand in terms of ship structural
complexity and comfort requirements. Moreover, even the most sophisticated theoretical investigations need to be complemented by
data (such as damping coefficients for the evaluation of the structural response) which can only be obtained through specific experi-

mental tests carried out on similar ships. This involves, in addition to the routine delivery trials, the assessment of the real
vibratory level of the ship in service by carrying out specific
complementary investigations, in order to improve the knowledge
necessary to deal with the vibration problem at the design stage
successfully. The ship response depends both on the magnitude of
the exciting forces and on the dynamic properties of the system at
the excitation frequency. The dynamic properties are easily identifiable at any point by examining the response to a given unit
excitation in the frequency range of interest. The frequency response thus obtained has peaks corresponding to the natural frequencies. Their amplitudes depend on the energy associated with the
point at the actual mode as well as on damping. Therefore, identifying the ship vibratory response implies knowing its modal parameters, i.e. natural frequencies, modes and damping. The identification of the modal parameters:

a) allows methodologies for predicting the natural frequencies and
   modes of a ship to be improved by refining the theoretical-
   experimental correlation;

b) provides useful information on the vibration behaviour of the
   ship for troubleshooting purpose;

c)allows the prediction of the structural response for a given
excitation to be verified, as it provides a correct definition
of the damping.

This last point is very important because the ship vibration response can be predicted if natural frequencies, modes, exciting
forces and damping are known. However, while the first three parameters can be successfully calculated, the last one can only be
estimated on the basis of statistical data from experimental tests
and modal parameter identification on similar ships. It is important to note that, for a suitable development of points a) and b),
it is necessary, after having identified frequencies and modes, to
know the mobility (response per unit excitation) at the most significant points. Damping is certainly, among the modal parameters,
the most difficult to evaluate and the most sensitive to the test
procedure, the accuracy of measurements and the effectiveness of
the analysis method. Various types of measurements enable identification of modal parameters, but the most commonly used and effective are certainly exciter tests. Currently, the hull damping is
modelled by combining all energy dissipation effects into a single
damping factor, although, physically, several dissipation mechanisms (structural, hydrodynamic and cargo damping) contribute in
different ways, depending on structure and cargo types and on the

frequency range under consideration. Moreover, the energy is not dissipated uniformly along the ship because of local structure and configurations, so that on real ships the damping coefficient also changes from point to point. Little reliable data concerning damping is available worldwide because of the complexity of this parameter and the uncertainties associated with its identification. In this section, some analysis procedures for the identification of the modal parameters are presented, together with numerical examples from which the degree of effectiveness and reliability of some procedures and problems arising from their application can be inferred. When considering the results, it must be remembered that the basis of the analysis consists of data obtained ,from experimental tests: therefore, identification procedures also include aspects related to test methodology, acquisition and treatment of the signal, all of which have a substantial effect on the quality of the final results. Several methods will be reviewed, ranging from simple ones (Response Curve, Phase Variation, Logarithmic Decrement) to others which are more sophisticated:Phase Resonance [1], Phase Separation [2],

Exponential Method [3], Maximum Entropy Method [4].

2 - Experimental Test Procedures

Identification of the modal parameters is carried out by analysing the vibratory response of the ship subjected to an excitation. The following types of excitation can be applied to the ship:

- steady state excitation;

- impulsive excitation;

- random or pseudo-random excitation.

To day the first method is the most widely used to identify modal parameters.

2.1 - Steady-State Excitation .

It is well-known that the transfer function of a linear system is obtained through harmonic analysis by applying sinusoidal signals of increasing frequency and calculating for each one the rate of the input-output Fourier Transform, normalized with respect to the input spectrum modulus. This procedure requires the frequency variation of the input signal to be very slow, so that steady-state excitation can be assumed. For correct determination of the transfer function, analysing the signal at each interesting frequency

takes a long time. Due to the fact that all signals at frequencies other than the input frequency are filtered out, steady- state testing has the best signal-to-noise ratio at the measurement frequency of all excitation techniques. This type of excitation is most commonly used with harmonic exciters.

2.2 - Impulsive Excitation

Impulsive excitation enables to determine the frequency response simply by exciting the ship structure with an impact load. This is possible since an impact is an approximation of an impulse function which contains energy at all frequencies. However, particular attention must be devoted to the selection of a proper impactor and signal processing technique. Impact tests are generally carried out using two types of actuators: wave impact (slamming and sweeping),and hammers. Several identifications were performed using wave impact excitation [5,6,14]. Hammers were extensively used in the past to excite the superstructures [24,25]. Recently a hammer was used successfully to excite the hull [26]. However, problems still exist for large ships with small response amplitude, where high-sensitivity detectors with low signal-to-noise ratio and proper analysis techniques are necessary to obtain satisfactory results. Impact tests also show implicit problems related to two characteristics of the structures, i.e. linearity and damping :

- since an impact has a very high ratio of peak to r.m.s. energy
  content, it tends to excite all the non-linearities in a system.
  For this reason, impacting does not work well on a non-linear
  system as the hull;

- the amount of damping in the system is also important: when
  there is too little damping, the response signal does not decay
  to zero within the duration of sampling, in severe leakage errors may occur. On the other hand, if the damping is high. noise
  becomes the problem as, while the response signal decays to zero
  rapidly after the start of sampling, any noise will be present
  throughout. To improve these conditions, proper signals processing techniques must be selected.

2.3 - Pseudo-Random Excitation

Pseudo random input testing has become a practical method of frequency response measurement since the development of the digital
Fourier analyser [15]. Using the fourier transform, periodic input
is not restricted to sinusoidal and can have almost any spectrum
content. The excitation signal is created in the frequency domain
and transformed to the time domain. It is always periodic within

the sample window, and therefore does not suffer from the leakage errors of pure randomness: this is one of the important advantages of pseudo-random excitation. The others are :

- it is fast;

- both the amplitude and frequency contents of the excitation signal can be precisely controlled;

- it has a low ratio of peak to r.m.s. energy;

- noncoherent noise can be conveniently averaged out.

However, for extremely environments, the process is not faster than swept sine;

- leakage errors are eliminated by using periodic input within the sample window of the Fourier analyser.

On the other hand, its disadvantages are :

- high sensitivity to rattle: loose components generate periodic, coherent noise which cannot be averaged out and appear as spikes on the frequency response measurement. This can cause difficulties in curve-fitting the data to extract modal parameters;

- the energy input at any frequency is small compared to swept-sine: the reason, of course, is that all frequencies are excited simultaneously.

No references have been found regarding relevant applications of the method to the identification of the modal parameters of the ship hull.

3 - Analysis Methods

The most commonly used or promising analysis procedures for the identification of the modal parameters will be described. We will consider in turn :

- Response Curve Method (RCM);

- Phase Variation Method (PVM);

- Logarithmic Decrement Method (LDM);

- Phase Separation Method (PSM);

- Exponential Method (EM);

- Phase Resonance Method (PRM);

- Maximum Entropy Method (MEM);

- Circle Fitting Parameter Estimation Method (CFPE);

- Analytical Identification Procedure (AIP).

Most applications of these procedures have as common data the recording of the response to a steady-state excitation. For the PSM, PRM and RCM, which work in the frequency domain, the response

operators are used, whereas for the EM, which work in the time
domain, the inverse Fourier transform of the response operator is
used when the method is applied to the analysis of exciter test
response. This is due to the fact that the EM works on limited time
series of a typical transient nature, and it can be shown that the
impulsive response of a system can be obtained by applying the
inverse Fourier transform to its response operator.

3.1 - The Response Curve Method

This method derives directly from the characteristic equation of
single degree of freedom system [5,6]. The absolute value of the
transfer function, defined here by the ratio of dynamic displaceme-
nt/static displacement in the frequency domain, can be written:

$$|H(\omega)| = \frac{1}{\sqrt{\left(1 - \frac{\omega^2}{\omega_0^2}\right)^2 + 4\zeta^2 \frac{\omega^2}{\omega_0^2}}} \qquad (1)$$

where ω_0 denotes the (circular) eigenfrequency of the undamped
system.

Taking into account the low damping values (i. e. $\zeta \ll 1$) one can
assume that the maximum amplitude corresponds to the resonance.

For $\omega = \omega_0$ one can write:

$$|H(\omega)|_{max} = |H(\omega_0)| = \frac{1}{2\zeta}$$

For slight variations around the resonance, it is possible to write:

$$\omega = \omega_0 + \Delta\omega \quad, \Delta\omega > 0$$

Substituing in (1) and dropping higher-order terms, one obtains:

$$|H(\omega_0 + \Delta\omega)| \approx \frac{1}{2\sqrt{\frac{\Delta\omega^2}{\omega_0^2} + \zeta^2}} \tag{2}$$

The amplitude near the resonance can be expressed as a fraction of the maximum, i.e.:

$$|H(\omega_0 + \Delta\omega)| = |H(\omega)|_{max} = \frac{1}{2\zeta} \qquad n > 1 \tag{3}$$

Combining the two formulae, one obtains:

$$\zeta = \frac{\Delta\omega}{\omega_0\sqrt{n^2 - 1}} \tag{4}$$

considering (-3 dB point) i.e. :

$$n = \sqrt{2} \rightarrow \zeta = \frac{\Delta\omega}{\omega_0} \tag{5}$$

Finally, considering two points respectively on the right and on the left of the peak, one obtains:

$$\zeta = \frac{\omega_2 - \omega_1}{2\omega_0} \tag{6}$$

where:

$$\omega_2 = \omega_0 + \Delta\omega$$

$$\omega_1 = \omega_0 - \Delta\omega$$

3.2 - The Phase Variation Method

This method relies on the determination of the tangent of the phase angle between the excitation and the response near a resonance frequency [5,16]. This analysis requires the resonance curve to be available in complex form. The damping ratio can be obtained as follows :

$$\zeta = \left| \frac{\omega_0^2 - \omega^2}{2\omega_0\omega} \tan\tau \right| \tag{7}$$

where:

ω_0 is the resonance circular frequency;

ω is a circular frequency close to ω_0 ;

$\tan\tau$ is the tangent of the phase angle between excitation and response at ω .

The main problem in using this method is that the phase angle curve exhibits a sharp drop as it approaches the resonance frequency.

3.3 - The Logarithmic Decrement Method

This method is based on the evaluation of the variation of the structural response caused by a sudden variation of the excitation [7]. The experimental tests are carried out by stopping the excitation and recording the amplitudes for a defined period of time [5]. The logarithmic decrement can be evaluated as follows:

$$\Delta = \frac{1}{P} \ln \frac{a_n}{a_n + P} \qquad (8)$$

where:

a_n - is the first amplitude observed;

P = is the number of observed periods.

The damping ratio can be easily obtained from the logarithmic decrement as follows:

$$\zeta = \frac{\Delta}{2\pi} \qquad (9)$$

The method is particularly interesting when the damping is very small and the structure is excited at a frequency corresponding to an uncoupled mode. At higher frequencies, however, a modulation can occur between the excited mode and the closest ones, which can introduce errors in the evaluation of the logarithmic decrement.

3.4 - The Phase Separation Method

This method [2,4,5,6] belongs to the normal mode testing category, and makes use of a modal transformation in terms of complex modes in order to linearize the transfer function matrix :

$$<H> = \sum_r \frac{\left(\Psi_r\right)\left(\Psi_r\right)^t}{\lambda_r - \mu} \tag{10}$$

where:

$\left(\Psi_r\right)$ = r-th mode of vibration:

normalized in such a manner that all the generalized masses are equal to 1

$\lambda_r = \overline{\omega}_r^2\left(1 + j\overline{y}_r\right)$

$\overline{y}_r, \overline{\omega}_r$ = loss factor, r-th natural frequency;

μ = square of the excitation frequency.

The advantage of this approach lies in the fact that, in addition

to mass and stiffness, the damping matrices are involved in the

modal transformation. The damping matrix is assumed to be the imag-

inary part of a complex stiffness matrix, (hysteretic damping).

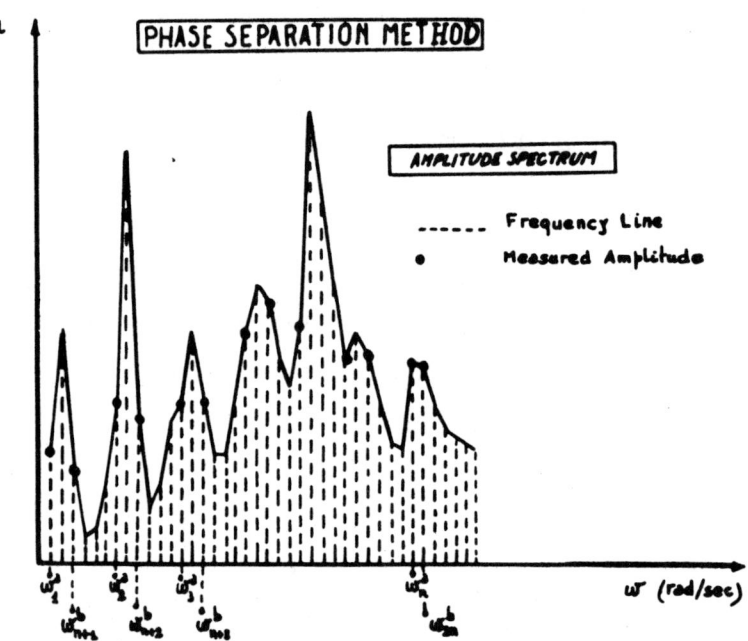

Figure 1 - Phase Separation Method:
Extraction of values from Transfer Function.

As is shown in Fig. 1, two sets of values (response versus frequen-

cy) are extracted from the transfer function by considering two

points on the left and on the right of each natural frequency in

the response diagram. Similar operations are performed on the

transfer functions of all the measurement points at the same frequency values ω^* and ω^+ .

Two sets of transfer functions are thus derived for the two sets of
frequency values. The modal parameters of the system can then be
identified by equating the solutions of the two sets and using a
polynomial matrix approach.

3.5 - The Exponential Method

The Exponential Method used at CETENA [3,8] derives from the optimization of an exponential algorithm developed for the first time,
by Spitznogle and Quazi [9] to analyze short, non-periodic damped
(i.e. transient) signals. The method can also be used to analyze
data from steady-state trials (response operator), as it can be
shown that the impulsive response in the time domain can be obtained by inverse Fourier transform of the response operator (frequency response function).

A series of numerical values is obtained by sampling the time
series, which is decomposed into M exponential functions of the
Laplace variable :

$$x(n,\Delta t) = \sum_{j-1}^{M} A_j e^{(s_j n\Delta t)} \qquad ;(n = 0,1,\ldots\ldots,2N-1)(11)$$

In most cases, M is unknown; an estimation of the solutions of the
system of linear equations can be made using the least-square
method. In developing the algorithm, an auto-regressive approach
has been selected. Assuming that the order of the exponential model
equals that of the auto-regressive one. the problem is now to
evaluate the order of the latter. The Akaike Information Criterion
(AIC) [10,11] has been used for the evaluation of the order and
coefficients of the auto-regressive model that best fits the starting time series.

Finally, the parameters typical of the harmonic components of the
original signal (complex amplitude, frequencies and damping) can be
derived via the resolution of the characteristic equation associated with the auto-regressive model and of a system of Van der Monde
linear equations.

3.6 - The Phase Resonance Method

This identification procedure [1,2] is also a "normal mode testing"
technique, but it is based on different assumption on the matrices
characterizing the system (stiffness, mass and damping), in order
to obtain a linerized form of the response operators. In particular, it is assumed that the damping matrix is proportional to the
mass and stiffness matrices (proportional damping). The mode shapes

are identified through a finite-difference approach on adjacent

spectral lines of the response operator.

This method usually gives reasonably good results when overlapping

between the modes is limited (low modes density). When the overlap-

ping is likely to cause significant errors, an iterative

least-square based estimation method, which tends to minimize a

suitable error function, can successfully be applied. A particular

version of this approach has been developed by Hewlett-Packard [6].

It uses the following linearized system response (vector) in the

frequency domain :

$$H(s) = \sum_{k=1}^{m} \frac{r_k}{2j(s-p_k)} + \frac{r_k^*}{2j(s-p_k^*)} \qquad (12)$$

where :

$s = +\sigma + j\omega =$ Laplace variable;

$p_k = -\sigma_k + j\omega_k =$ Pole of Transf. Func.;

$r_k =$ Complex residue of mode shape ;

$(.)_k^* =$ Complex conjugate of $(.)_k$

The iterative algorithm uses a least-square estimation method,

based on the minimization of the following error terms :

$$\Xi = \sum_{i=1}^{N} \left(H_i H \left(\omega_i \right) \right)^2 \tag{13}$$

where :

H_i = measured system response vector at frequency i ;

$H(\omega_i)$ = analytical system response vector at frequency i ;

N = number of spectral lines of the response vector .

For each iteration of the algorithm, all modal parameters (frequency, damping and mode shapes) are newly estimated, in such a manner
that gradually the error between the measured response operator and
the calculated one, can be reduced.

3.7 - The Maximum Entropy Method

This method seems to be promising analysis technique for ship vibration measurements. It is based on an extension of the auto-correlation function, which gives the required frequency resolution
[4]. Statistics on the reliability of the estimation of modal parameters are also possible.

3.8 - Circle-Fitting Parameter Estimation

The circle-fitting method [27,28] was originally developed by Kennedy and Pancu for system with hysteretic damping. It is also valid for system with general viscous damping, and further extensions also allow the estimation of complex modal displacements, yielding complex mode shapes. In practice, the method works as follows :

- the number of modes is estimated by visual inspection of the set of response curve ;

- the damped natural frequencies are defined at the local maxima of the response curves ;

- using a least square approximation method, a circle is fitted through points in the vicinity of the damped natural frequency ;

- the damping ratio and the modal displacement are defined in amplitude and phase by the position and dimension (diameter) of the circle .

3.9 - Analytical Identification Procedure

The System Identification Procedure was described in [12]. A 3 - D Finite Element Model of the hull and superstructures was correlated with the response measured with an exciter test. The calculated responses at the main resonant frequencies were modified to obtain the best possible selection of damping values. In essence, the damping determined in this way is not a "true" but an "equivalent" damping, which is to some extent a function of the analytical model and possibly of the condensation technique used .

It must be remembered that the analytical model, no matter how detailed, represents the total dynamic behaviour only to a limited extent . Nevertheless, the derived results were in good agreement with normally accepted damping values for hull and superstructure .

4 - Damping Identification Literature

Up to now very few recent literature concerning damping identification is available . The following presents a review of identified damping value with, whenever possible, indication of the identification procedure used .

The Germanisch Lloyd performed extensive testing on ships of varying sizes and with different loading condition [16] . Responses to both propeller and exciter are measured, and the modal parameters (frequency and damping) are identified .

The damping coefficients for hull and superstructures are identified both by "response curve" and "phase variation" methods .

The hull damping identified by the response curve method from pro-

peller excited vibrations, presents the following features :

- for vertical vibration, damping values decrease as the frequency

  increases. An example of identified damping pattern is shown in

  Fig. 2, where values are obtained from different measured points

  on different ships : the damping values show a sharp drop at the

  first vibration modes.

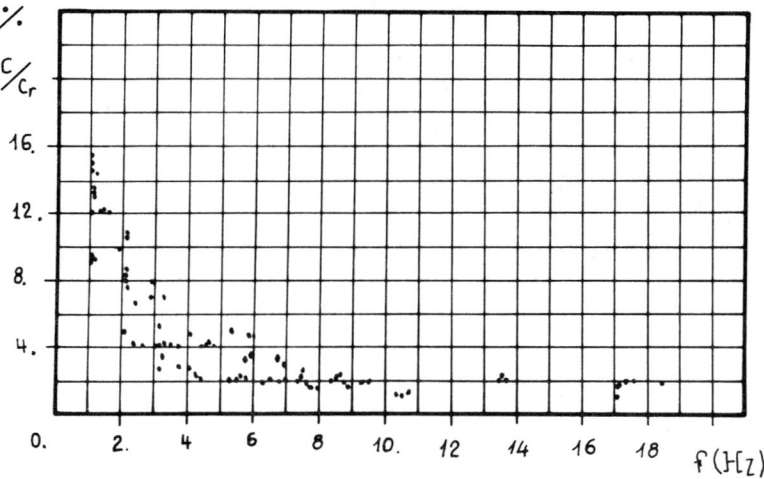

Figure 2 - Damping in different points for several ships ([16]) .

- The damping does not depend only on the mode at the current
frequency : coupled modes leads to higher damping values.

Table 1 shows hull damping values for various types of ships calculated by the response curve method applied to propeller excited
vibrations . Damping values for the first hull modes seem to be too
high . Results from exciter tests performed at quay show damping
values relatively high in the 7 - 12 Hz range . The results of
these tests, carried out with a main interest in superstrutture
vibration, are mainly analysed using the phase variation method.

Table 1 - HULL DAMPING VALUES ([16]).		
Ship type	Frequency Range	Critical Damping %
LPG Tanker	0.91-10.50	9.1-2.0
LPG Tanker	0.91-4.02	7.6-2.6
Product Carrier	1.07-7.78	9.4-1.5
RO-RO	1.17-18.40	13.9-1.8
RO-RO	1.07-17.00	14.1-1.0
RO-RO	1.07-21.30	14.2-0.9
Container	1.14-10.55	9.1-1.1

Another set of damping values is identified in [5] . Various ships
at sea and at quay were tested . Mostly exciter tests are performed
at quay, and the vibration response obtained is analysed using the
response curve method . The response of ships at sea was analysed

mainly with the logarithmic decrement method .

Table 2 shows some of the hull damping values reported in the paper. Values marked with an asterisk were obtained using the response curve method .

Table 2 - HULL DAMPING VALUE ([5])			
Ship Type	Frequency Range	Critical Damping %	Notes
Tanker	3.8 - 13.3	0.9 - 1.9	-
Tanker	3.8 - 13.3	1.1 - 2.5	-
Tanker	5.0 - 11.5	0.7 - 1.3	Full load
Tanker	5.3 - 12.8	0.8 - 2.9	Ballast
Tanker	.47 - 13.3	0.9 - 2.48	B.Sh.W.
Tanker	7.0 - 12.3	1.0 - 1.9	-
L.N.G.	4.5 - 6.7	1.46 - 1.7	-
Cont.S.	2.5 - 15.7	1.85 - 3.7	-

Reference [14] presents a survey of damping values for various ships in different loading conditions . The logarithmic decrement method is used in the identification . Some of the results are shown in Table 3 .

Table 3 - HULL DAMPING VALUES ([14])			
Ship Type	Condition	Critical Damping %	Reference
Cargo ship	Part Load	0.3	[18]
Cargo ship	Full Load	0.3	[18]
Cargo ship	Ballast	0.58	[19]
Cargo ship	Loaded	0.62	[19]
Cargo ship	Part Load	1.0	[20]
Cargo ship	Full Load	1.13	[20]
Cargo ship	Light	0.38	[22]
Cargo ship	Deep	0.41	[22]
Cargo ship	Light	0.7	[23]
Cargo ship	Deep	1.8	[23]
Container	Normal Draft	0.7	[21]
Container	Deep Draft	1.0	[21]

An extensive description of the identification procedures applied to four ships is described in [6] . Due to the particular configuration of one of the ships (twin-skeg), tests with impulsive excitation due to stern-wave impact were performed in addition to the usual steady-state exciter tests .

The paper discusses the respective advantages and drawbacks of both types of excitation, and their consequences on the damping evaluation . Four identification methods are discussed :

- the Phase Separation Method (PSM) ;

- the Exponential Method (EM) ;

- the Phase Resonance Method (PRM) ;

- the Response Curve Method (RCM) .

The basic concepts are reviewed for each method. Two applications are given which consist of simulating experimental vibration tests with the results of a theoretical model obtained by structural analysis . Two other examples are quoted, concerning the totalling four ships .

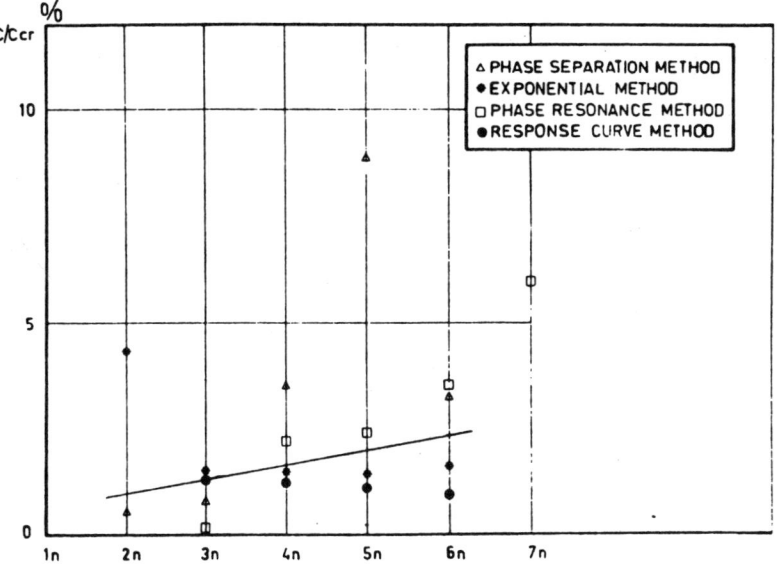

Figure 3 - Damping Identified by Different Analysis Methods for a
Car Ferry (from [6])

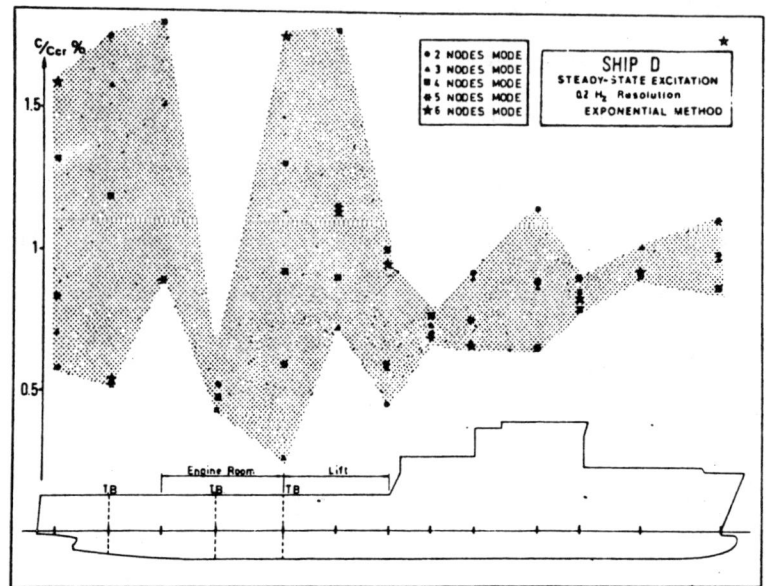

Figure 4 - Variation of Damping along the Ship for Different Vibration Modes (from [6])

For each case, the applicability of each method is discussed and

the main results are shown . Various aspects are investigated, such

as :

- behaviour of the methods at various modes (Fig. 3);

- damping pattern for individual modes along the ships from

   measurements taken at various locations (Fig. 4) .

The main conclusion are :

- the natural frequencies are identified with good accuracy by all
  methods;

- the mode shapes are identified with less, although acceptable
  accuracy;

- the damping identification is good in the simulation models, no
  matter which method is used . On the other hand, notable decrease in accuracy takes place for the experimental models;

- the damping values increase slightly with the frequency;

- coupling between hull and subsystems leads to higher damping
  values.

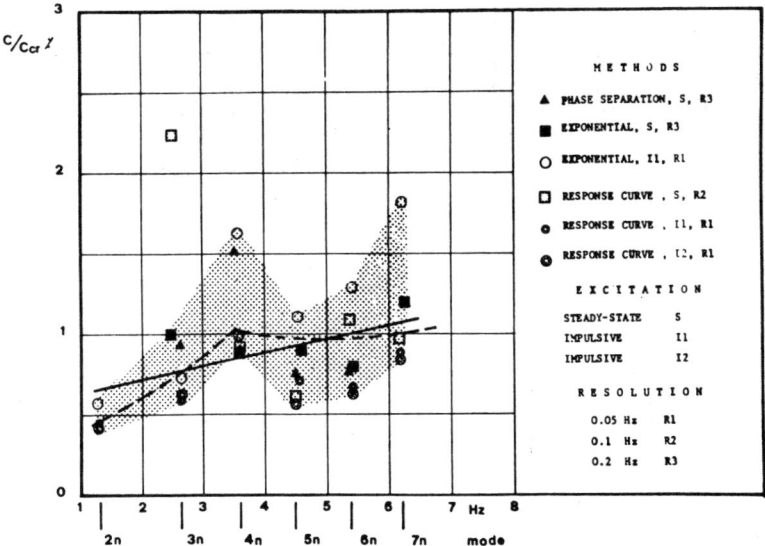

Figure 5 - Damping Identification from Impulsive and Exciter Tests

In Fig. 5, the damping pattern shows two peaks which correspond to
coupled hull - double bottom modes.

Results of damping identification obtained by an analytical procedure described in [12,17] are shown in Fig. 6 and in Table 4 .

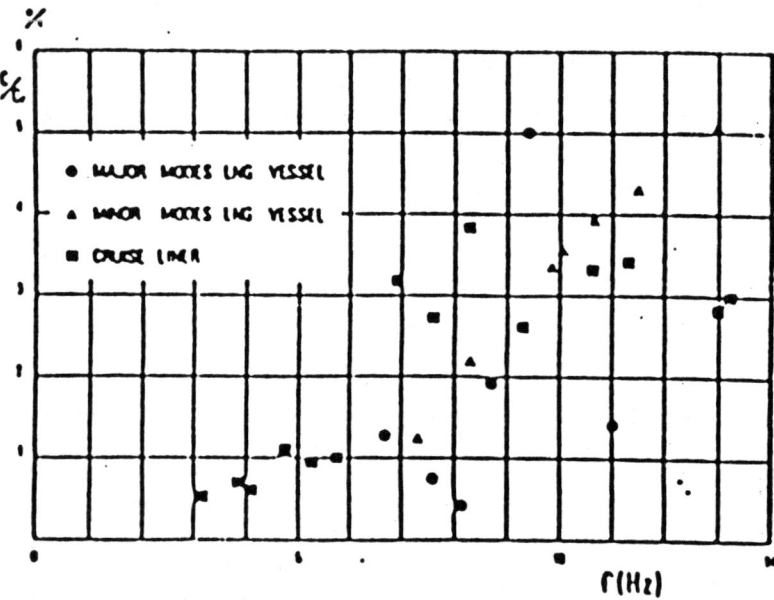

Figure 6 - Damping (from [12],[17])

For LNG carriers, distinction is made between major and minor
modes . All the damping values are between 0.4 and 2 % of critical
for the major modes . Minor modes show higher values characterized
by a more regular pattern .

TABLE 4 - Damping Ident. [12] [17]			
Ship	Frequency Range	Critical Damping (%)	Notes
LNG Vessels	7.0 - 10.9 —	0.39 - 1.90 1.25 - 3.87	Hi. Modes Lw. Modes
Cruise Liners	3.20 - 5.75 6.90 -13.20	0.4 - .1.1 2.6 - 3.8	1st Fr.Rg 2nd Fr.Rg

5 - REFERENCES

1. Formenti D. : "Analytical and Experimental Modal Analysis"
    Ph.D.Thesis, Leuven, 1977.

2. Natke H.G. : "Survey of European Ground and Flight Vibration
    Test Methods ",Proc. of Aerospace Engineering and Manufactur-
    ing Meeting, S.Diego, Dec.1979.

3. Maltese R. : "Optimization of a Complex Exponential Algorithm
    for Representation and Spectral Analysis of Time-Limited Sig-
    nals",Cetena Rep. 1860, Mar.1983 (in Italian).

4. Campbell R. B., Vandiver J. K.: "The Determination of Modal
    Damping Ratio from Maximum Entropy Spectral Estimates", ASME
    Winter Annual Meeting, Paper no.80, WA/DSC-29,Chicago,
    Ill.,Nov. 1980.

5. Volcy G.C., Baudin M., Morel P. : "L'Amortissement dans les
    Vibrations des Navires",Nouveautés Techniques Maritimes,1978.

6. Camisetti C.,Maltese R., Scavia F.: "Identification Proce-
    dures of the Modal Parameters from Full-Scale Measurements;
    Some Aspects of the Application to Ship Vibration", Paper
    no.19, Int. Symp. On Ship Vibrations, Genoa, May 1984.

7. Angelov I. A. : "Rechnerische Bestimmung der Dampfungshoef-
    fizienten fur Vertikale Schiffskörper Schwingmiflu anf der
    Grumdlage von Messergebnissen", Schiffbauforschung, June 1978.

8. Box G. E. P. : "Statistical Model for Forecastings and Con-
    trol", Holden Day, S. Francisco, 1980.

9. Quazi A. H., Spitznogle F. R. : "Representation and Analysis
    of Time-Limited · Signals Using a Complex Exponential Algo-
    rithm",J.A.S.A. Vol.47 no 5 1970.

10. Akaike H. :   "A Bayesian Extention of Minimum AIC Procedure
    of Autoregressive Model Fitting", Biometrika, Vol.66, no.2,
    pp. 237-242, 1979.

11. Akaike H.,Kitagawa G.: "A Procedure for Modeling of
    Non-stationary Time Series", Annals of the Institute of Sta-
    tistical Mathematics, Vol.30 no.2 1978.

12. Armand J. L.,Orsero P. : "Analitycal Identification of Damp-
    ing in Ship Vibrations from Full-Scale Measurements", Paper
    no.14, RINA Symp. On Propeller Induced Ship Vibration, Lon-
    don, Dec. 1979.

13. Camisetti C., et. al. : "Propeller-Induced Vibration -- the
    State of the Art of Research and Current Practice in The
    Italian Shipbuilding Industry", Paper no.2, Int. Symp. On
    Ship Vibration, Genoa, May 1984.

14. Betts C. V., Bishop R., Price W. G. : "A Survey of Internal
    Hull Damping", Trans. RINA, Vol.118, 1976.

15. Brown D., Carbon G., Ramsey K.   :   "Survey of Excitation
    Techniques Applicable to the Testing of Automotive Struc-
    tures", Int. Automotive Engineering Congress, Detroit, 1977.

16. Kaube K. et al:"Zulässige Schwingungen an Bord von Schiffen",
    BMFT, Forschungsvorhaben MTH 0058 Germanisch Lloyd's Report
    STB-465, 1977-79.

17.   Grass F.,Mavrakakis Y.,Orsero P.,Paradis A.:   "Prevention   of
      Vibration Aboard   a Cruise Liner Calculations, Measurements
      and Correlation", Paper no.10, Int. Symp. on Ship Vibrations,
      Genoa May 1984.

18.   Lockwood T. : "Vibration of Ships", Trans. RINA, Vol.72,
      pp.162-196, 1930.

19.   Aertssen G., De Lembre R. : "Calculation and Measurements of
      Vertical and Horizontal Vibration Frequencies of a Large
      Ore-Carrier", Trans. NECIES, Vol.86, pp. 9-12, 1970.

20.   Aertssen G., De Lembre R. : "A Survey of Vibration Damping
      Factors Found from Slamming Experiments of Four Ships",
      Trans. NECIES, Vol.87, pp. 83-86, 1970.

21.   Aertssen G., De Lembre R. : "Hull Flexural Vibrations of the
      Container Ship Dart Europe",Trans. NECIES, Vol.90, pp. 19-26,
      1974.

22.   Johnson A. :   "Vibration Tests of an All-Welded and an
      All-Riveted 10000 Ton Dry Cargo Ship", Trans. NECIES, Vol.67,
      pp. 205-276, 1951.

23.   McGoldrick R., Russo V. C. :"Hull Vibration Investigation on
      SS Gopher Mariner", Trans. SNAME, Vol.63, pp. 436-494, 1955.

24.   Skear K. T. : "Vibration Parameters", Paper no. 8, Seminar on
      Ship Vibration, June 1977.

25.   Ramsey K. :   "Effective Measurements for Structural Dynam-
      ics Testing", J. of Sound and Vibration, Apr. 1976.

26.   Hara T., Ohtaka K., Kagawa K.: "On the Application of the
      impulse Test Method to Ship Vibration", 3rd Int. Congress on
      Marine Technology, Session 8, Athens, 1984.

27.   Ward G., Willshare G. T.: "Propeller-Excited Vibration with
      Particular Reference To Full-Scale Measurements", Trans. RINA
      Vol. 18, 1976.

28.   Fujii K., Tanida K.: "Measurements of Ship Vibration and An-
      ti-Vibration Design", IFIP Computer Application in the Au-
      tomation of Shipyard Operation and Ship Design 4, North Hol-
      land Publ. Co., 1982.

DETERMINATION OF FREQUENCY RESPONSES
BY MEANS OF PSEUDORANDOM SIGNALS

A. Lingener
University of Technology Otto von Guericke, Magdeburg, G.D.R.

1. SUMMARY

System identification demands in each case a suitable excitation. Independent of the application of one of the numerous system indentification methods so much mechanical energy has to be fed into the system, that an evaluable reaction of the system under investigation occurs. The aim of identification is to provide an experimentally secured model describing the dynamic behaviour.

The usefulness of advanced methods is essentially based on the application of wide-band exciting signals in connection with direct computer coupling.

The methods in frequency domain which have proved a success start from an experimental determination of the elements of the frequency response matrix.

The frequency responses are either used for direct assessment of the dynamic behaviour or are taken as fundamentals for further investigations for example modal analysis.

2. REQUIREMENTS OF EXCITING SIGNALS

Mechanical structures are excited by: periodic signals (sinusoidal excitation) transient signals (chirp, impact, step) random signals (natural random, periodic pseudorandom).

Fig.1 shows the different kinds of exciting signals and their corresponding spectra. Each kind of excitation has its specific advantages and disadvantages.[1,2]

With regard to a wide-band excitation which should be as wide as possible random excitations are most suitable. Also because of the linearity of the system, which has always to be assumed, a random excitation is to be preferred to impact and step, as smaller displacements are introduced into the system. An essential advantage is that nonlinear disturbances may be abolished by averaging. If required, by reasons of higher accuracy to excite a system reproducibly in a defined frequency range, then there exists nearly no other possibility of excitation than by artificially generated pseudorandom signals.

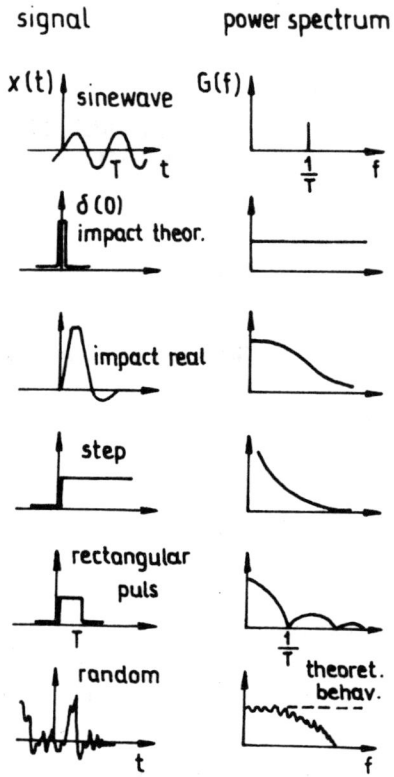

Fig.1. Different types of exciting signals and their power spectra

3. GENERATION OF PSEUDORANDOM SIGNALS

For the generation of pseudorandoms with defined qualities some older methods are known, working without Fourier-transformation.[3,4] Methods based on FFT - algorithms, however are to be preferred, as they can use the FFT-algorithms neccessary for signal- and system analysis anyway. Periodic pseudorandom signals may be generated according to the following principle.

The outset is a discrete spectrum U(k) consisting of N complex amplitudes. The absolute values U_k of this spectrum are given point by point
and the phase angles are added in the form of independent random numbers
equipartitioned in /0,2π/. The transformation of this spectrum (together with its conjugated complex) into the time domain by FFT-routines
yields 2N real values of one sample function of the random signal. The
principle of this method is shown in fig. 2.

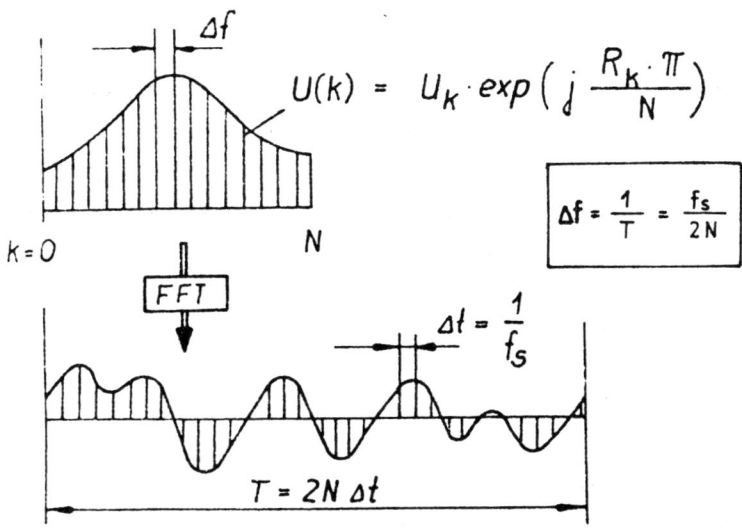

Fig.2. Generation of periodic random signals

We can see, that after fixing the sampling frequency of the signal to
be generated the distance Δf of the spectral lines depends on the length
N of the original spectrum. A higher resolution in frequency domain is
possible only by increasing the number N of samples in the original
spectrum. For different reasons, such as core storage capacity, accuracy
of calculations, resolution capability of ADC or DAC, respectively, the
length N is not expandable to any number. Usual mini- and microcomputers
process a number of N = 1024 values. Consequently, the possible resolution Δf of the generated pseudorandom signals is limited by

$$f = f_N /1024$$

with f_N = 0,5 f_s the Nyquist-frequency.
The time function generated in this way is periodical with

$$T = 2N \Delta t.$$

The calculations on a computer use a storage saving algorithm, which calculates a complex sequence of values. The real and imaginary parts of this sequence form the real sequence required [5].

4. EXPERIMENTAL DETERMINATION OF FREQUENCY RESPONSES

The discrete values a(n) of the time function a(t), gained by inverse Fourier-transformation from a predetermined spectrum, are read out with a fixed sampling frequency. Via a DAC they control an exciter, which itself excites the mechanical system (fig. 3). Exciting force and system respon- se are measured and read in the computer again via ADC. The computer calculates auto- and cross power spectra, frequency responses and coher- ence functions.Because of periodicity of the pseudorandoms leakage eff- ects do not appear in spectral analysis. Averages of spectra, gained by different pseudorandom signals with different phase angles but the same amplitudes, are used for weakly disturbed systems only for the determina- tion of the coherence function.

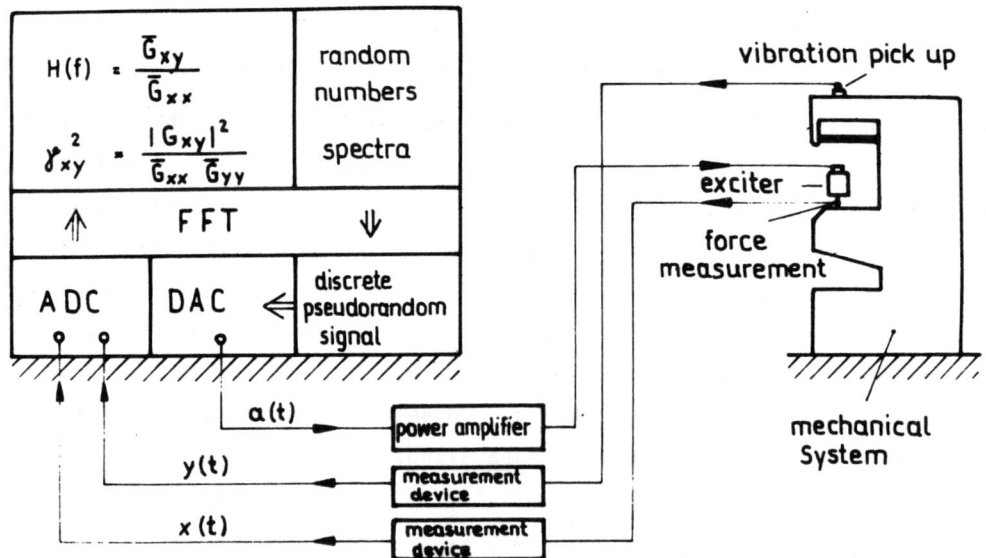

Fig.3. Measurement setup for determining mechanical frequency responses

5. MEASUREMENTS OF FREQUENCY RESPONSES BY MEANS OF PERIODIC NARROW-BAND-PSEUDORANDOM

The procedure described so far produces exciting signals with a frequency resolution $\Delta f = f_s/2N$. However, in many cases, after a basic measurement a higher resolution is neccessary (for instance if narrow neighbouring eigenfrequencies occur). A compression of the bandwidth without increasing transformation length now is possible by the inversion of the common Zoom-transformation.[6]

The basic principle consists in what follows: The whole range of the spectrum is divided into a number of blocks of N spectral lines each. After this, one block representing the frequency range of interest is zoomed in the time domain.

In this way, periodic narrow-band pseudorandom signals may be generated with a bandwidth of $\Delta f = f_s/2 N$. As a restriction it has to be supposed, that only this one block contains spectral lines. The basic idea of this method is, similary as in zoom-FFT a shifting of the frequency range of interest into the origin of frequency axis and the application of the inverse FFT.[7] After system excitation and measurements the sampled time signal has to be Fourier-transformed again into frequency domain.

The principal steps avoiding complicated formulae[5,7] (see appendix) are shown in fig.4 for the case of real Zoom-FFT. Predetermined are $N/2 = 512$ absolute values of an amplitude spectrum, fig.4(1) in the frequency range $/f_1, f_2/$ with a frequency bandwidth of $\Delta f = (f_2-f_1)/N$. A pseudorandom signal and the amplitude spectrum of the response of a system, excited by this signal are to be determined.

Before applying the FFT - algorithm β subspectra $\bar{Z}_i$ in $/0, f_1-f_2/$ are calculated from the predetermined spectrum completed by combination with random phase angles $R_{k\beta}$, $k = 0,....,511$ and their conjugate complex values repeated periodically, fig.4(2) and transformations of 1024 complex spectral lines each according to the rules of inverse Zoom-FFT[7] into the time domain are carried out. Another direction of the inclined straight line represents the corresponding conjugate complex spectrum. β is supposed to be a power of 2. The β discrete time signals $\bar{z}_i$ generated

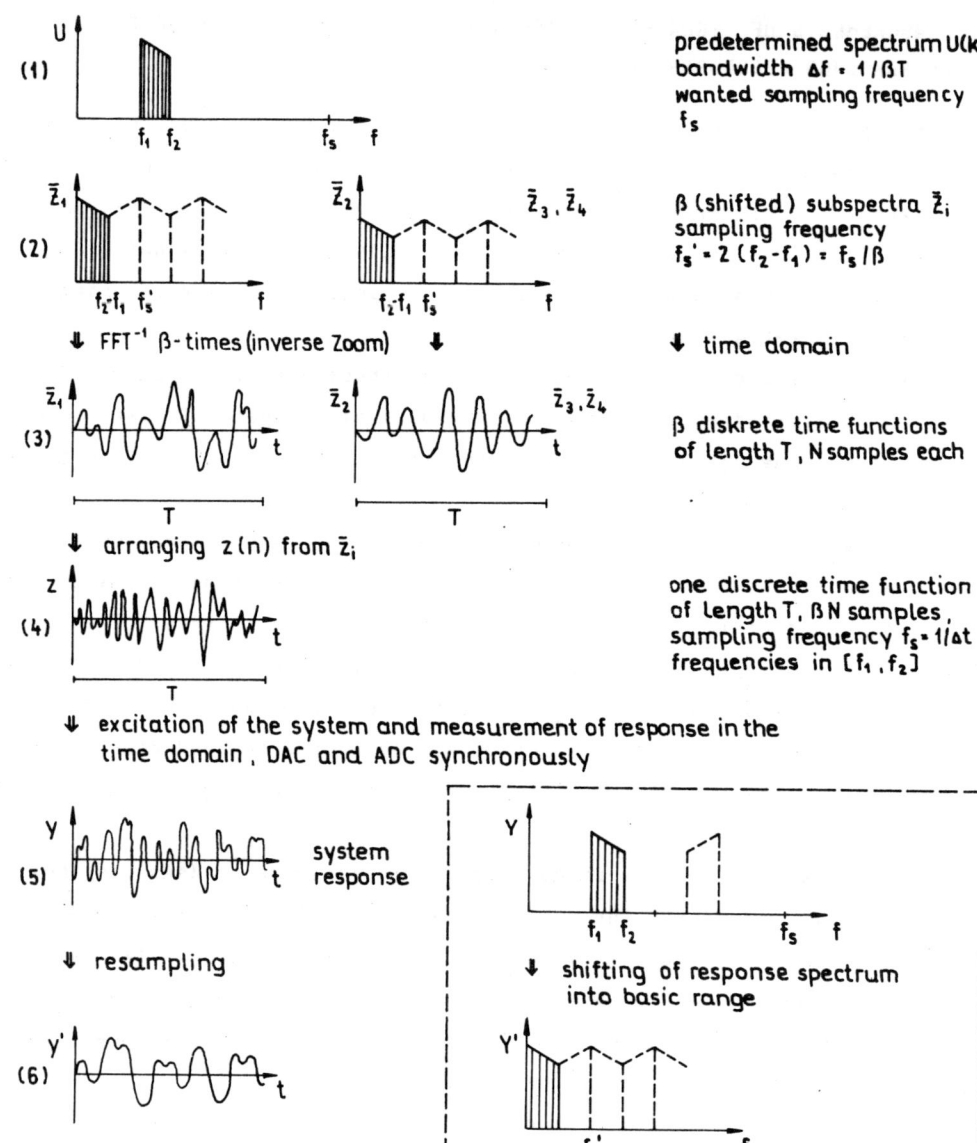

Fig. 4. Signal flow during measurement of frequency responses by means of periodic narrowband pseudorandom signals

in this way are sampled with a frequency $f_s^{'} = 2(f_2-f_1)$ and their lenghth is $T = 2N/f_s$ each fig.4(3).

These β time series $\bar{z}_i$ are fitted into each other and one time function of length T with βN samples is arranged. This procedure corresponds to an increase of the sampling frequency to $f_s = \beta f_s^{'}$, in fig.4(4) shown for $\beta = 4$.

After DC-conversion the system is excited by this signal and the response is measured as indicated in fig.3. The response signal invol-ves only frequencies in the range $/f_1, f_2/$. Its spectrum would take the form as shown in fig.4(5).

To find out this spectrum a modified zoom transformation is now applied. After AD-conversion at first the sampling frequency is dimished from f_s to f_s again by omitting $\beta - 1$ of β successive values (resampling). In order to apply FFT at the same time the discrete time signal has to be shifted again to the frequency range $/0, f_2-f_1 /$ by multiplying with $\exp(-2\pi jf_1 t)$, fig.4(6). But as β and N are both powers of 2, applies $\exp(-2\pi jf_1 t) = 1$ and multiplication is not neccessary.

The resampled discrete time function is now Fourier-transformed (only one transformation) and as result we find the desired spectrum of the system response in the frequency range $/f_1, f_2/$ with a bandwidth of Δf fig.4(7). Without the procedure described here we would have to have used generally a sampling frequency $2f_2$ and would have obtained a spec-trum with the bandwidth $f_2/N \gg \Delta f$.

6. INFLUENCE OF AD-CONVERSION

In applications of simulated pseudorandom signals to test rigs[8] an analog signal, which may be represented by a staircase curve

$$x(t) = \sum_{n=0}^{N-1} x(n) \Delta t\, b(t-n\Delta t)$$

with

$$b(t) = \begin{cases} 1/\Delta t \text{ for } -\Delta t/2 \leq t \leq \Delta t/2 \\ 0 \text{ otherwise} \end{cases}$$

is formed by DAC, using a discrete signal

$$x_d(t) = \Delta t \sum_{n=0}^{N-1} x(n)\, \delta(t-n\Delta t).$$

As   x(t)   may   be   thought of as   convolution

$$x(t) = x_d(t) * b(t) = \int_{-\infty}^{\infty} x_d(\tau) b(t-\tau)d\tau$$

the amplitude spectral density of   x(t) follows because of   the

convolution theorem as

$$X(f) = x_d(f) B(f)$$

Here $X_d(f)$ is the predetermined amplitude spectral density and

$$B(f) = \frac{f_s}{\pi f} \sin \frac{\pi f}{f_s}$$

the amplitude spectrum of rectangular pulse.

Forming   the   cross   power   spectra both amplitude spectra $X_d(f)$ and

$Y_d(f)$ are multiplied with the factor B(f) so that for the spectra,   used

for determination of frequency responses applies

$$G_{xy}(f) = G_{xyd}(f) B^2(f) \quad B(f) = G_{xyd}(f)|B(f)|^2$$

The function $|B(f)|^2$ is presented in fig 5.

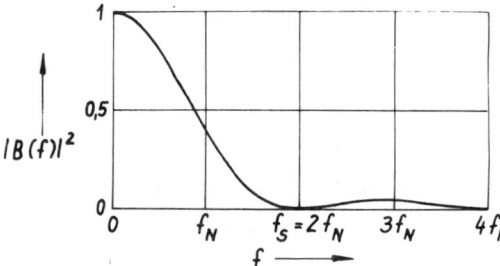

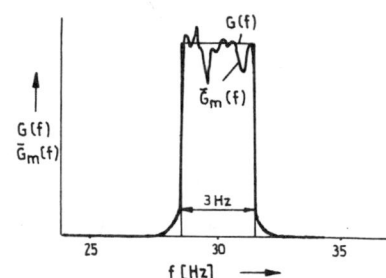

Fig.5. Influence of digital-analog   Fig.6. Result of spectral analysis
       conversion   of a simulated narrow band
                                                 noise

In   order   to compensate for this influence the   predetermined   power

spectral   density   of   the signal to be simulated   is   multiplied   before

applying FFT with $|B(f)|^{-2}$.

The simulated signal in the frequency range above the Nyquist   frequ-

ency   contains periodic shares caused by discretisation as well as shares

caused by the staircase curve.   These shares have to be abolished by   low

pass   filtering.   By reasons of the non ideal filter characteristics   the

application   of a higher Nyquist frequency than is neccessary is   recomm-

ended. Good experiences were made with a Nyquist frequency wich is double

the size as neccessary. In
such a case the filter characteristics between $f_N/2$ and
$3f_N/2$ excerts no influence on
the result. In fig. 6 is shown
the measured spectrum $G_m(f)$ of
narrow-band noise with a constant spectrum $G(f)$ in the
frequency range /28.5 - 31.5
Hz/ generated according to
this principle after 50 averages (without synchronous sampling).

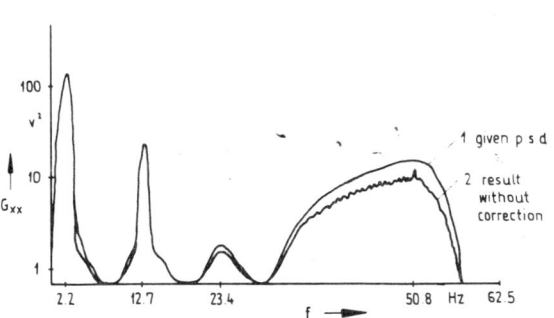

The quantitative influence
of the correction factor can
be recognised from fig. 7 in
which the result of a simulation of a predetermined spectrum with several peaks was
analysed with and without correction factor. Applying pseudorandom signals for the de-

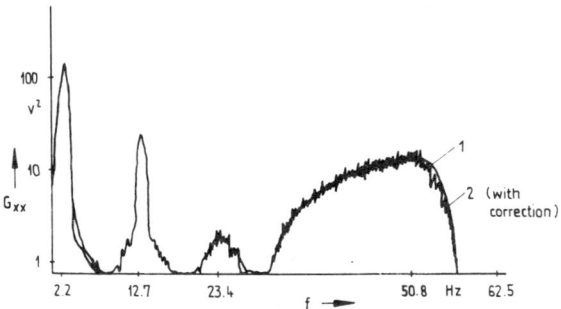

Fig.7. Power densitity spectrum with
several peaks without and with
correction (50 averages)

termination of frequency responses indeed does not demand any correction,
as input and output signals are falsified in the same way.

7. TECHNOLOGICAL IMPLEMENTATION

In practical applications of this method the exciting signals are
generated in advance to the measurement and (if possible) stored on a
data disk. In all cases several independent signal sections are generated. In the case of field measurements, if no suitable computer is available, the pseudorandom signals have to be stored in advance on a data
recording medium or generated by means of a special signal generator
directly in the field. In the same way, the measured signals of exciting

forces and system responses have to be stored too.

After data processing the frequency response may be displayed in any
suitable form or used for further processing.

8. EXAMPLES

8.1. Resolution of a frequency response with a quarter of the bandwidth

A frequency response of a mechanical system according to fig. 8 was
measured by means of a pseudorandom excitation in a frequency range from
10 to 1024 Hz with a bandwidth of f = 2 Hz. The fig. shows the corresponding Nyquist plot. With the aim of an improved resolution of the
frequency range 512 - 763 Hz a pseudorandom signal was generated according to the method described in paragraph 5 and the system was excited
again.
Fig. 9 shows the resulting cut out of the Nyquist plot with a frequency
resolution of 0,5 Hz.

In both cases were carried out 4 averages. The time for measurement
was for the basic range 10 - 1024 Hz, 4 sec (1 sec for each particular
signal). With the zoom - factor β = 4 the measuring time increases
to the fourfold. The calculation of the Nyquist plot in fig. 8 takes
20 sec. with 4 averages.

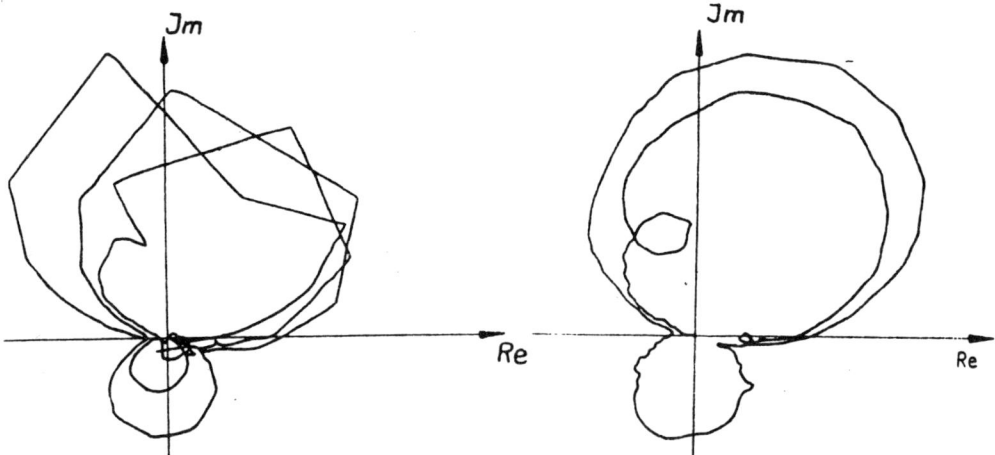

Fig.8.Frequency response 10-1024 Hz   Fig.9.Freqency response 512-768 Hz
    bandwidth 2 Hz   bandwidth 0.5 Hz

8.2. Random fatigue tests with simulated operational loads

To carry out fatigue tests the predetermination of operational loads is neccessary. For this aim, often measured loads or their spectra are taken as a basis. The method of simulation of pseudorandom signals described here allows a real time simulation of predetermined spectra. A specially prepared microcomputer calculates single independent sections of periodic pseudorandoms. These signals control via DAC an electric servohydraulic device.

While one section is put out, the computer is internally calculating the next section as a digital signal etc. Such kind of application was elaborated for a tractor factory. The upper limiting frequency in this application was 128 Hz.

9. REFERENCES

1. Natke, H.G.   Transiente Anregungen von mechanischen Schwingungen in der Versuchstechnik. Technisches Messen tm, 52 H.11, p.393-398, 1985

2. Lingener, A. Analysis of Mechanical Systems Excited by Random Vibrations. Proceedings of the IUTAM-Symposium Random Vibrations and Reliability Frankfurt/Oder 1982 Akademie-Verlag Berlin, p. 173 - 194, 1983

3. Anderson, G.B./Finnie, B.W./Roberts, G.T.   Pseudo-Random and Random Test Signals HEWLETT-PACKARD-JOURNAL, Vol. 19, Nr. 1, 1967

4. Shinozuka, M.   Simulation of multivariable and multidimensional random processes. J. Acoust. Soc.Am. 49 , p. 357 - 368, 1971

5. Wahl, F. Ein effektives Verfahren zur experimentellen Ermittlung von Systemkennfunktionen mit Hilfe von Pseudozufallssignalen. Technische Mechanik 3 , H.1 p. 11 - 17, 1982

6. Thrane, N.   Zoom-FFT Bruel & Kjaer Technical Review Nr. 2, 1980

7. Wahl, F. Die bandselektive Fouriertransformation und ihre Anwendung zur Simulation von Schmalbandprozessen Technische Mechanik 3, H.4 p. 23 - 28, 1982

8. Sperling, L. Zur Simulation zufaelliger Belastungen mittels Fourierreihen nach der Spektraldichte. Tagungsberichte Festkoerpermechanik, Dynamik und Getriebetechnik, Band C p. XLIV/1 - XLIV/15 Fachbuchverlag Leipzig 1985

10. APPENDIX

10.1. Zoom-FFT

Zoom-FFT transforms a real times series of βN samples by means of β common FFT (N samples). Predetermined is a real time series of N samples

$$z(n) = [\ z(0),\ z(1),\ldots,\ z(\ N-1)].$$

The expected spectrum (βN samples, conjugated complex) is

$$Z(k) = \frac{1}{\beta N} \sum_{n=0}^{\beta N-1} z(n)\ \exp\ (\ -j\ \frac{k2\pi n}{\beta N}\ ) \qquad\qquad k=0,1,\ldots,\beta N-1$$

$Z(k)$ may be formally written as

$$Z(k) = \frac{1}{\beta} \sum_{i=0}^{\beta-1} \bar{Z}_i\ (k)\ \exp\ (\ -j\ \frac{k2\pi i}{\beta N}\ ) \qquad\qquad k=0,1,\ldots,\beta N-1$$

with subspectra

$$\bar{Z}_i\ (k) = \frac{1}{N} \sum_{n=0}^{N-1} z_i(n)\ \exp\ (\ -j\ \frac{k2\pi n}{N}\ ), \qquad k=0,1,\ldots,N-1 \qquad (1)$$

calculated from subseries (N samples)

$$z_i(n) = [\ z(i),\ z(i+\beta),\ z(i+2\beta),\ldots,\ z(i+(N-1)\beta)] \qquad i=0,1,\ldots,\beta-1$$

$\bar{Z}_i(k)$ are periodically with k=N, i.e. $Z_i(k)$ may be calculated by common FFT (N samples). For the p-th partial spectrum $Z(pN+r)$ p=0,1,..., $\beta-1$ yields:

$$Z_p = Z\ (pN+r) = \sum_{i=0}^{\beta-1} \exp\ (-j\ \frac{p2\pi i}{\beta}\ )\ \bar{Z}_i(r)\ \alpha_i\ (r) \qquad \begin{array}{l} p=0,1,\ldots,\beta-1 \\ r=0,1,\ldots,N-1 \end{array} \qquad (2)$$

with $\alpha_i\ (r) = \exp\ (\ -j\ \frac{r2\pi i}{N}\ )$.

Putting

$$Z(r) = [\ Z_0,\ Z_1,\ldots,\ Z_{\beta-1}\ ]^T \qquad \text{vector of partial spectra}$$

$$\bar{Z}(r) = [\ \bar{Z}_0,\ \bar{Z}_1,\ldots,\ \bar{Z}_{\beta-1}\ ]^T \qquad \text{vector of subspectra}$$

$$A(r) = \begin{bmatrix} \ddots & & \\ & \alpha_i(r) & \\ & & \ddots \end{bmatrix}$$ diagonal matrix of $\alpha_i(r)$, $i=0,1,\ldots,\beta-1$

$$K(r) = [\ exp\ (\ -j\ \frac{2\pi i r}{\beta}\ )\ ]$$

β,β-matrix of complex unit vectors

$i, r = 0,1,\ldots,\beta-1$

we can write (2) in matrix presentation

$$\beta Z(r) = K(r)\ A(r)\ \overline{Z}(r)\ . \tag{3}$$

For K and A yield:

$$K^{-1} = \frac{1}{\beta}\ K^* \qquad\qquad A^{-1} = A^*\ . \tag{4}$$

10.2. Complex Zoom-FFT

FFT allows processing of complex data. For this reason from a real time series u_0, $u_1,\ldots,u_{2\beta N-1}$ with $2\beta N$ samples a complex time series is arranged. samples with even numbers become real parts $x(n)$, $n=0,1\ldots,\beta N-1$, samples with odd numbers become imaginary parts $y(n)$, $n=0,1,\ldots,\beta N-1$:

$$z(n) = x(n) + j\ y(n).$$

The formulae of real Zoom-FFT are valid and lead to a spectrum $Z(k)= X(k) + j\ Y(k)$. The spectrum $U(k)$ of the real time series consists of β partial spectra and results from the partial spectra of $Z(k)$ as follows:

$$4\ U\ (pN + r) = Z\ (pN + r) + Z^*\ (\beta N - pN - r)$$

$$- j\ [\ Z(\ pN + r\ ) - Z^*\ (\beta N - pN - r)]\ exp\ (-j\ \frac{(pN + r)\pi}{\beta N}\ )$$

$$r=0,1,\ldots,\ N-1 \qquad\qquad p=0,1,\ldots,\beta-1 \tag{5}$$

The factor $exp\ (\ -j\ \frac{(pN+r)\pi}{\beta N}\ )$ corrects the shifting in time domain of the odd samples u_{2k+1} by Δt.

10.3. Inverse Zoom-FFT

Starting from βN complex spectral lines $U(k)$, $k=0,1,\ldots,\beta N-1$ inverse Zoom-FFT yields $2\beta N$ real time samples $u(n)$, $n=0,1,\ldots,\ 2\beta N-1$ by the following steps:

1. Calculation of βN complex spectral lines $Z(k)$ in β blocks of partial spectra of the complex time series $z(n)$ according to the reversal

formula of (5):

$$Z \ (pN + r) = U \ (pN + r) + U^* \ (\beta N - pN - r)$$

$$+ \ j \ [\ U \ (pN + r) - U^* \ (\beta N - pN - r)] \ exp \ (\ +j \ \frac{(pN + r)\pi}{\beta \ N})$$

$$r = 0, 1, \ldots, \ N-1$$
$$p = 0, 1, \ldots, \ \beta-1$$

2. Calculation of β complex subspectra $\bar{Z}_i(r)$ according to the reversal formula of (3) and with (4)

$$\bar{Z} \ (r) = A^* \ (r) \ K^* \ (r) \ Z \ (r) \qquad\qquad r = 0, 1, \ldots, N-1$$

$$\bar{Z}_i \ (r) = \alpha_i^* \ (r) \sum_{p=0}^{\beta-1} Z \ (pN+r) \ exp \ (+ \ \frac{p2\pi i}{\beta}), \qquad r = 0, 1, \ldots, N-1 \qquad (6)$$

3. FFT^{-1} (β-times) of (1) yields β complex subseries $\bar{z}_i(n)$:

$$\bar{z}_i \ (n) = \sum_{r=0}^{N-1} \bar{Z}_i \ (r) \ exp \ (+ \ j \ \frac{n2\pi r}{N}) \qquad \begin{matrix} n = 0, 1, \ldots, N-1 \\ i = 0, 1, \ldots, \beta-1 \end{matrix}$$

4. Arranging the complex time series z(n), n=0,1,..., N-1

$$z_0 \ (0), z_1 \ (0), \ldots, z_{-1} \ (0), z_0 \ (1), z_1 \ (1), \ldots, z_{-1} \ (1),$$

$$z_0 \ (2), \ldots \ \ldots, z_{-1} \ (N-2), z_0 \ (N-1), \ldots, z_{-1} \ (N-1)$$

5. Arranging the real time series u(n), n=0,1,..., $2\beta N-1$.

10.4. Simplification of Inverse Zoom-FFT in the case
      of narrowband pseudorandom

In the case of narrowband pseudorandom is $U = U_k (k) \neq 0$ only in the p-th of the β frequency intervals (fig. 10).
Introducing new indices r and m and separating real and imaginary parts in (6) result

$$\bar{Z}_i \ (r) = U_r \ (cos \ \varphi_1 - sin \ \varphi_2) + U_m \ (cos \ \varphi_3 + sin \ \varphi_4)$$

$$+ \ j \ [\ U_r \ (sin \ \varphi_1 + cos \ \varphi_2) + U_m \ (sin \ \varphi_3 - cos \ \varphi_4)]$$

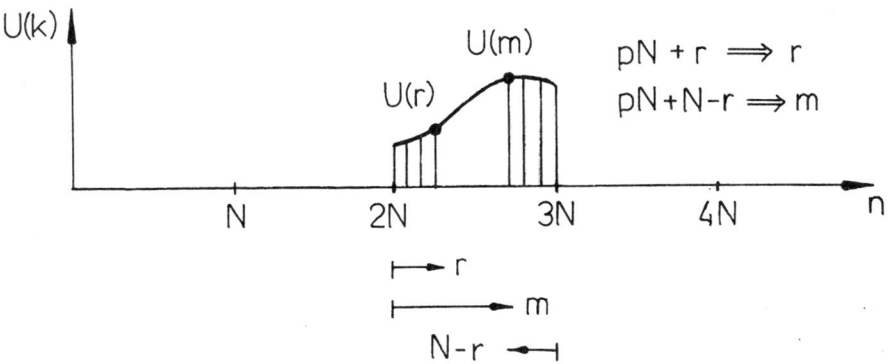

Fig.10. Spectrum of narrowband pseudorandom and new indices ($p=2$, $\beta=4$)

$$\bar{Z}_i(m) = U_r(\cos\varphi_1 + \sin\varphi_2) + U_m(\cos\varphi_3 - \sin\varphi_4)$$

$$+ j \,[\, U_r(-\sin\varphi_1 + \cos\varphi_2) + U_m(-\sin\varphi_3 - \cos\varphi_4)\,]$$

with

$$\varphi_1 = \left[\frac{(pN+r)2i + R_r}{2\beta}\right]\frac{2\pi}{N} \qquad \varphi_3 = \left[\frac{((\beta-p)N-m)2i - R_m}{2\beta}\right]\frac{2\pi}{N} \qquad r=1,2,\ldots,N/2$$

$$\varphi_2 = \left[\frac{(pN+r)(2i+m)+R_r}{2\beta}\right]\frac{2\pi}{N} \qquad \varphi_4 = \left[\frac{((\beta-p)N-m)(2i+m)-R_m}{2\beta}\right]\frac{2\pi}{N} \qquad \begin{array}{l} m=N-r \\[4pt] i=0,1,\ldots,\beta-1 \end{array}$$

If β is a power of 2, then $\varphi_1,\ldots,\varphi_4$ may be calculated without multiplications only by additions and shifting of dual numbers and the calculation time essentially only depends on the time for $(\beta+1)$ FFT^{-1}.

APPLICATION OF MODAL ANALYSIS TO LINEAR ELASTIC MECHANICAL SYSTEMS BY THE EXAMPLE OF A RIBBED PLATE

A. Lingener
University of Technology Otto von Guericke, Magdeburg, GDR

1. SHORT DESCRIPTION OF THE METHOD APPLIED

The method of modal analysis used here assumes the structure under investigation to be treated as a discrete linear nonproportionally viscously damped system with n degrees of freedom and symmetric system matrices.[1,2]

The dynamic behaviour of such a system is described by a linear time-invariant system of differential equations

$$M \ddot{y}(t) + D \dot{y}(t) + C y(t) = f(t) \tag{1}$$

with

    M   – mass matrix (n x n)

    D   – damping matrix

    C   – stiffness matrix

    y(t) – columm vector of n independent coordinates of the system

    f(t) – columm vector of exciting forces acting at the n masses

Solving the set of n differential equations (1) we find 2n conjugated complex eigenvalues and eigenvectors.

With

$$f_{dr} = \omega_{dr}/2 \qquad \text{- r-th resonance frequency}$$

m_r - modal mass of the r-th mode

δ_r - modal damping of the r-th mode (decay constant)

c_r - modal stiffness of the r-th mode

x_r - complex elements of the r-th mode,

we have a relationship between the elements of frequency response matrix $H(f) = (H_{kl}(f))$ and the modal quantities as follows:

$$H_{kl}(f) = \sum_{r=1}^{n}\left\{\frac{x_{rk}\,x_{rl}\,/c_r}{2[(1-f/f_{dr})+j\delta_r/2\pi f_{dr}]} + \frac{x_{kr}^{*}\,x_{rl}^{*}\,/c_r}{2[(1+f/f_{dr})-j\delta_r/2\pi f_{dr}]}\right\} \qquad (2)$$

In each case, in the vicinity of the r-th resonance dominates the modal frequency response belonging to f_{dr}, which may be approximated by a circle.

Directly in resonance, $f = f_{dr}$, applies

$$H_{kl}(f_{dr}) = -j\,\frac{\pi f_{dr}}{c_r\delta_r}\,x_{rk}\,x_{rl} + R_{kl}(f_{dr}) = D_{kl}(f_{dr}) + R_{kl}(f_{dr}) \qquad (3)$$

Here D_{kl} is a complex quantity. Its absolute value is the diameter of the approximating circle. For fixed k and fixed r apart from a constant $D_{kl}(f_{dr})$ represents the component x_{rl} of the modal vector x_r. R_{kl} describes the influence of the other modes, not being in resonance.

2. PRINCIPLE OF COMPUTER-AIDED MODAL ANALYSIS

Investigating multi-degree-of-freedom systems the measured data have to be processed by numerical methods of data processing. Here the parameter values, gained from the measured frequency responses by following the methods for single-degree-of-freedom systems are used as starting estimates of curve-fitting methods.

We start from equation (2) which has to be modified accordingly.

When checking real mechanical systems in each case a limited frequency range [f_A, f_B] is investigated only. For this reason, eq. (2) is dissected

$$H_{kl}(f) \approx - \frac{1}{M_{kl}(2\pi f)^2} + \sum_{r=1}^{m} \left\{ - - - - - - - \right\} + S_{kl} \qquad (4)$$

In this equations means

    M_{kl} - the residual mass, describing the influence of resonances in

               the frequency range [0, f_A] on the range [f_A, f_B]

    S_{kl} - the residual stiffnes, describing the influence of resonances

               in the frequency range [f_B, ∞] on the range [f_A, f_B]

    m   - the number of resonances in [f_A, f_B], m < n.

    Dividing the quantities $D_{kl}(f_{dr})\delta_r$ into real and imaginary parts, we find

$$\delta_r D_{kl}(f_{dr}) = -j \frac{f_{dr}}{c_r} x_{rk} x_{rl} = u_{klr} + jv_{klr} \qquad (5)$$

and introducing this into eq.(4), the well known basic relation of modal analysis results^1 :

$$H_{kl}(f) = - \frac{1}{M_{kl}(2\pi f)^2} + \sum_{r=1}^{m} \left\{ \frac{u_{klr} + jv_{klr}}{\delta_r + j2\pi(f-f_{dr})} + \frac{u_{klr} - jv_{klr}}{\delta_r + j2\pi(f+f_{dr})} \right\} + S_{kl} \qquad (6)$$

By this formula it is possible to describe approximately the systems behaviour in the frequency range [f_A, f_B] with a small number of degrees of freedom. This is a decisive advantage of the modal view.

    The basic idea of computer aided modal analysis consists of an approximation (curve fitting) of the analytical term to the measured data as good as possible. This curve fitting is carried out by means of least square methods.

    The nonlinear equations, resulting from the demand $|H(f)|^2 \longrightarrow$ Min are solved iteratively. With the correction values, obtained from these equations the starting estimates are improved.After obtaining a predetermined limit of the sum of squared deviations the calculation is stopped.

The programme package modal analysis, developed at the Technical^3
University of Magdeburg works according to the following principle:

1. Input of the number of degrees of freedom to be analysed

2. Input of starting estimates of δ_r and f_{dr} (for instance taken from the imaginary part of $H_{kl}(f)$)

3. Calculation of starting estimates of u_{klr}, v_{klr} and the residues M_{kl}, S_{kl} by means of a linear set of equations

4. Iterative determination of 4n modal quantities u_{klr}, v_{klr}, δ_r, f_{dr} and of the two residuals M_{kl} and S_{kl}

5. Calculation of $H_{kl}(f)$ according to eq. (3) and determination of a set of corrections

The method is applicable up to n_{max} = 17. This limitation depends on the capacity of the primary store of the computer.

3. MODAL ANALYSIS OF A BED PLATE OF A PRECISION INSTRUMENT

3.1. Description and aim

The subject under investigation is shown in fig 1. The aim was to determine eigenfrequencies and eigenmodes in a frequency range 10-1000 Hz on a prototype. The results were meant to provide vibration-free opera-

Fig.1. Photo of the bed plate

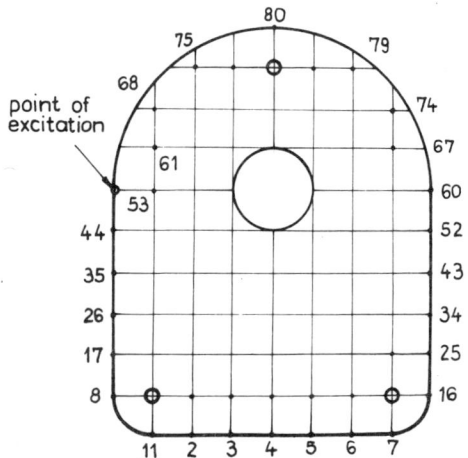

Fig.2. Measurement points on the plate (80 points)

tion. Those modes at which a deformation of the plate itself occured, were of essential interest, as the rigid body motions, depending on the supporting conditions, were not regarded of interest.

The investigations were carried out theoretically by calculating eigenfrequencies and modes by means of FEM[4] and experimental-ly by means of vibration measurements. The FEM calculations were carried out without damping. 80 points according to fig. 2 were defined as points of measurement. During the experimental investigations the plate was supported on a soft three - point rubber spring supporting as well as on a foam rubber mat.

3.2. Measurement engineering problems

The experimental investigations were carried out by means of a pseudo-random excitation and subsequent modal analysis (phase·separation tech-nique). The plate was excited by an electrodynamic exciter in the fre-quency range 10 - 1024 Hz.

The eigenfrequencies and some eigenmodes were measured for compari-tative purposes by harmonic excitation, too. These last mentioned mea-surements were carried out with a frequency - selective measuring method (Vibroport). The time expense in this case was very high, 60 - 90 min per mode.

The excitation by pseudorandom signals was carried out as described in the previous paper: There was a direct coupling between computer and test rig. An intermediate storage is not neccessary. In each case the gene-rated 4 independent sections of the signal were fed via a power ampli-fier into the exciter. The exciting force at a fixed point and accelera-tion at one of the 80 measurement points were measured and read in the computer again and stored. The time effort including the calculation of four frequency responces, averaging and coherence funktion was about 1 min per measuring point. As the rigid body frequencies were not of interest and because of the bad coherence at low frequencies the fre-quency range was limited to > 50 Hz by high - pass filtering.

The point of excitation, point 53, was chosen, as no nodal line is run-ning through this point (except rigid body vibrations).The evaluable fre-quency range was limited to 1024 Hz by the sampling frequency of 2048 Hz.

3.3. Processing of measured data by phase separation technique

From the measured time functions the averaged cross power spectra $\bar{G}_{xiF}(f)$ between exciting force $F(t)$ and acceleration $x_i(t)$ of the measuring point and the auto-power-spectra $\bar{G}_{FF}$ were calculated and from these spectra 80 frequency responses of dynamic flexibility

$$H_{i53}(f) = -\bar{G}_{xiF}(f)/4\pi^2 f^2 \bar{G}_{FF}(f) \qquad i=1,\ldots,80$$

According to the 4 independent measurements at each measuring point in each case four frequency responses were averaged.

These averaged frequency responses were plotted partly (as Nyquist − plots, fig. 3a as example and printing lists) and the starting estimates for the modal analysis routines were taken from the plots. Because of the relative insensitivity of the method in view of differing starting estimates of damping in each case the same starting value of δ_r was predetermined (table 1). From the 80 frequency responses, obtained by curve fitting (fig. 3) the complex eigenmodes can be deducted (fig. 5 and 6).

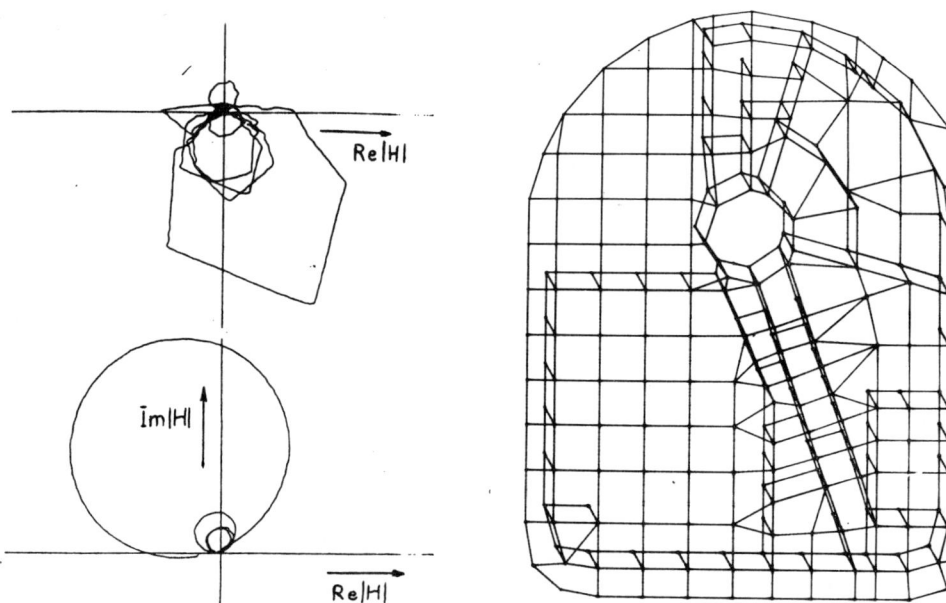

Fig.3. Frequency response H_{753} of the plate   Fig.4. FEM-mesh
      a) measured (inertance)
      b) analytical (receptance)

3.4. Results

The frequency responses, measured at the measuring point 53 contained 6 utilizable eigenfrequencies $f_1, \ldots, f_6$. From these f_1 = 63 Hz is a rigid-body frequency. A comparison calculation had proved, that the 6 measured eigenfrequencies comprised all possible eigenfrequencies in the range of interest. The starting estimate of damping was uniformly defined by a half-width of the resonance peaks of $f_H = \delta/2\pi = 2$Hz. From this value result different damping ratios $\vartheta_i = \delta/\omega_{oi}$. The results of modal analysis are summarized in table 1.

The calculation by means of finite element method based on the mesh in fig.4 was carried out by means of the programme package COSAR.[4] The spatial calculation model consists of 267 elements and 785 nodes. The calculations were made with different stiffness parameters of the supports. The results are summarized in table 2. The deviations in % refer to the measured values.

Table 1. Starting estimates and modal parameters of the plate under investigation. Frequency response H_{7053}, 4 averages

	f_i (Hz)	ϑ_i (δ_i=12,57/s)	u_i (mm/Ns)	v_i (mm/Ns)
starting estimates	f_1 = 63 f_2 =246 f_3 =394 f_4 =590 f_5 =666 f_6 =748	ϑ_1 =0,032 ϑ_2 =0,0081 ϑ_3 =0,0051 ϑ_4 =0,0034 ϑ_5 =0,0030 ϑ_6 =0,0028		
modal parameters after 7 iterations	f_1 = 62,32 f_2 =242,12 f_3 =394,20 f_4 =589,95 f_5 =667,35 f_6 =746,56	ϑ_1 =0,0793 ϑ_2 =0,0299 ϑ_3 =0,0055 ϑ_4 =0,0044 ϑ_5 =0,0075 ϑ_6 =0,0026	u_1 =-,51E-4 u_2 =-,42E-2 u_3 =-,21E-2 u_4 =-,60E-4 u_5 =-,93E-4 u_6 =-,17E-4	v_1 =-,79E-4 v_2 =-,89E-2 v_3 =+,74E-2 v_4 =+,17E-2 v_5 =+,21E-2 v_6 =+,86E-3

Table 2   Calculated and measured eigenfrequencies

	soft support $c_x = c_y = c_z = 0,5$ N/mm		nearly real support $c_x = c_y = 150$ N/mm $c_z = 850$ N/mm		measured values
f_i	Hz	deviation %	Hz	deviation %	Hz
1	–		24.1	-18.9	29.7
2	–		25.2	-19.2	31.2
3	–		34.5	-16.3	41.2
4	–		54.0	-14.3	63.3
5	–		61.9	- 6.2	66.0
6	–		86.7	-12.6	99.2
7	212.7	-5.0	231.9	- 4.2	242.1
8	403.2	2.1	407.7	2.4	394.2
9	546.0	-7.5	547.3	- 7.2	590.0
10	682.6	2.0	685.9	2.6	667.4
11	811.8	8.4	815.8	9.6	746.6
12	1114.6	2.9	1116.0	3.1	1082
13	1258.7	2.8	1260.9	2.7	1228
14	1330.1	1.1	1333.3	1.4	1315
15	1446.9	3.9	1452.1	4.1	1395

The nodal lines of the eigenmodes, determined by modal analysis and by FEM calculation (without damping) are compared in fig. 5. Because of (small) nonproportional damping, the question here concerns lines with "minimum vibration maxima". Fig. 6 examplifies two eigenmodes with a spatial presentation of the amplitudes.

3.5. Investigations of accuracy

The evaluation of the frequency responses in the whole frequency range, in our case with 6 eigenfrequencies takes a high expenditure of time. Therefore, with this example investigations were made with the aim of reducing the number of degrees of freedom taken into consideration in each case. In the borderline case this means to analyse each eigen-

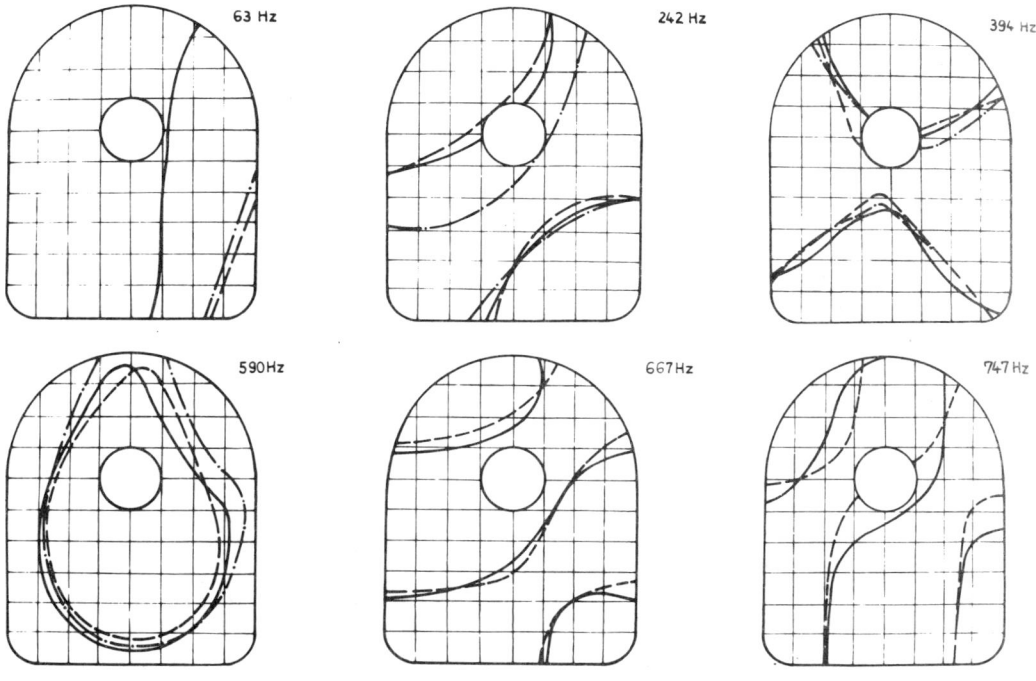

Fig.5. Nodal lines of different
modes
——— modal analysis
—·—·—harmonic excitation
————FEM (COSAR)

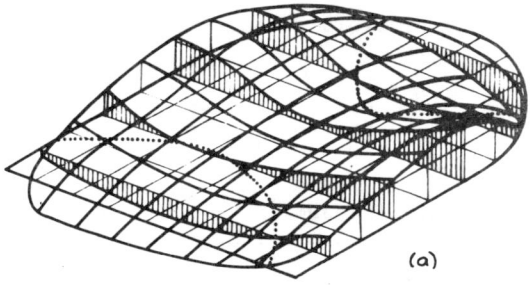

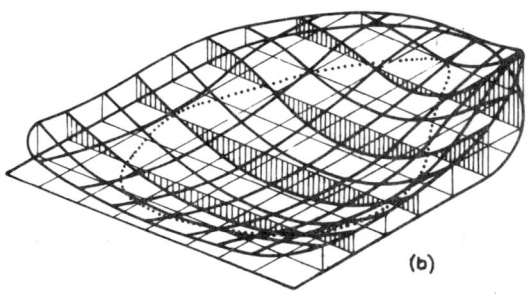

Fig.6. Measured modes of
394 Hz(a) and 590 Hz(b)

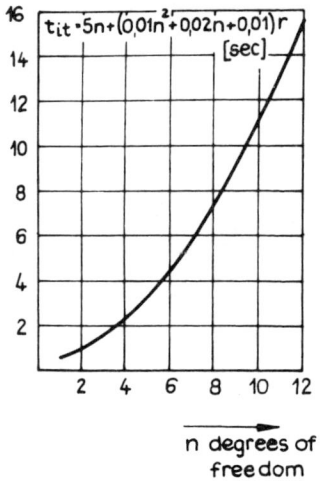

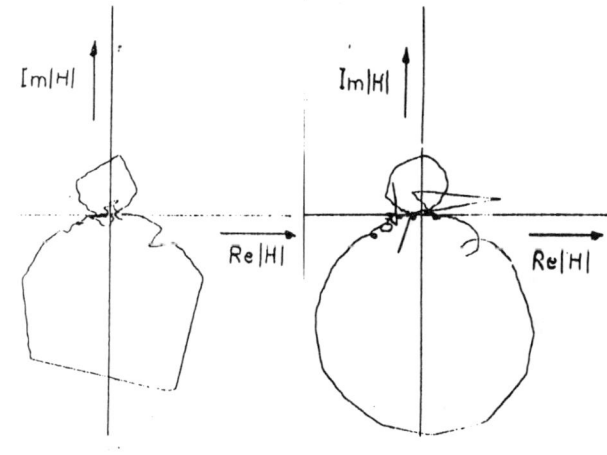

Fig.7. Computation time for
       one iteration
       (r=500 frequency lines)

Fig.8. Frequency response H$_{5853}$ in the
       range 512-768 Hz, 2 Hz(a) and
       0.5 Hz(b) frequency resolution

frequency separately, according to the single-degree-of-freedom method.

Fig. 7 shows the saving of time for evaluation of one (averaged) frequency response. The deviation of modal parameters due to this reduction was negligibly small (table 3).

Table 3:   Modal parameters calculated under conditions of
           different effective degrees of freedom, f = 394 Hz
           measuring point 70

degree of freedom	f	δ	u	v
5 (7 iterations)	394.20	13.57	-.214E-2	.737E-2
4 (5 iterations)	394.24	13.26	-.197E-2	.725E-2
2 (5 iterations)	394.17	13.57	-.226E-2	.736E-2
1 (5 iterations)	394.21	13.13	-.207E-2	.718E-2

The influence of a modification of the truncation constant for the iterations was investigated in a similar way. The truncation constant, i.e. the ratio of standard deviations (the sum of the remaining error squares) of the measured and analytical frequency response of successive

iteration steps in the complex plane, is predetermined in the programme
by 0,95. Dimishing this value to 0,5 did not change the modal parameters
in any way (table 4). The reason for this behaviour can be seen in the
quality of the frequency responses measured by means of pseudorandom
signals and in great distances between resonances.

Table 4:   Modal parameters calculated with different truncation
             constants. f = 667 Hz, n   =4 degrees of freedom.
                                     eff

truncation constant	f	f_H	u	v
0.95 (5 iterations)	667.37	4.62	-.824E-4	.193E-2
0.80 (5 iterations)	667.37	4.62	-.824E-4	.193E-2
0.50 (4 iterations)	667.37	4.62	-.821E-4	.193E-2
0.40 (4 iterations)	667.37	4.62	-.821E-4	.193E-2

The frequency bandwidth of the underlying spectra was 2 Hz, resulting
from a frequency range up to 1024 Hz with 512 discrete values in the
spectrum. A higher accuracy due to a higher frequency resolution is
possible by using narrow-band pseudorandom and zoom transformation in
the frequency ranges of interest. The results of the frequency resolu-
tions 2 Hz and 0,5 Hz, respectively, are compared in fig. 8. The Nyquist
plot is essentially improved. This method, proving its worth in the case
of narrowly neighbouring eigenfrequencies but demanding a longer calculation time, did not lead to improved results in the present case.

4. CONCLUSIONS

The results underline that the method of experimental modal analysis
is an effective means for determining the dynamic qualities of complicated structures . The effort of a single investigation of such a kind of
structures is essentally smaller than a calculation by FEM. On the basis
of a verified calculation model, the latter results in more advantageous
possibilities of investigating different variants. .
The eigenmodes point out that the plate is relatively soft in the

right upper part. The coincidence of the eigenmodes of interest is satisfactory. The eigenfrequencies show deviations up to 10 %. Here investigations aiming at a systematic improvement of the calculation model are
neccessary.

5. REFERENCES

1. Ewins, D.J. Modal Testing: Theory and Practice.
   Research Studies Press Ltd., John Wiley & Sons 1984

2. Natke, H.G. Einfuehrung in die Theorie und Praxis der Zeitreihen- und
   Modalanalyse. Vieweg Verlag Braunschweig; Wiesbaden 1983

3. Vasel, T. Die Modalanalyse mechanischer Schwingungssysteme mit dem
   Programmsystem ASAM. Wiss. Z. Techn. Hochsch. Magdeburg 30
   H. 7, S.102-105 (1986)

4. Gabbert, U., Berger, H., Zehn, M., Fels, D. Universelles FEM-Pro-
   grammsystem COSAR - Uebersicht ueber den nachnutzbaren Leistungsum-
   fang. Maschinenbautechnik 34, H. 8 S. 352 - 356 (1985)

ESTIMATION OF DYNAMIC SPRING AND
DAMPING PARAMETERS OF THE SUPPORTS OF A NUCLEAR REACTOR
BY MEANS OF AN ADAPTIVE METHOD IN TIME DOMAIN

A. Lingener
University of Technology Otto von Guericke, Magdeburg, GDR
S. Doege
Technical University, Dresden, G.D.R.

1. INTRODUCTION

In connection with the construction of powerful nuclear power stations in the GDR a dynamic model of the reactor should be derived, suitable for assessing measured data for diagnostics purposes. Therefore it was neccessary to find a structural setup and confirm and determine the parameters of the model.

Measurements could be made at two reactors during assembling in different assembly states.
The task involved the development of the method including programming as well as carrying out measurements and evidence of efficiency of the method. At the beginning, there was no experience with mechanical systems of such large dimensions (more than 200 t).

2. DESCRIPTION OF THE SYSTEM

A principal scheme of the reactor vessel is shown in fig. 1. The vessel with a total mass of more than 200t is supported below the pipe

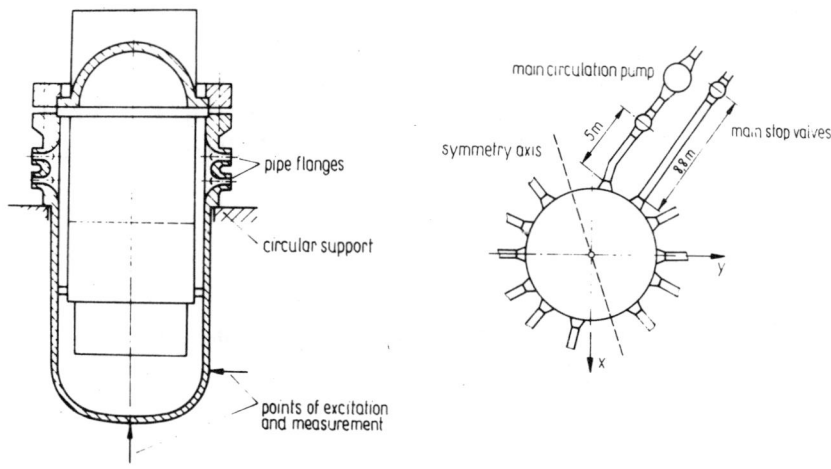

Fig.1. Principal scheme of the reactor vessel

flanges on a circular support. Its centre of gravity is close to the ring
plane. The twelve pipes leading to the six heat exchangers excert an
influence on the supporting stiffness with the consequence that the
support of the reactor vessel is not rotationally symmetrical.

3. METHOD OF SOLUTION [1,2]

    From preliminary investigations such as statical loading, frequency
analysis of response functions with a very good acccuracy the supposition
of a rigid body model could be confirmed . A system of such large dimensions like a reactor vessel can be excited practically only by means of
a step excitation realized by fracture elements. The vessel was excited
in horizontal and vertical direction at two different points (fig. 1).
The displacements were measured at the same points.

    With the aim to determine the starting estimates a rotationally
symmetrical model was assumed. Based on this two possible models a four-
and a three-point supported vessel were investigated.Such models allow

due to decoupling the differential equations a simple
parameter estimation.

The estimation of spring
and damping parameters is
carried out by means of an
adaptive method, comparing
the response function of a
mathematical model with the
measured data.

The resulting unsym-
metrically supported rigid-
body-model is used later as a
basis for interpreting vibration measurements during the
operation of the reactor
vessel.[3]

4. THE METHOD OF PARAMETER ADJUSTMENT

The method of parameter
adjustment used here compares
and minizes the sum of squares of differences between
discrete values of the
measured time signal and
corresponding values of a
computational model. The basic
principle is shown in fig. 2.
The differential equation
system of a general rigid-
body-model (fig. 3) consists
of 6 coupled differential
equations. x, y, z at the same

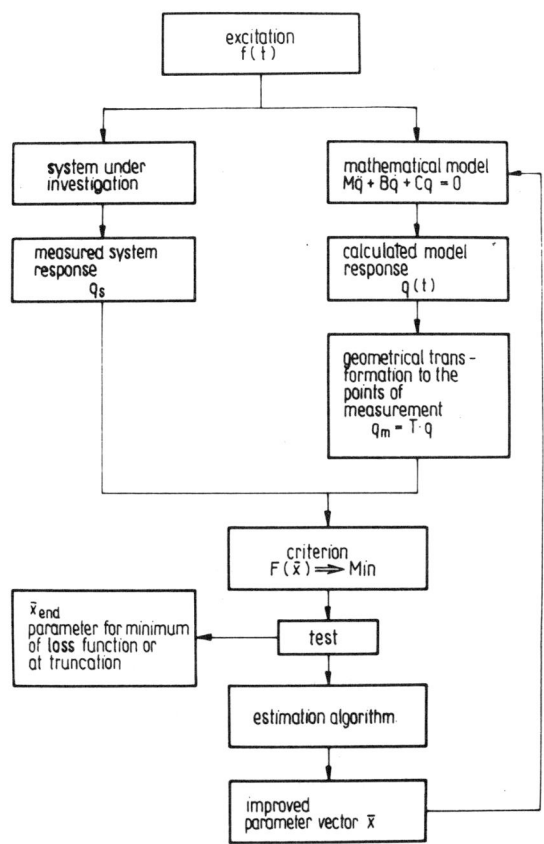

Fig.2. Principle of the adaptive method for the adjustment of the model parameters

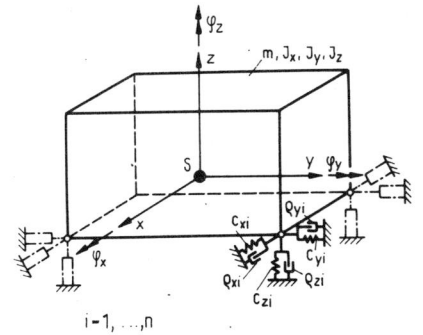

Fig.3. General rigid body model

time are the principal axes of inertia.

Let φ_x, φ_y, φ_z the rotational angles related to these axes. Then the differential equations with the vector of coordinates $q = (x, y, z, \varphi_x, \varphi_y, \varphi_z)^T$ are as follows:

$$M\ddot{q} + B\dot{q} + Cq = f(t) \tag{1}$$

In this equation the mass matrix M is diagonal, the stiffnes matrix C is symmetrical as follows:

$$
C = \begin{bmatrix}
\Sigma c_{xi} & 0 & 0 & 0 & \Sigma c_{xi} l_{xzi} & -\Sigma c_{xi} l_{xyi} \\
 & \Sigma c_{yi} & 0 & -\Sigma c_{yi} l_{yzi} & 0 & \Sigma c_{yi} l_{yxi} \\
 & & \Sigma c_{zi} & \Sigma c_{zi} l_{zyi} & -\Sigma c_{zi} l_{zxi} & 0 \\
 & & & \Sigma c_{zi} l_{zyi}^2 + \Sigma c_{yi} l_{yzi}^2 & -\Sigma c_{zi} l_{zyi} l_{zxi} & -\Sigma c_{yi} l_{yzi} l_{yxi} \\
 & & & & \Sigma c_{zi} l_{zxi}^2 + \Sigma c_{xi} l_{xzi}^2 & -\Sigma c_{xi} l_{xyi} l_{xzi} \\
 & & & & & \Sigma c_{yi} l_{yxi}^2 + \Sigma c_{xi} l_{xyi}^2
\end{bmatrix}
$$

i marks the spring (i = 1, 2, 3 or i = 1, ...,4), the l are the distances between the points of spring application and the centre of gravity. The damping matrix is set up analogously. The exciting vector depends on the point of the transient excitation.

The solution vector q(t) of the mathematical model has to be transformed by a geometrical transformation to the measuring points of the real system (the vessel). As a result we find the calculated model response q_m. The measured response of the vessel is q_s. The solution of (1) was carried out by means of recursive numerical integration. The time step 0.008 s of integration coincides with the time step of discretisation of the measured time function. The differences between the discrete values of q_m and q_s form a vector $\Delta q = q_m - q_s$, containing as much elements as discrete values result from measurement and calculation respectively.

As loss functions F(x) of minimization were used two functions:

1. Weighted least squares (weighting matrix W, later W=I)

$$F(\bar{x}) = \Delta q^T W \Delta q$$

2. The product of a weighting vector W with the vector $|\Delta q| = |q_m - q_s|$:

$$F(\bar{x}) = W|\Delta q|$$

$\bar{x} = (x_1, x_2, \ldots, x_n)$ is the vector of the variables of the model. From 4 springs (with 3 parameters each), 4 dampers and 6 mass or inertia result 30 parameters as a whole.

As the coordinates q_M of the mathematical model depend in a nonlinear way on the parameters of the model a nonlinear adjustment problem results.

The numerical processing uses a gradient method, introduced by Zettl.[5]

Additionally to the amount of the loss function the coincidence of eigen-frequencies of the model and the measured values was checked for supporting structural suppositions and assessment of the result of parameter adjustment. For this purpose, available programmes were used.[6]

The adjustment algorithm demanded and made it possible to take into account numerous computational specialities, allowing an economical treatment of the problem.

5. DETERMINATION OF START VALUES

Before starting the calculations, the parameter vector $\bar{x}$, being expected to converge towards the solution, has to be occupied with start values.

a) Spring parameters from a test using free vibrations

Assuming a symmetrical model and that mass parameters and eigenfre-quencies are known, the spring parameters can be determined easily under special conditions.

For instance, under condition of double symmetry e.g. related to the xz and yz-plane, the equations (1) are decoupled into two independent sys-tems of equations, each consisting of two equations, and two totally decoupled equations. The corresponding determinants are

$$
\begin{vmatrix}
\Sigma c_{xi} - m\omega^2 & \Sigma c_{xi} l_{xzi} \\
\Sigma c_{xi} l_{xzi} & \Sigma c_{zi} l^2_{zxi} + \Sigma c_{xi} l^2_{xzi} - J_y \omega^2
\end{vmatrix} = 0 \qquad (2)
$$

$$
\begin{vmatrix}
\Sigma c_{yi} - m\omega^2 & -\Sigma c_{yi} l_{yzi} \\
\Sigma c_{yi} l_{yzi} & \Sigma c_{zi} l^2_{zyi} + \Sigma c_{yi} l^2_{yzi} - J_x \omega^2
\end{vmatrix} = 0 \qquad (3)
$$

$$
\Sigma c_{zi} - m\omega^2 = 0 \qquad (4)
$$

$$
\Sigma c_{yi} l^2_{yxi} + \Sigma c_{xi} l^2_{xyi} - J_z \omega^2 = 0 \qquad (5)
$$

The eigenfrequencies which are neccessary for the determination of the spring parameters, may be derived from a spectral analysis of the measured time functions.

The question of associating of the measured eigenfrequencies to those of a symmetrical model in such a case cannot be decided unambiguously, such that more than one possibility of the start vector could result. Under these suppositions from the equations (2)...(5) may be derived simple formulae for the spring parameters from which the numerical values of a symmetrical model can be determined.

b) Spring parameters from static investigations

Under the supposition of a linear system and if a sufficient number of static measurements is available the spring parameters may be determined directly from the stiffness matrix C in eq. (1) An indirect determination of static spring parameters beginning with a start vector by adapting it to the measured values by means of a similar method as in the case of the dynamical model is possible. Corresponding programmes are available.[6]

c) Damping parameters

From experience follows that the determination of start values for damping parameters is difficult. As our problem could not be treated

without damping, the value resulting from the application of the well-
known formulae of damping determination of one-degree-of-freedom-systems
to the test using free vibrations was uniformly supposed.

6. PROBLEMS OF MEASUREMENTS

The force, causing the displacement
of the vessel is applied by a hydraulic cylinder. The force is
applied by means of a notched greycast iron bolt, breaking under a
load of maximum 300kN. In this way,
under horizontal loading displacements of about 0,5 mm were pro-
duced, under vertical loading only
about 0,1 mm. The loading device is
shown in fig. 4. During the free
vibrations after the breaking of
the greycast iron bolt measurements
were made at different points of
the vessel in horizontal and verti-

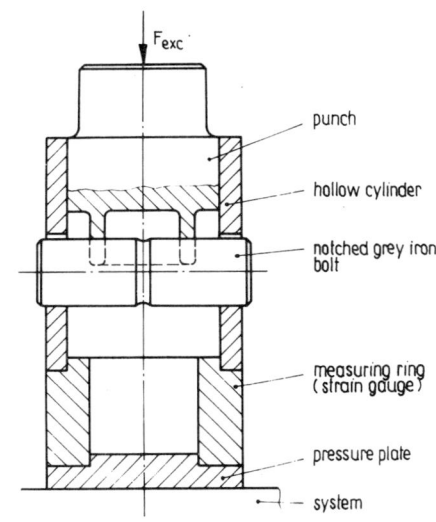

Fig.4. Loading device

cal direction. Displacement was fixed as measuring quantity. The measurement was carried out with non-contacting inductive sensors, measuring at
the same time the static displacement. Parallel acceleration measurements
were accomplished by means of piezoelectric transducers. All measuring
signals, including the force were stored as analogue signals on a tape.

The signals were sampled with 125Hz sampling frequency, as the resulting upper limiting frequency was high enough. The highest frequency of
interest of the vessel was nearing 50Hz. The positions of force application and measuring points are given in fig. 1.

The measuring programme included a great number of measurements at
two reactor vessels during assembly. In this way, the influence of different installations and the water level could be determined. As a whole,
were carried out 38 different measurements.

7. RESULTS OF MEASUREMENTS

Principally during each test two
time functions were measured:
The displacement q_o in the direction of force and perpendicular to
that direction q_{90}. Examples of
measured time functions are presented in fig. 5. From the measured
time functions were derived amplitude spectra, the characteristic
shape of which is presented in
fig. 6.

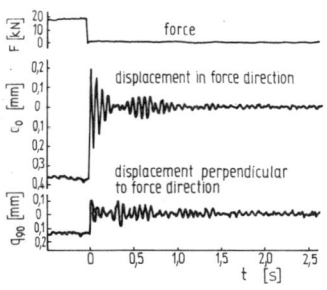

Fig.5. Signals of force and
displacement caused by
horizontal force

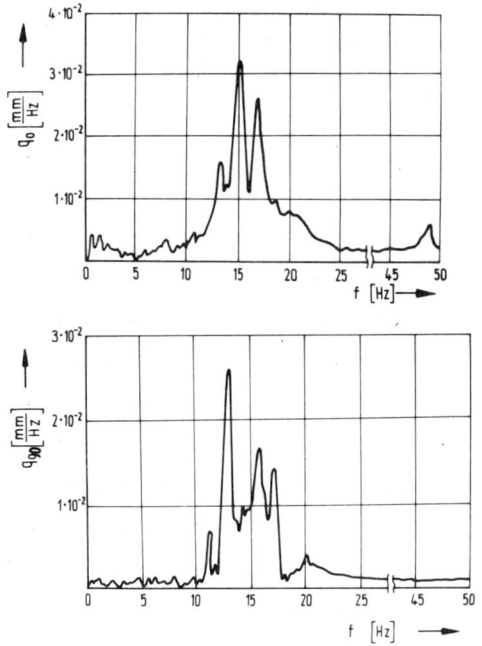

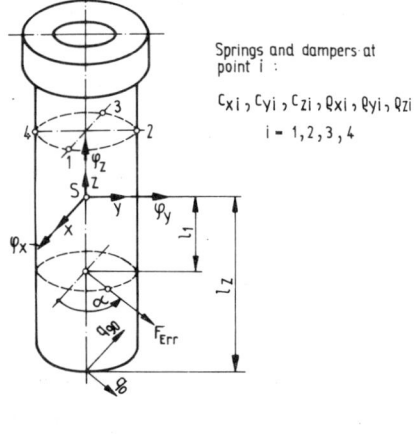

Springs and dampers at
point i :

$c_{xi}, c_{yi}, c_{zi}, \varrho_{xi}, \varrho_{yi}, \varrho_{zi}$

$i = 1,2,3,4$

Fig.6. Absolute value of amplitude
spectrum of measured
displacements

Fig.7. Rigid body model with
4 supporting points

8. PARAMETER ADJUSTMENT

The method is presented for the case of a horizontal displacement of the empty vessel. The model with 4 supporting points (decoupling in the case of symmetry) is shown in fig. 7. If the dominating frequencies in the spectra (fig. 6.) are taken as eigenfrequencies of a symmetrically supported rigid-body-model, than the equations (2) - (5) lead to a start-model 1 (horizontal motion). The numerical values of the parameters are given in table 1.

Using these values of the parameters, the time-dependent displacements of the different models 1 - 7 can be calculated.

The result (fig. 8) shows that the measured time function is represented very imperfectly by the start-model 1.

Model 2 resulted from the deliberation that in the vertical measurements the eigenfrequencies of vertical vibrations shall dominate. Now for economical reasons, in the next step the static displacement of the models 1 and 2 were adjusted, using the method mentioned in paragraph 5b.

The resulting models are 3 and 4. These models describe the real behaviour much better than model 1 (fig. 8), but not good enough as the springs remain symmetrically.

To demonstrate the advantage of this procedure, the model 1 was immediately adapted to the measured function. The result, model 5, is nearly the same (fig. 8) but the calculation time was 36 times more than for the models 3 and 4. Finally, the damping parameters were adjusted. From model 4 resulted model 6, from model 5 - model 7 (fig. 9)

These models guarantee a sufficiently exact description of the measured time function and good coincidence of eigenfrequencies. In all cases $F(\bar{x}) = W|\Delta q|$ with $W = I$ was used as aim functional.

An adjustment with the least square function was carried out too. The results, which are not presented in this paper were not better than with the method used here.

In consequence of the incomplete identification (only one input-output relation) the measured beating is not described by the models (fig.9). All models represent only the function q_o as the essential vibrational characteristics during the operation of the vessel. Table 2 contains the

Table 1. Numerical parameter values of different starting and improved models

model	1	2	3	4	5	6	7
c_{x1} [10^8N/m]	4,9	2,7	5,0965	3,6151	2,1945	2,4144	2,0179
c_{x2} "	4,9	2,7	5,0965	3,6151	3,4078	2,3329	2,5087
c_{x3} "	4,9	2,7	5,0965	3,6151	1,8612	2,2863	1,7996
c_{x4} "	4,9	2,7	5,0965	3,6151	2,6828	2,271	2,682
c_{y1} "	3,55	5,5	3,9195	6,276	9,1253	6,0063	10,052
c_{y2} "	3,55	5,5	3,9195	6,276	5,7531	6,0692	5,972
c_{y3} "	3,55	5,5	3,9195	6,276	5,0949	5,8557	5,387
c_{y4} "	3,55	5,5	3,9195	6,276	6,508	6,0692	7,0393
c_{z1} "	30,5	4,9	39,859	31,156	36,438	36,017	35,503
c_{z2} "	30,5	4,9	47,417	51,418	53,065	58,108	56,606
c_{z3} "	30,5	4,9	39,859	31,156	36,212	36,017	36,911
c_{z4} "	30,5	4,9	47,417	51,418	53,06	58,11	54,531
ϱ_{x1} [10^6Ns/m]						1,6584	1,3841
ϱ_{x2} "		for all models				1,6768	1,3615
ϱ_{x3} "						1,6507	1,3374
ϱ_{x4} "		m = 2.1 10^5kg				1,6602	1,3447
ϱ_{y1} "						1,2879	1,3592
ϱ_{y2} "		$J_x = J_y = 3.3217\ 10^6$ kgm^2				1,2879	1,3207
ϱ_{y3} "						1,2556	1,3611
ϱ_{y4} "		$J_z = 7.074\ 10^5$ kgm^2				1,2556	1,3611
ϱ_{z1} "						1,2333	1,3487
ς_{z2} "		for all models 1-5 uniformly				1,2871	1,3062
ϱ_{z3} "						1,2333	1,3512
ϱ_{z4} "		$Q = 1.285\ 10^6$ Ns/m				1,2871	1,2961

Table 2: Eigenfrequencies of different rigid-body-models and residual errors

model	f 1 [Hz]	f 2 [Hz]	f 3 [Hz]	f 4 [Hz]	f 5 [Hz]	f 6 [Hz]	Y min $\frac{mm}{100\,sample}$
1	11,48	12,00	15,01	15,04	16,92	38,31	18,63
2	5,04	5,07	11,87	14,78	15,25	16,89	232,03
3	12,87	13,31	15,51	17,60	17,81	45,84	6,66
4	11,59	14,97	15,20	16,26	19,99	44,59	5,07
5	10,12	14,76	14,79	15,29	20,34	46,40	3,97
6	10,12	14,76	14,79	15,29	20,34	47,61	3,65
7	9,97	14,91	15,38	16,35	21,48	47,01	3,38
meas.	(7,80)	13,18	15,38	16,11	19,80	(49,80)	

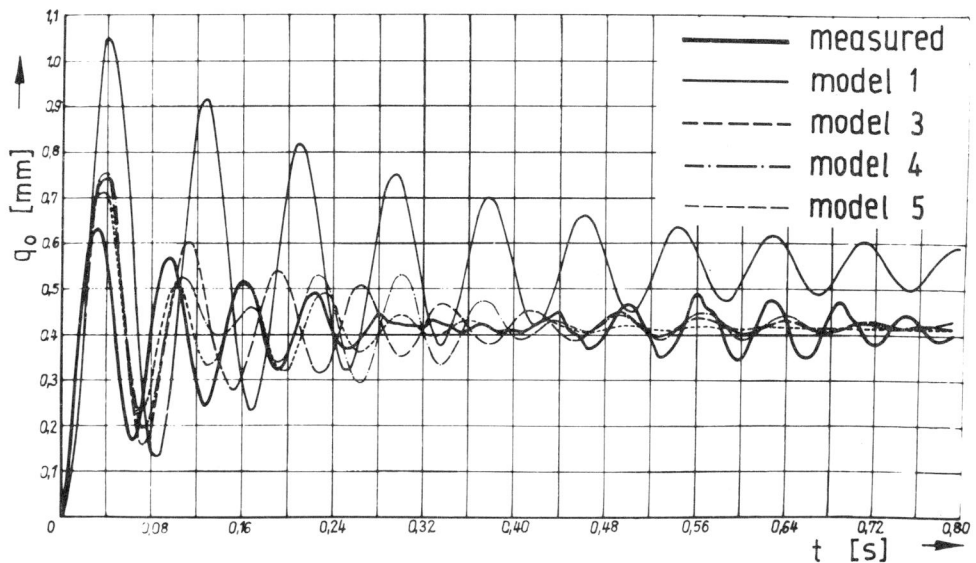

Fig.8. Measured and calculated displacements for the models 1,3,4,5

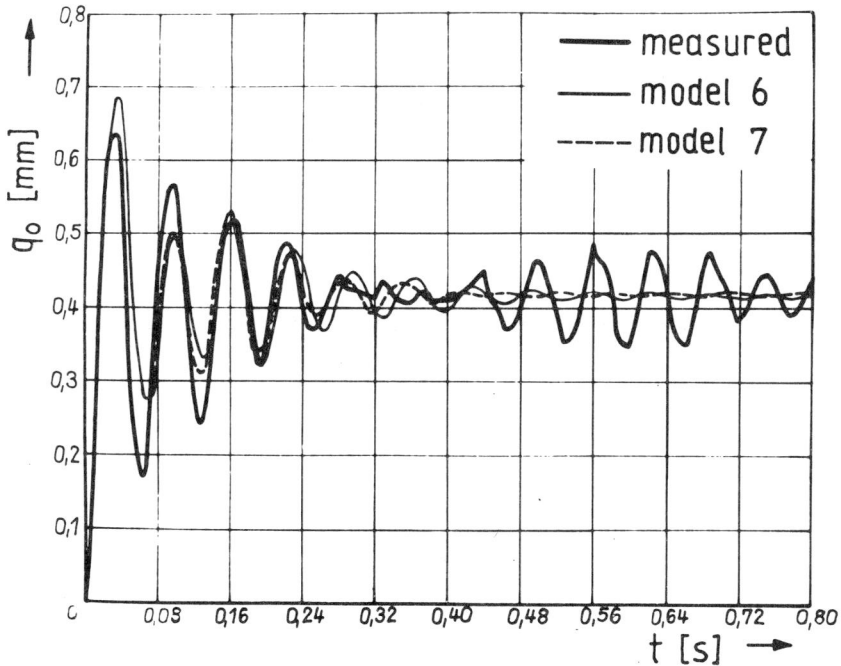

Fig.9. Displacements measured and calculated for the models 6 and 7

eigenfrequencies of the different models and the residual error related
to the measured values and 100 samples i.e. 0,8 s.

Analogous investigations were made for the case of three supporting
points. Furthermore, the completely assembled vessel, ready for operation was investigated. The different assembly states differed mainly by
their eigenfrequencies.

Finally, to show the usefulness of the rigid-body model an extended
model with 30 degrees of freedom was calculated taking into account the
bending stiffnes of the mounted pipes connecting the vessel with the main
circulation pumps and the main stop valves. These results emphasized that
the eigenfrequencies in the frequency range between 7 and 20 Hz are not
changed essentially.

But the last and deciding criterion had to be the verification of
the rigid-body-model during the operation of the reactor vessel.

9. USEFULNESS OF THE RIGID-BODY-MODEL

The results of identification and parameter determination are the
basis for measurements and their evaluation during operation. A reliable
model allows a safe interpretation of measurements and a corresponding
diagnosis.

Table 3 compares the eigenfrequencies of the model with those resulting
from frequency analysis of the vessel vibration during operation.

Table 3: Eigenfrequencies of an adapted rigid-body-model and
         frequencies during operation [Hz]

	f_1	f_2	f_3		f_4	f_5		f_6
theoretical	6,9	8,3	11,3		16,2	18,7		47,9
experiment.		8,9	11,2	12,7	15,0	18,9	24,6	

These results allow the following conclusions.

- The rigid-body-model is sufficient for the description of reactor vessel vibrations as a whole.

- The frequency 24.6 Hz corresponds to the operational speed of the main circulation pumps and is no eigenfrequency.

- The frequency 11.2 Hz is that with the greatest share of vibrational power; it represents a mode, which is fixed by the symmetry plane of the whole primary circuit. From this follows, that the mounted pipes are of essential influence on the stiffness of the supports.

- The frequency 12.7 Hz cannot be associated to a vessel frequency. Further investigations have shown, that this frequency occurs above all at the flange of the vessel and that the coherence with other measuring points is very low.

It is supposed, that the frequency 12.7 Hz concernes the so-called reactor shaft vibrations [7] which cannot be described by a rigid-body-model. Here an extended model would be required, which takes into consideration the internal structure of the vessel.

10. REFERENCES

1. Doege, S. Bestimmung der dynamischen Parameter von Starrkoerpersystemen mit Hilfe der Methode der adaptiven Modellfindung. Maschinenbautechnik 30 S. 399-402 (1981)

2. Doege, S. Parameterbestimmung fuer elastisch gelagerte Starrkoerper mit Hilfe eines adaptiven Modellfindungsverfahrens. Dissertation TU Dresden 1983

3. Grunwald, G. u.a. Schwingungsuntersuchungen am Reaktordruckgefaess WWER-440. Kernenergie 28, 1, S. 18-26 (1985)

4. Trujillo, D.M. The Direct Numerical Integration of Linear Matrix Differential Equations Using Pade Approximation. Internat. J. Num. Math. Engng. 9 (1975)

5. Zettl, G. Ein Verfahren zum Minimieren einer Funktion bei eingeschraenktem Variationsbereich der Parameter Numerische Mathematik 15 (1970)

6. Hardtke, H.J. Programmsystem SYSKEN, TU Dresden, Sektion 13 Programmbibliothek

7. Bastl, W. Measuring and Analysis Methods applied to On-Line Vibration and Noise Monitoring in PWR Power Plants. Nucl. Engng. Design 28 S. 377 (1974)

OPTIMIZATION OF A REDUCED TORSIONAL MODEL USING A PARAMETER IDENTIFICATION PROCEDURE

P. Schwibinger, R. Nordmann
University of Kaiserslautern, Kaiserslautern, F.R.G.

INTRODUCTION

Large steam turbine-generators in operation may be stimulated to torsional vibrations by dynamic moments at the generator due to electrical system transients, Fig. 1. The induced torsional stresses have grown growing attention over the past few years /1-3/.

For the solution of the torsional vibration problem it is essential to find an appropriate torsional model for the turbine-generator shaft. A common approach is to model the torsional system by the finite element method.

For some applications it is desirable to have a torsional model with a reduced number of degrees of freedom (DOF), which reproduces the finite element model only in the lower eigenfrequencies and modes. One example, where a reduced torsional model is necessary, are torsional monitoring systems for the supervision of turbogenerators in operation.

A reduced torsional model is also required for the analysis of certain
network disturbances, where the nonlinear coupling between the mechanical (= torsional vibration system, Fig. 1) and the electrical system
(= generator and network) must be considered in the torsional analysis.
Therefore this paper describes a method how to find a most accurate reduced torsional model with discrete masses and springs from the finite
element model.

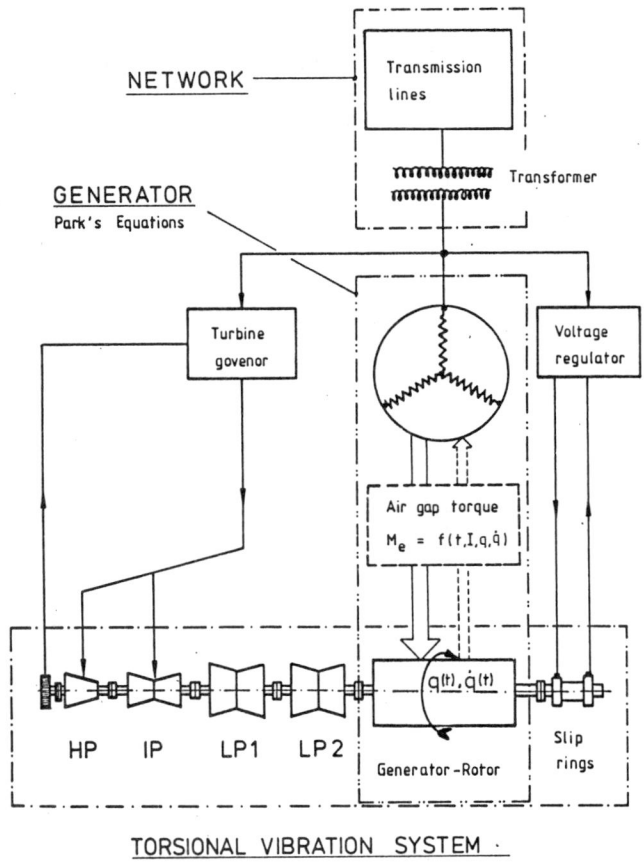

Fig. 1   Subsystems for calculating the effects of electrical system
         faults

FINITE ELEMENT MODEL

Modelling of Turbogenerator Shaft. In the first step we transform
the real system in an equivalent mechanical model. Therefore the turbine-generator shaft is subdivided in N-1 finite torsional elements,
Fig. 2. We consider small - that means linearized - torsional vibrations
about a stationary rotation of the shaft.

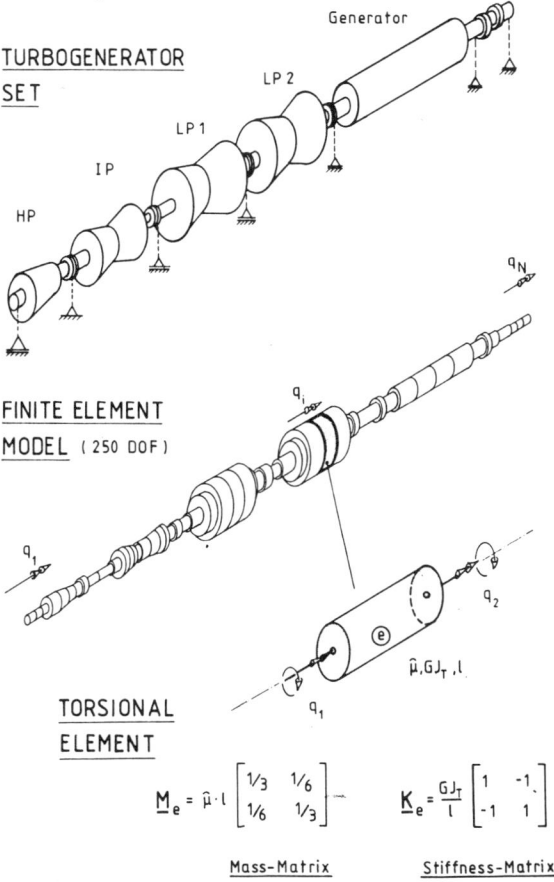

$$\underline{M}_e = \hat{\mu} \cdot l \begin{bmatrix} 1/3 & 1/6 \\ 1/6 & 1/3 \end{bmatrix} \qquad \underline{K}_e = \frac{GJ_T}{l} \begin{bmatrix} 1 & -1 \\ -1 & 1 \end{bmatrix}$$

Mass-Matrix   Stiffness-Matrix

Fig. 2 Transformation of the turbogenerator in a mechanical model

By using the principle of virtual work we can describe the dynamic
behavior of our model mathematically; we get the mathematical model.

$$\delta W = \underbrace{\sum_{e=1}^{N-1} \delta W_i^e}_{\text{virtual work}} + \underbrace{\sum_{e=1}^{N-1} \delta W_T^e}_{\text{virtual work}} + \underbrace{\sum_{e=1}^{N-1} \delta W_F^e}_{\text{virtual work}} = 0 \qquad (1)$$

of elastical moments of inertia moments of external moments

If we approximate the unknown torsional displacement in an element with
static deflection functions, we obtain a 2x2 mass- and stiffness-matrix,
which describe its dynamic behavior. Each element has two nodes with one
angular DOF each. Superposition of the element matrices yields the mass-
and stiffness-matrix for the complete system $\underline{M}$ and $\underline{K}$. The damping matrix
is assumed as proportional to $\underline{M}$ respectively $\underline{K}$.

$$\underline{D} = \alpha \, \underline{M} + \beta \, \underline{K} \qquad (2)$$

Modal Parameters of the Torsional Problem. The equations of motion are solved for the natural vibrations, we get form the homogeneous
equations of motion.

$$\underline{M} \, \ddot{\underline{q}} + \underline{K} \, \underline{q} = \underline{0} \qquad (3)$$

Assuming a solution of the form

$$\underline{q}(t) = \underline{\varphi} \, e^{\lambda t} \qquad (4)$$

we obtain the eigenvalue problem

$$(\underline{K} - \bar{\lambda}^F \underline{M}) \, \underline{\varphi} = \underline{0} \qquad \text{with } \bar{\lambda}^F = -\lambda^2 \qquad (5)$$

Eigenvalues $\bar{\lambda}_i^F$ Modes $\underline{\varphi}_i$

REDUCTION OF THE FINITE ELEMENT MODEL

If we discretisize e.g. a 600 MW turbine-generator in about 250 torsional elements the described finite element model is a very good model for the dynamical behavior of the shaft. Why do we need an additional reduced model?

A reduced model for the turbogenerator is needed for the analysis of electrical transients, in which the coupling of the torsional shaft model with the electrical network is remarkable /1,3/. Because the model of the electrical system contributes additional variables to the problem and makes it highly nonlinear, a reduced torsional model is essential to limit the number of DOF for the coupled electrical-mechanical model. Another example where a reduced torsional model is often used are torsional monitoring system for the supervision of turbogenerators in operation.

Reduction Algorithm. There exist different methods to reduce the systems DOF; e.g. static-, dynamic- and modal-condensation. Especially for torsional chains a very effective reduction algorithm can be developed using an electrical analogous model for the shaft train /4-6/. To apply the reduction algorithm to the finite element model, in a first step we have to transform it to a lumped mass model, Fig. 3. This is done by an equal distribution of the mass of an element to its nodes. In a second step we reduce this large lumped mass model with an electrical-mechanical analogy method to a reduced discrete mass model, we'll call further on reduced model.

This reduced model differs in its eigenfrequencies from the accurate finite element model. Therefore it is our goal to improve the reduced torsional modell by fitting it to the finite element model with many DOF. The fitted reduced model we'll call further on improved reduced model.

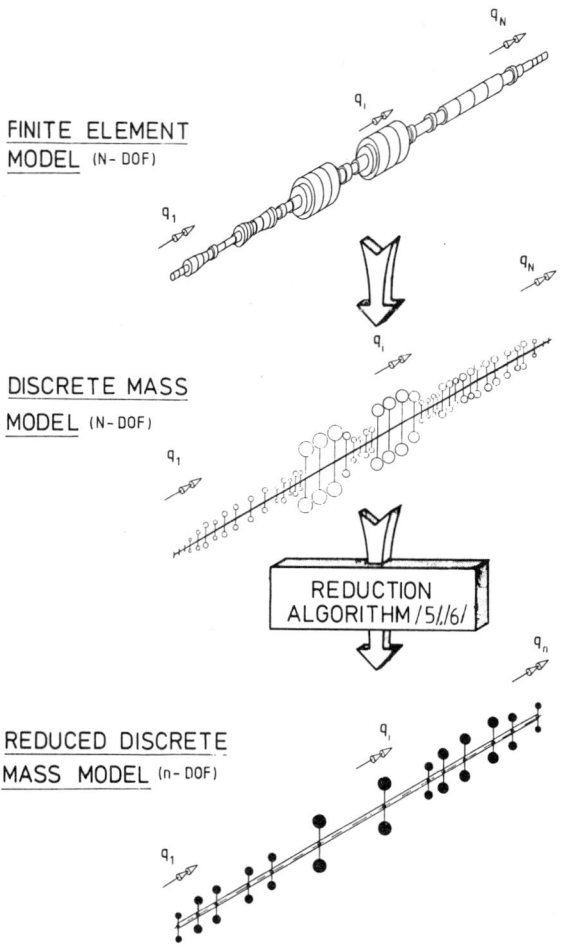

FINITE ELEMENT
MODEL (N- DOF)

DISCRETE MASS
MODEL (N- DOF)

REDUCTION
ALGORITHM /5/,/6/

REDUCED DISCRETE
MASS MODEL (n- DOF)

Fig. 3 Reduction of the finite element model of a turbogenerator
        to a discrete mass model

Improvement of the Reduced Model.

    Basic Idea. The fitting of the reduced model is performed by an
indirect parameter identification method /7,8/. Fig. 4 shows the basic
idea: On the one side there is the finite element model with its eigen-
values $\bar{\lambda}_i^F$. On the other side the reduced model yields the eigenvalues $\bar{\lambda}_i$.
In general they differ from the solutions of the finite element model.
The goal is to improve the parameters of the reduced model (e.g. its

stiffnesses) in a way, that they correspond better with the solutions
of the finite element model. The applied algorithm improves the parameters by minimizing the difference between the eigenvalues of the reduced model and the finite element model iteratively.

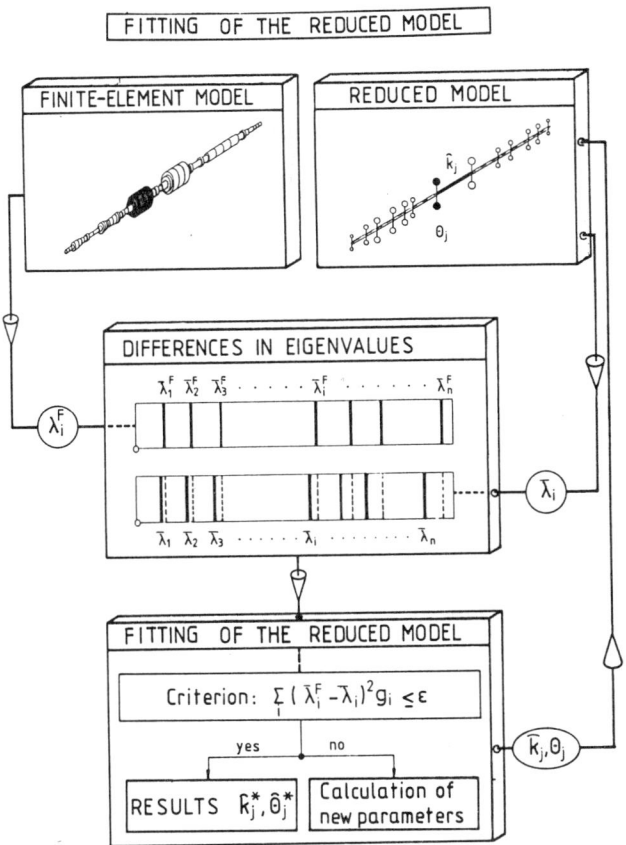

Fig. 4   Fitting of the reduced model by means of the eigenvalues

Because the masses and its distribution along the shaft of the reduced model looks physically sensible compared to the turbine-generator
shaft, the mass matrix $\underline{M}$ of the reduced model is supposed to be correct.
Damping is not taken into account for the fitting process. In the calculations for the described torsional system damping is introduced by

'modal damping factors' (2) after the reduction and fitting process.
The identification procedure can therefore be limited to the torsional
stiffnesses arranged in the stiffness matrix $\underline{K}$.

Derivation of the Identification Algorithm. We partition the
stiffness matrix $\underline{K}$ in submatrices $\underline{K}_j$ with the corrective factors a_j.

$$\underline{K} = \sum_{j=1}^{m} a_j \underline{K}_j \tag{6}$$

For the torsional chain, we take preferably the springs between the individual rotational masses as substructures. Fig. 5 shows two reasonable
possibilities to define substructures in the reduced model: In one case
every individual spring forms an independent substructure; in the other
case two adjacent springs are summarized in one submatrix. The following
example of application will show, which kind of substructuring proves
to be the better one.

The eigenvalue problem of the reduced model yields its eigenvalues

$$(\underline{K} - \overline{\lambda}\,\underline{M})\,\underline{\varphi} = \underline{0} \tag{7}$$

$\overline{\lambda}_i$ and the corresponding eigenvectors $\underline{\varphi}_i$. If we normalize the generalized mass equal one and partition the stiffness matrix (6), we get from
(7) the eigenvalues as a function of the corrective factors a_j.

$$\overline{\lambda}_i = \underline{\varphi}_i^T \left(\sum_{j=1}^{m} a_j \underline{K}_j \right) \underline{\varphi}_i \tag{8}$$

It is our goal to determine the corrective factors minimizing the difference ('residuals') between the eigenvalues $\overline{\lambda}_i^F$ of the finite element
model and the corresponding eigenvalues $\overline{\lambda}_i$ of the reduced model.

$$v_i = \overline{\lambda}_i^F - \overline{\lambda}_i = \overline{\lambda}_i^F - \underline{\varphi}_i^T \left(\sum_j a_j \underline{K}_j \right) \underline{\varphi}_i \tag{9}$$

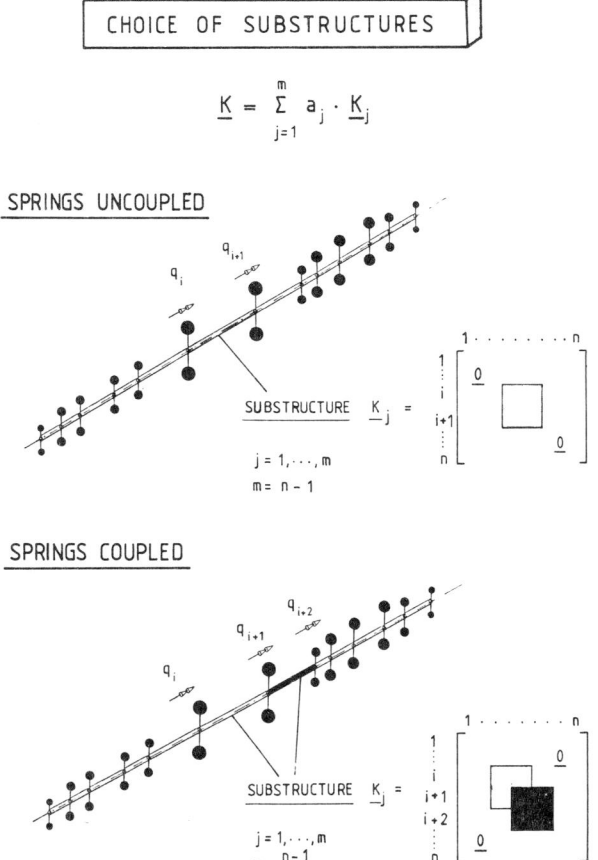

Fig. 5 Choice of substructures for the fitting process

Considering several eigenvalues, the residuals can be arranged in the vector $\underline{v}$ and the corrective factors in the vector $\underline{a}$. The least square method leads with the weight matrix $\underline{W}$ to an equation system, which can be solved iteratively for the unknown corrective factors $\underline{a}$.

$$D\underline{v}^T \, \underline{W} \, D\underline{v} \, \underline{a} \;\; = \;\; D\underline{v}^T \underline{W} \, \underline{b} \tag{10}$$

$D\underline{v}$ is the sensitivity matrix with the generalized stiffnesses of the submatrices.

$$Dv_{ij} = \underline{\varphi}_i^T \underline{K}_j \underline{\varphi}_i \tag{11}$$

The partial derivative

$$-\frac{\partial v_i}{\partial a_j} = \frac{\partial \bar{\lambda}_i}{\partial a_j} = \underline{\varphi}_i^T \underline{K}_j \underline{\varphi}_i \tag{12}$$

points out the sensitivity of the eigenvalue $\bar{\lambda}_i$ with regard to the sub-matrix $\underline{K}_j$. Fig. 6 shows the first row of the sensitivity matrix Dv: the influence coefficients of the first eigenvalue to the torsional stiff-nesses of the shaft. In this case every torsional spring between two masses is considered as one independent substructure. Obviously the first eigenvalue is strongly influenced by the coupling stiffness bet-ween the two low pressure turbines (LP1-LP2) while the outer springs have nearly no effect. This becomes clear with the associated eigenvec-tor, which has its main deflection (max. gradient) in the middle of the shaft.

The vector $\underline{b}$ contains the eigenvalues of the finite element model.

$$b_i = \bar{\lambda}_i^F \tag{13}$$

As (11) depends on the eigenvalues of the fitted model, (16) must be solved together with the eigenvalue problem (7). The procedure is sum-marized in the scheme of Fig. 7.

Analysis Results and Discussion. The torsional finite element mo-del of a 600 MW turbogenerator with 250 DOF is reduced to a 13 DOF dis-crete mass model by the electrical-mechanical analogy method. The fit-ting procedure of Fig. 7 is then applied to this reduced model to im-prove its torsional stiffnesses.

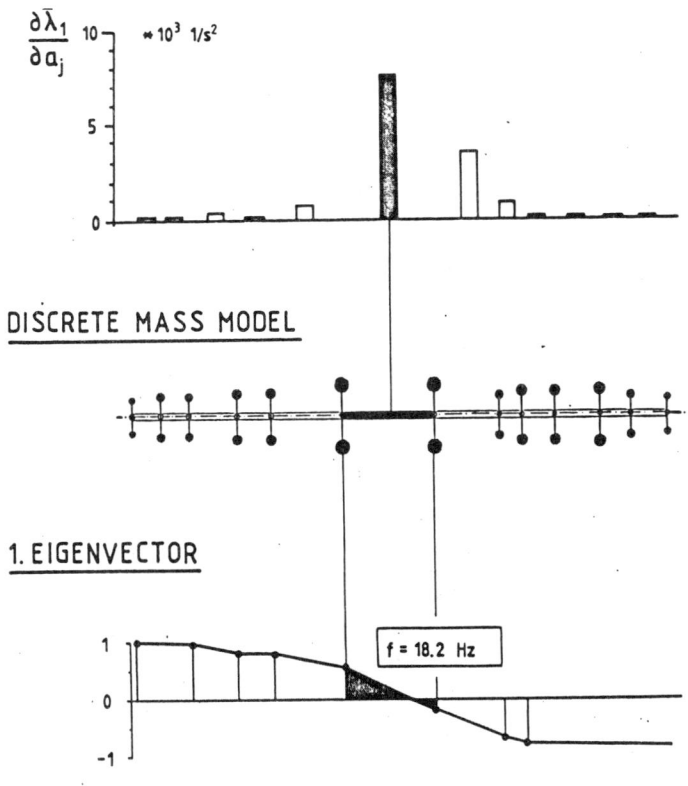

RELATIVE SENSITIVITIES

of 1.eigenvalue to torsional stiffnesses

$\frac{\partial \bar{\lambda}_1}{\partial a_j}$ $\approx 10^3 \, 1/s^2$

DISCRETE MASS MODEL

1. EIGENVECTOR

f = 18.2 Hz

Fig. 6   Relative sensitivities of first eigenvalue to change
of torsional stiffnesses

Modal Parameters. Fig. 8 shows that the eigenfrequencies of the
reduced model are all to low compared to the finite element model. The
error increases in the higher frequencies, e.g. is about 18 % in the
eleventh eigenfrequency. It has to be noted that the quality of the reduced model is not the best possible by the reduction algorithm with
the electrical-mechanical analogy method. It is nevertheless used to
show the efficiency of the fitting procedure.

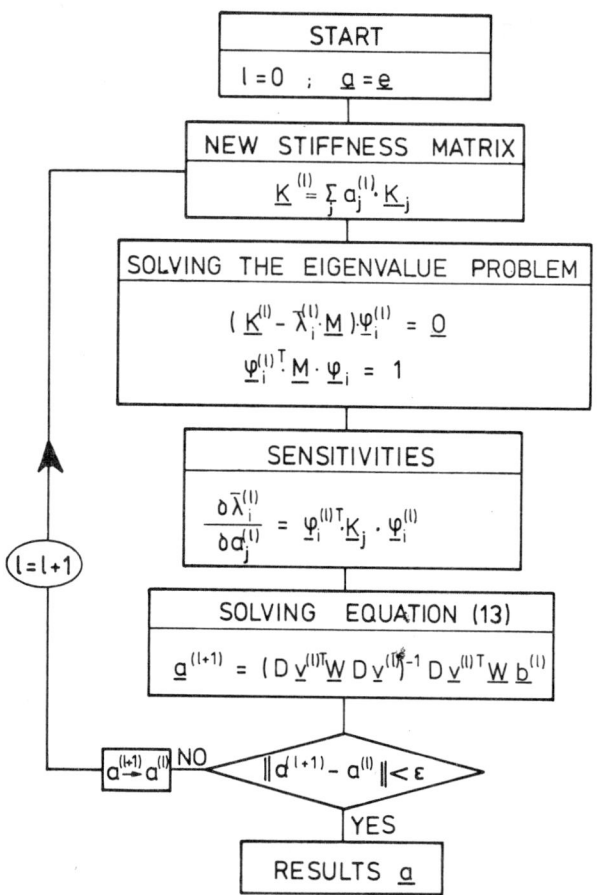

Fig. 7 Scheme of the fitting algorithm

To improve particularly the more important lower eigenfrequencies, the following weight matrix $\underline{W}$ is introduced.

$$\underline{W} = \text{diag}\left[w_i\right] \quad \text{with} \quad w_i = (1/\bar{\lambda}_i^F)^4 \tag{14}$$

Besides the weight matrix the choice of the submatrices $\underline{K}_j$ in (6), Fig. 5, has a strong influence on the improvement of the reduced model. If all torsional springs form independent substructures, we found this lead to divergence of the fitting algorithm for systems with more than

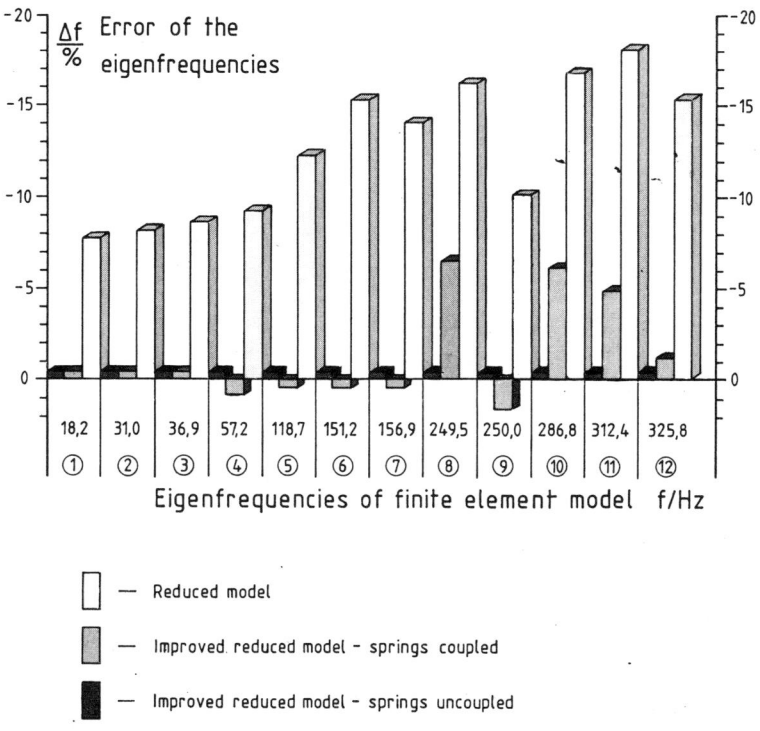

Fig. 8   Errors in eigenfrequencies of the reduced and the improved
         reduced model compared with the finite element model

six DOF. To solve the problem the total stiffness matrix was subdivided
in different substructures, whereas one substructure summarizes two or
several springs. If two adjacent springs are summarized in one substructure, as it is shown in Fig. 5, convergence of the fitting algorithm is
achieved in all calculations. Fig. 8 shows the errors in the eigenfrequencies after the fitting process with two coupled springs as one substructure. The eigenfrequencies in the lower modes (1-200 Hz) almost
coincide with the finite element solutions, while the errors in the
higher modes are still up to 7 %. When we fit this reduced model a second time, all springs may form an independent substructure and the algorithm converges and yields excellent results in the eigenfrequencies,
Fig. 8. After this second step all twelve eigenfrequencies of the im-

proved reduced model almost coincide with the solutions of the finite element model.

The results presented next are all calculated by this twice improved reduced model.

The introduced algorithm, Fig. 7, corrects simultaneously with the fitting of the eigenfrequencies via the eigenvalue problem also the eigenvectors of the reduced model. The amplitudes and zero passages of the improved reduced model represent the modes of the finite element very good, Fig. 9.

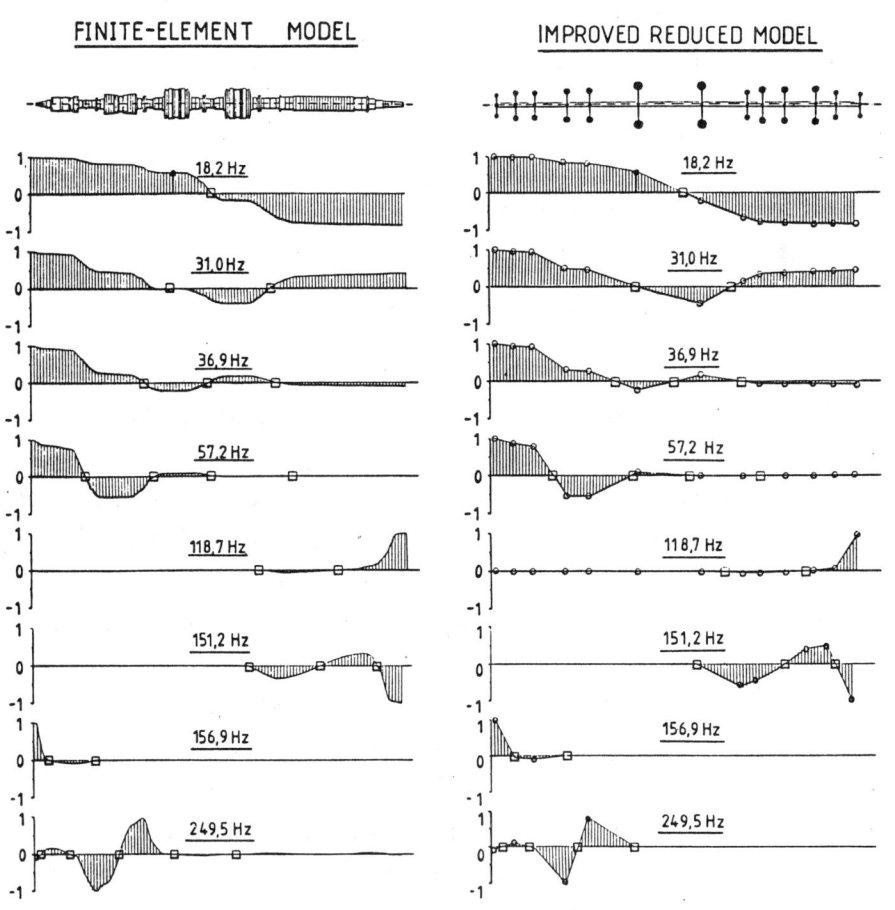

Fig. 9 Modes of the finite element and the improved reduced model

<u>Transient Vibrations</u>. The practician will also ask: How good is
the improved reduced model for the calculation of elastic stresses at
the couplings due to electrical excitation moments?

Fig. 10 shows the time history of the elastic moment in the coupling between the low pressure turbine and the generator. The maximum

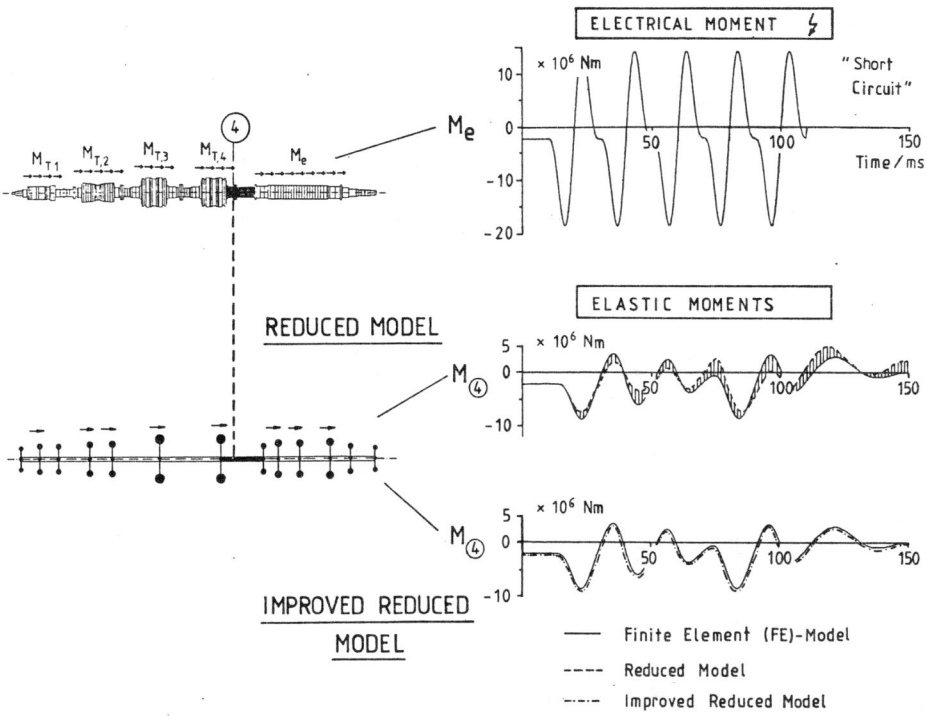

Fig. 10 Elastic moment in the LP2-Generator coupling due to a short-
circuit for the finite element model, the reduced model and
the improved reduced model

elastic moment occurs at this coupling, which is also the critical cross
section in our case. The system is excited by steam forces at the turbines and the electrical moment due to a short circuit.

We can see, that the elastic moment in the coupling of the reduced

model differs in amplitude and phase from the solution of the finite
element model, while the improved reduced model yields considerable better results.

In Fig. 11 the maximum amplitudes of the coupling moments of the
finite element model are compared with the solutions of the reduced
resp. the improved reduced model. The reduced model yields elastic moments, which are up to 17 % to low whereas the error of the fitted model is only 2 %.

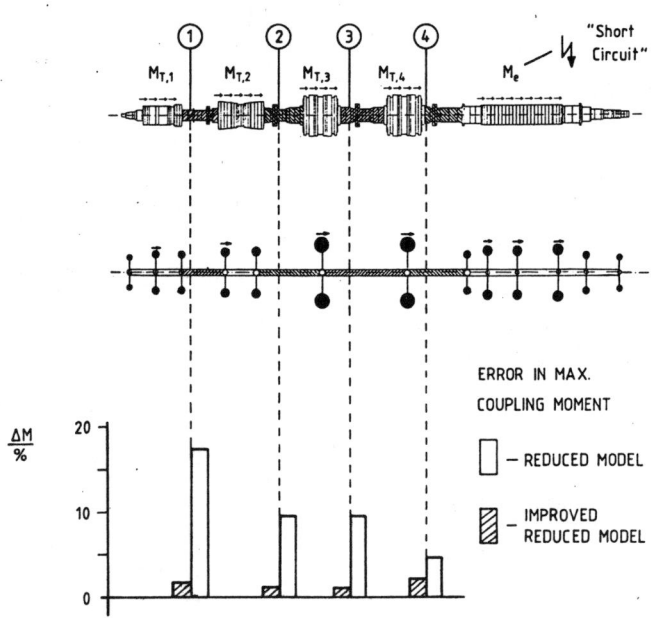

Fig. 11   Error in the max. coupling moment due to a short-circuit for
          the reduced and the improved reduced model

CONCLUSIONS

To calculate the torsional vibrations of turbine-generator sets
the dynamic behavior of the shaft system can be modelled by the finite
element method. The result are the linear equations of motion. The solution for the natural vibrations yields the eigenfrequencies and modes.

For a coupled electrical-mechanical torsional analysis and torsional monitoring systems a reduced torsional model for the shaft system
is required. Therefore an algorithm is developed to reduce the finite-
element model with many DOF to a discrete mass model with up to 20 DOF.
The reduced torsional model is optimized by fitting its eigenfrequencies
to the lower spectrum of a large finite element model. Therefore a parameter identification algorithm is used. The described procedure proves
in practice to yield good results concerning

- the eigenfrequencies and modes
- the coupling moments due to electrical system transients.

REFERENCES

1. Berger, H., Kulig, T.S.: Simulation Models for Calculating the
    Torsional Vibrations of Large Turbine-Generator Units after Elec-
    trical System Faults, Siemens Forschungs- und Entwicklungsberichte,
    Band 10, 1981, Nr. 4.

2. Schwibinger, P.: 'Torsionsschwingungen von Turbogruppen und ihre
    Kopplung mit den Biegeschwingungen bei Getriebemaschinen', PhD-
    Thesis, Univ. Kaiserslautern, 1986.

3. IEEE Comitee Report: 'A Bibliography for the Study of Subsynchronous
    Resonance between Rotating Machines and Power Systems', IEEE Trans.,
    Vol. PAS-95, 1976.

4. Schwibinger, P.; Feng, T.; Nordmann, R.: 'Improvement of a Reduced Torsional Model by means of Parameter Identification', Intl. Conf. on Rotordynamics - IFToMM and JSME, Tokyo, Sept. 14-17, 1986.

5. Bethke, J.: 'Reduktion der Freiheitsgrade von kettenförmigen Torsionsschwingungssystemen mit elektrischen Analogiemodellen', Master Thesis, Univ. Kaiserslautern, 1985 (unpublished).

6. Di, Phan Nguyen: 'Beitrag zur Reduktion diskreter Schwingerketten auf ein Minimalmodell', PhD-Thesis, Univ. Dresden, 1974.

7. Zimmermann, H., Collmann, D., Natke, H.G.: Erfahrungen zur Korrektur des Rechenmodells mit gemessenen Eigenfrequenzen am Beispiel des Verkehrsflugzeuges VFW 614, Zeitschrift für Flugwissenschaften und Weltraumforschung 1 (1977), Heft 4.

8. Natke, H.G.: Einführung in Theorie und Praxis der Zeitreihen- und Modalanalyse, Vieweg-Verlag, Braunschweig, Wiesbaden, 1983.

IDENTIFICATION OF STIFFNESS, DAMPING AND INERTIA
COEFFICIENTS OF ANNULAR TURBULENT SEALS

R. Nordmann
University of Kaiserslautern, Kaiserslautern, F.R.G.

Introduction

An important assumption for the reliability of high speed centri-
fugal pumps is a good rotordynamic behavior. Connected to this problem
hydraulic forces acting on the rotor are of major importance. It is
well known that neck ring seals as well as interstage seals (Fig. 1)
may have a large influence on the bending vibrations of a pump rotor.
Besides their designed function of reducing the leakage flow between
the impeller outlet and inlet or between two adjacent pump stages, the
contactless seals have the potential to develop significant forces.
This type of forces, created by lateral rotor vibrations can be described by a linear model with stiffness, damping and inertia coefficients.

If contactless seal elements are used in a turbopump the fluid-
mechanical interactions have to be considered when predicting the vibration behavior of the pump rotor in the design process. However,

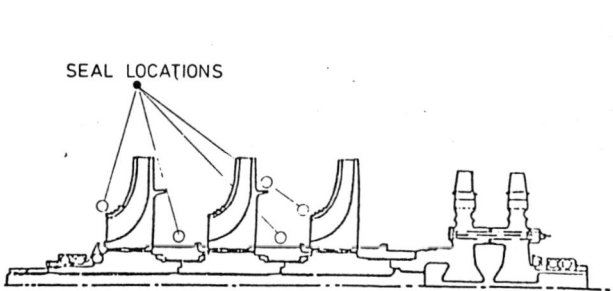

TURBOPUMP   TURBINE

SEAL LOCATIONS

Fig. 1   Seals of a High-Pressure-Turbopump Rotor in
         Aerospace Engineering

there is often a uncertainty, concerning the data for the dynamic coefficients. Up to now the stiffness and damping characteristics of seals
are not very well known and there is a need for additional research in
this area. This is particularly the case for grooved seals, which are
very common in practice. Different research projects have been started
to investigate the dynamics of seals by theoretical models as well as
experimental procedures. The following chapter presents a possible model, based on a bulk flow theory and describes an experimental procedure to identify the stiffness, damping and inertia coefficients.

Modelling of Annular Seals with Turbulent Flow Conditions

Seal model.   To explain the seal model, we consider a very simple
geometrical form, consisting of a cylindrical shaft with circular cross
section, surrounded by a cylindrical housing (Fig. 2). This annular
seal seperates the two chambers with pressure p_1 and pressure p_2, respectively. The pressure difference is $\Delta p = p_1 - p_2$. Caused by this pressure difference there is a leakage flow in axial direction, which is
always almost a turbulent flow with average velocity V. A velocity in
circumferential direction is superimposed due to the rotation of the
shaft with angular of velocity Ω. In order to obtain the governing

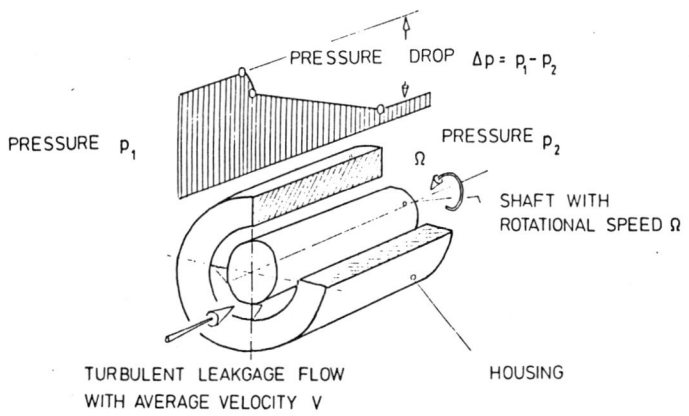

PRESSURE   DROP $\Delta p = p_1 - p_2$

PRESSURE p_1

PRESSURE p_2

Ω

SHAFT WITH
ROTATIONAL SPEED Ω

TURBULENT LEAKGAGE FLOW
WITH AVERAGE VELOCITY V

HOUSING

Fig. 2   Modelling of an Annular Seal with Turbulent Flow

equations for the presented seal we assume pure translational movements
of the shaft in radial direction.

To derive the pressure around the shaft and then the force-motion-
relationships for the vibrating rotor we are using a bulk flow model,
which was originally derived by Hirs /1/. Childs /2/ introduced this
bulk flow theory for seal elements. The first basic idea of this theory
is, that the fluid velocity distributions in radial direction are substituted by average velocities. For a fluid element between the rotor
and stator surface (Fig. 3), located at the axial coordinate Z, and the
circumferential coordinate θ, the average axial velocity is $U_z(Z, \theta, t)$
and the average circumferential velocity is $U_\theta(Z, \theta, t)$. The corresponding pressure for this location is $p(Z, \theta, t)$ and the seal radial
clearance $H(\theta, t)$. The shaft circumferential velocity is $U = R\,\Omega$.

The second basic assumption in Hirs theory is his empirical finding, that the relationship between the shear stress at the wall and the
mean velocity of the bulk flow - relative to the wall - can be expressed by well known formulas. With the bulk flow velocity $V_R^2 = \{(U_\theta - R\Omega)^2$
$+ U_z^2\}$ relative to the rotor surface we obtain the wall shear stress
at the rotor

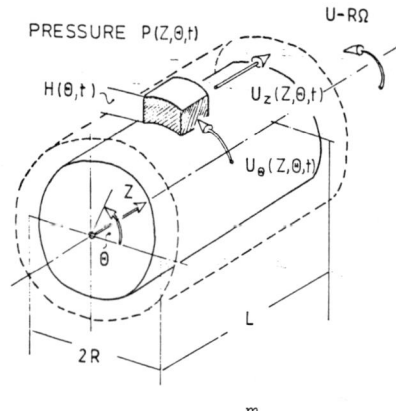

PRESSURE P(Z,θ,t)

U-RΩ

H(θ,t)

$U_z(Z,\theta,t)$

$U_\theta(Z,\theta,t)$

Z

θ

L

2R

Fig. 3   Velocities, Pressure and
Radial Clearance for a
Fluid Element

$$\tau_R = n_o \left(\frac{2H \, V_R}{\nu} \right)^{m_o} \frac{\rho}{2} V_R^2 \tag{1}$$

n_o, m_o are empirical turbulence coefficients, ρ is the fluid density
and ν the kinematic viscosity of the fluid. In a similar way V_S
$V_S^2 = \{U_\theta^2 + U_z^2\}$ is the bulk flow velocity relative to the stator surface and the corresponding stator shear stress is

$$\tau_S = n_o \left(\frac{2H \, V_S}{\nu} \right)^{m_o} \frac{\rho}{2} V_S^2 \tag{2}$$

Formulas (1) and (2) correlate the shear stresses τ_R, τ_S with the Reynolds numbers, which are defined in parantheses.

We can now derive the two momentum equations, expressing the "equilibrium" for the fluid element in axial and circumferential direction
(Fig. 4). If we introduce the shear forces at the walls, the pressure
forces and the inertia forces we end up with the two equilibrium equations (3) and (4), which are shown in Fig. 5 together with the continuity equation (5)

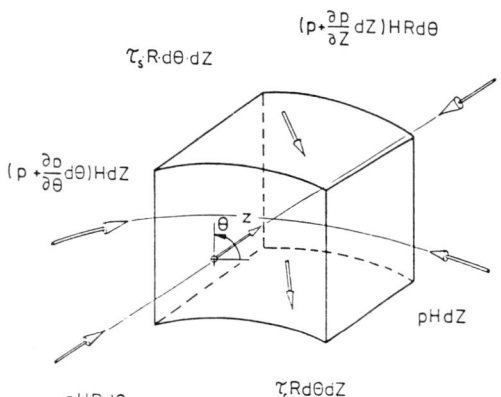

Fig. 4   Pressure and Shear Forces acting on a Fluid Element

Axial Momentum Equation

$$-\frac{H^2}{\eta U}\frac{\partial P}{\partial Z} = \frac{n_o}{2} R_\theta^{1+m_o} \{U_Z(U_\theta^2 + U_Z^2)^{\frac{1+m_o}{2}} + U_Z(U_\theta-1)^2 + U_Z^2{}^{\frac{1+m_o}{2}}\}$$

$$+ R_\theta \{\frac{H}{U}\frac{\partial U_Z}{\partial t} + \frac{HU_\theta}{\partial R}\frac{\partial U_Z}{\partial \theta} + H\ U_Z\frac{\partial U_Z}{\partial Z}\}$$

(3)

Circumferential Momentum Equation

$$-\frac{H^2}{\eta U}\frac{1}{R}\frac{\partial P}{\partial \theta} = \frac{n_o}{2} R_\theta^{1+m_o} \{U_\theta(U_\theta^2 + U_Z^2)^{\frac{1+m_o}{2}} +(U_\theta-1)^2((U_\theta-1)^2 + U_Z^2)^{\frac{1+m_o}{2}}\}$$

$$+ R_\theta \{\frac{H}{U}\frac{\partial U_\theta}{\partial t} + \frac{\partial U_\theta}{R}\frac{\partial U_\theta}{\partial \theta} + H\ U_Z\frac{\partial U_\theta}{\partial Z}\}$$

(4)

Continuity Equation

$$H\frac{\partial U_Z}{\partial Z} + \frac{1}{R}\frac{\partial}{\partial \theta}(H\ U_\theta) + \frac{1}{R\Omega}\frac{\partial H}{\partial t} = 0$$

(5)

Reynolds number in circumferential direction

$$R_\theta = \frac{HU}{\nu} = \frac{\rho HU}{\eta}$$

Fig. 5   Governing Equations for a Fluid Element

Perturbation Analysis. With the assumption of small motions of the shaft about a centered position we can expand the equations (3), (4), (5) by means of a perturbation analysis /2/

$$H(\theta,t) = H_0 + \varepsilon H_1 \quad ; \quad H_0 = \Delta R$$

$$p(Z,\theta,t) = p_0 + \varepsilon p_1$$

$$U_Z(Z,\theta,t) = U_{Z0} + \varepsilon U_{Z1} \quad ; \quad U_{Z0} = V \tag{6}$$

$$U_\theta(Z,\theta,t) = U_{\theta0} + \varepsilon U_{\theta1}$$

ε is a perturbation parameter. H_0, p_0, U_{Z0}, $U_{\theta0}$ are the quantities for the zero eccentricity flow condition (centered position of the shaft). H_1, p_1, U_{Z1}, $U_{\theta1}$ correspond to the flow conditions for small shaft motions.

If we introduce (6) into the governing equations (3), (4), (5) we obtain zeroth-order and first order perturbation equations. The substitution and the solution procedures of the remaining equations are very extensive. They are described in more detail in /2/ and /3/. In this presentation we discuss only the essential results.

The solution of the zeroth-order equations ($\varepsilon = 0$, shaft without radial motion) defines the steady state leakage or the pressure drop Δp in axial direction (Fig. 6)

$$\Delta p = p_1 - p_2 = \{1 + \xi + 2\sigma\} \frac{\rho}{2} V^2 \tag{7}$$

The pressure drop Δp is proportional to ρ and the squared velocity V
and consists of three parts. The first one shows the change of pressure
energy to kinetic energy. The second part points out the pressure loss
at the seal entrance and the third part expresses the pressure drop

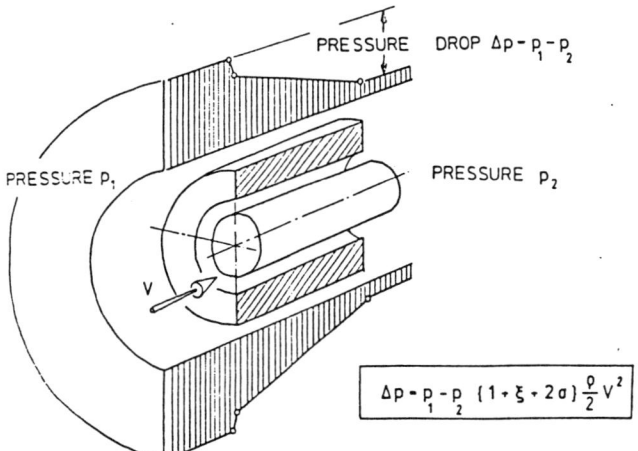

PRESSURE DROP $\Delta p = p_1 - p_2$

PRESSURE p_1 PRESSURE p_2

V

$$\Delta p = p_1 - p_2 \ (1 + \xi + 2\sigma)\frac{\rho}{2}V^2$$

Fig. 6   Pressure Drop in Axial Direction

along the seal, caused by friction. The seal behaves like a hydrostatic
bearing. A static displacement of the shaft in radial direction causes
a restoring force and the fluid acts like a spring. The second important
result of the zeroth-order equation is the development of the circumferential velocity of a fluid element proceeding axially along the seal.
This velocity influences the cross coupled stiffness coefficients.

The first order equations describe the pressure and flow quantities
due to a small shaft motion $H(\theta,t)$ about the centered position. These
equations can be solved numerically. When we introduce a circular harmonic orbit as a special motion, we can express $H(\theta,t)$ in terms of this
motion. By the further assumption of a harmonic pressure and velocity
distribution in circumferential direction, the first order equations can
be reduced to a system of three coupled complex ordinary differential
equations for the unknowns U_{z1}, $U_{\theta 1}$, p_1 /2/, /3/. From the pressure field
solution the reaction forces acting on the rotor due to the circular

shaft motion have to be determined by integration of the pressure along the seal and in circumferential direction. Finally the force motion relationship is established. The dynamic system of a seal can be modeled by a linear system with stiffness, damping and inertia terms, if small movements about the seal centre are assumed (Fig. 7).

$$
\begin{vmatrix} F_y \\ F_z \end{vmatrix} = \begin{vmatrix} m_{yy} & \\ & m_{zz} \end{vmatrix} \begin{vmatrix} \ddot{y} \\ \ddot{z} \end{vmatrix} + \begin{vmatrix} c_{yy} & c_{yz} \\ c_{zy} & c_{zz} \end{vmatrix} \begin{vmatrix} \dot{y} \\ \dot{z} \end{vmatrix} + \begin{vmatrix} k_{yy} & k_{yz} \\ k_{zy} & k_{zz} \end{vmatrix} \begin{vmatrix} y \\ z \end{vmatrix} \qquad (8)
$$

In general a numerical procedure is needed to calculate the dynamic coefficients of equation (8). For the special case of a short seal a solution in an analytical form is possible /4/, /5/.

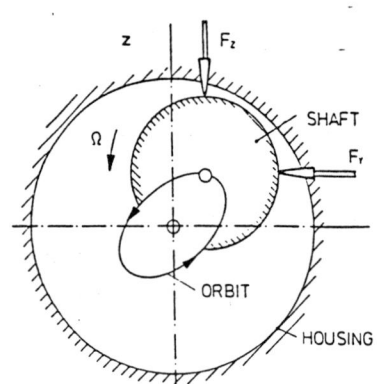

Fig. 7   Dynamic Seal Forces caused by Small Shaft Motions about a Centered Position

The main diagonal elements in each of the matrices are equal and the cross-coupled terms are opposite in sign. The coefficients are mainly dependent on the pressure drop, the average axial velocity V, the rotational speed Ω of the shaft the seal geometry (seal length L and Radius R) and on some quantities characterizing the friction in a seal. It is important to note, that the cross coupled stiffnesses $k_{yz} = - k_{zy}$ are strongly influenced by the rotational speed Ω and the fluid entry swirl, which is the circumferential velocity of the fluid at the seal

entrance. This effect may cause serious instability problems in high
speed rotating machinery, when the cross coupled stiffness terms become
dominant.

In one of his first publications about seals, Black /6/ has derived
stiffness, damping and inertia coefficients for short seals. Fig. 8
presents this coefficients in dependence of the most important influence
parameters Δp, V, Ω, T = L/V and some friction coefficients μ_0, μ_1, μ_2.
In Blacks derivatives the entry swirl was assumed to be half of the cir-

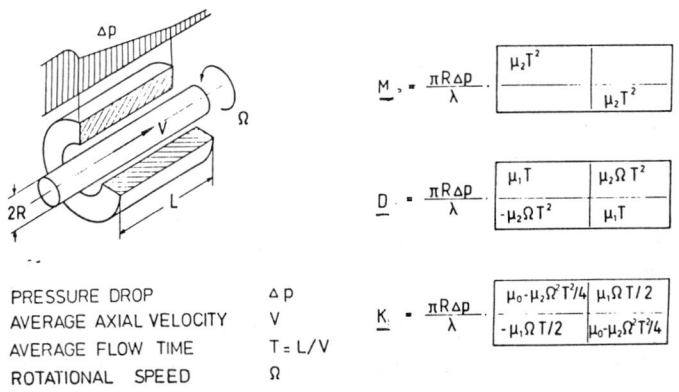

$$\underline{M} = \frac{\pi R \Delta p}{\lambda} \cdot \begin{array}{|c|c|} \hline \mu_2 T^2 & \\ \hline & \mu_2 T^2 \\ \hline \end{array}$$

$$\underline{D} = \frac{\pi R \Delta p}{\lambda} \cdot \begin{array}{|c|c|} \hline \mu_1 T & \mu_2 \Omega T^2 \\ \hline -\mu_2 \Omega T^2 & \mu_1 T \\ \hline \end{array}$$

PRESSURE DROP	Δp
AVERAGE AXIAL VELOCITY	V
AVERAGE FLOW TIME	T = L/V
ROTATIONAL SPEED	Ω

$$\underline{K} = \frac{\pi R \Delta p}{\lambda} \cdot \begin{array}{|c|c|} \hline \mu_0 - \mu_2 \Omega^2 T^2/4 & \mu_1 \Omega T/2 \\ \hline -\mu_1 \Omega T/2 & \mu_0 - \mu_2 \Omega^2 T^2/4 \\ \hline \end{array}$$

Fig. 8 Dynamic Coefficients of a Short Seal /6/

cumferential velocity of the shaft $R\Omega/2$.

With the à priori knowledge about the seal dynamic coefficients we
have valuable informations about the structure of the model (linearity,
order of the model, skewsymmetric etc.), which will be used in the following parameter identification procedure.

Identification of the Dynamic Seal Coefficients

Most of the experimental methods to determine dynamic coefficients
of bearings or seals are working with test forces (input signals) and

are measuring the relative displacement between shaft and housing (output signals). The unknown seal parameters can then be calculated by
means of input-output relations in the time domain or in the frequency
domain (Fig. 9).

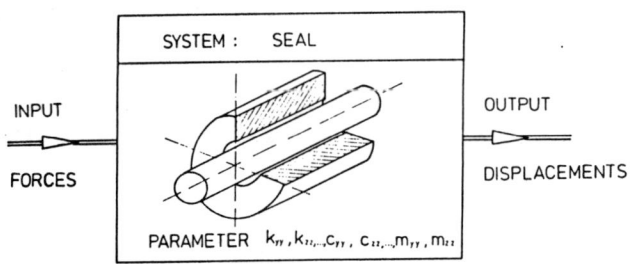

IDENTIFICATION PROBLEM :
TO FIND THE PARAMETERS OF THE SEAL
FROM INPUT-OUTPUT- RELATIONS

Fig. 9 Identification of the Parameters of a Seal

The basic steps of the applied identification procedure are pointed
out in Fig. 10. Measurements are carried out at a test rig, that consists of a very stiff rotating shaft and an elastically mounted rigid
housing with two symmetric seals between shaft and housing. Water flows
axially across the two seals in opposite directions, while the shaft is
running with a fixed rotational speed Ω. The housing is excited by a
test force and the system response is measured as a radial motion between the seal surfaces. From measured input and output time signals
frequency response functions can be determined by signal processing.

Corresponding to the seal test rig a linear mechanical model ex-
cists, consisting of the rigid mass (housing) and the stiffness and
damping coefficients of the seals. The analytical frequency response
functions of the model depend on the unknown seal parameters, which
have to be determined.

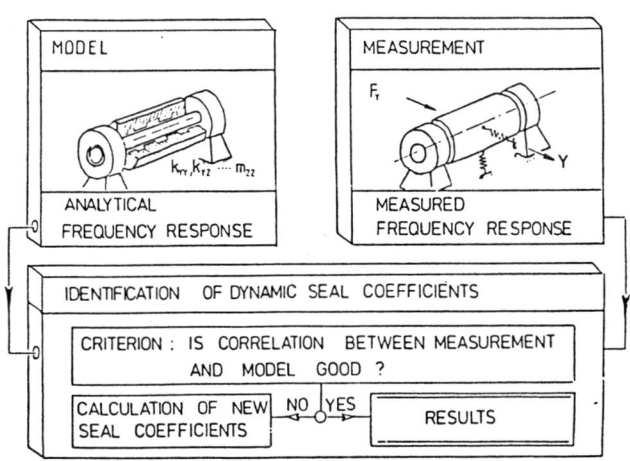

Fig. 10   Basic Steps of the Identification Procedure

The dynamic seal coefficients are estimated by a linear procedure and uses the measured mobility frequency response function and the analytical functions of the model. The main steps of the identification are following in more detail.

Mechanical model.   Fig. 11 shows the mechanical model with a rigidly supported very stiff shaft the rigid mass m of the housing and the stiffness and damping coefficients elements corresponding to the seals and the flexible springs supporting the casing. If test forces are acting in the center of the housing, the system responds only with translatory motions in the two directions y and z. The equations of motion for the model

$$
\begin{vmatrix} m + 2m_{yy} & \\ & m + 2m_{zz} \end{vmatrix}
\begin{vmatrix} \ddot{y} \\ \ddot{z} \end{vmatrix}
+2
\begin{vmatrix} c_{yy} & c_{yz} \\ c_{zy} & c_{zz} \end{vmatrix}
\begin{vmatrix} \dot{y} \\ \dot{z} \end{vmatrix}
+2
\begin{vmatrix} k_{yy} & k_{yz} \\ k_{zy} & k_{zz} \end{vmatrix}
\begin{vmatrix} y \\ z \end{vmatrix}
=
\begin{vmatrix} F_y \\ F_z \end{vmatrix}
$$

(9)

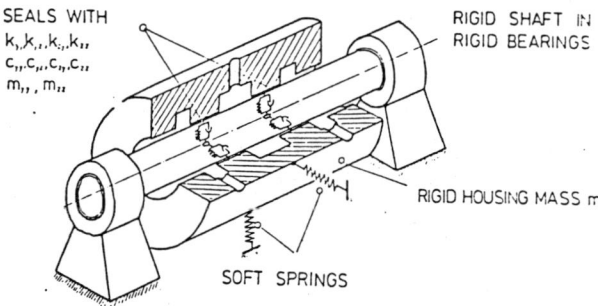

Fig. 11 Mechanical Model of the Seal Test Rig

describe the equilibrium of the inertia forces (housing), the seal
forces and exciter forces.

    For the considered two degree of freedom test rig system a total
of four stiffness frequency functions as well as four flexibility frequency functions can be derived. They depend on the seal coefficients
and the exciter frequency ω, as well. The exciter frequency ω is usually
different from the frequency of rotation Ω. Fig. 12 points out the mathematical expressions of the two types of frequency response functions
$\underset{\sim}{H}^*(\omega)$ and $\underset{\sim}{K}^*(\omega)$. By inversion of the frequency response $\underset{\sim}{H}(\omega)$ we obtain the
stiffness response $\underset{\sim}{K}^*(\omega)$. Both functions are used in the identification
procedure.

    <u>Measurements of the Frequency Response Functions $\underset{\sim}{H}^*(\omega)$.</u> Up to now
we have considered frequency response functions of the model. For the
determination of the frequency response functions from measured input
and output time data, we take advantage of the fact that the ratio of
the Fourier transformed signals is equal to the frequency response. Due
to this possibilities excitation signals with broadband character in
the frequency domain (impact, random etc.) can also be applied.

COMPLEX FLEXIBILITY FREQUENCY RESPONSE $H_{ik}(\omega)$

$$\underline{H}^{*} = \begin{vmatrix} H_{yy}(\omega) & H_{yz}(\omega) \\ H_{zy}(\omega) & H_{zz}(\omega) \end{vmatrix} = \frac{1}{4\Delta} \begin{vmatrix} k_{zz}-\omega^{2}(\frac{m}{2}+m_{zz})+i\omega c_{zz} & -(k_{yz}+i\omega c_{yz}) \\ -(k_{zy}+i\omega c_{zy}) & k_{yy}-\omega^{2}(\frac{m}{2}+m_{yy})i\omega c_{yy} \end{vmatrix}$$

$$\Delta = (k_{yy}-\omega^{2}(\frac{m}{2}+m_{yy})+i\omega c_{yy})(k_{zz}-\omega^{2}(\frac{m}{2}+m_{zz})+i\omega c_{zz})-(k_{yz}-i\omega c_{yz})(k_{zy}+i\omega c_{zy})$$

COMPLEX STIFFNESS FREQUENCY RESPONSE $K_{ik}(\omega)$

$$\underline{K}^{*} = \begin{vmatrix} K_{yy}(\omega) & K_{yz}(\omega) \\ K_{zy}(\omega) & K_{zz}(\omega) \end{vmatrix} = 2 \begin{vmatrix} k_{yy}-\omega^{2}(\frac{m}{2}+m_{yy})+i\omega c_{yy} & k_{yz}+i\omega c_{yz} \\ k_{zy}+i\omega c_{zy} & k_{zz}-\omega^{2}(\frac{m}{2}+m_{zz})+i\omega c_{zz} \end{vmatrix}$$

Fig. 12   Frequency Response Functions $H(\omega)$ and $K(\omega)$
          of the Mechanical Model

The force and response signals are measured in the time domain, transformed to the frequency domain by means of Fast Fourier Transformation
and the quotient is calculated (Fig. 13). This procedure is executed by
efficient two channel Fourier Analyzers. Fig. 14 shows in principal the
measurement equipments. A hammer was used in this case to excite the
housing by an impulse force (see also Fig. 13). By this excitation the
signal contains energy in a desired frequency range, which can be influenced by the hammer mass, the flexibility of the impact cap and the
impact velocity. The relative displacements between housing and shaft
are measured with displacement pick ups. The time signals are amplified,
digitized by the analyzer and the frequency response functions are cal-

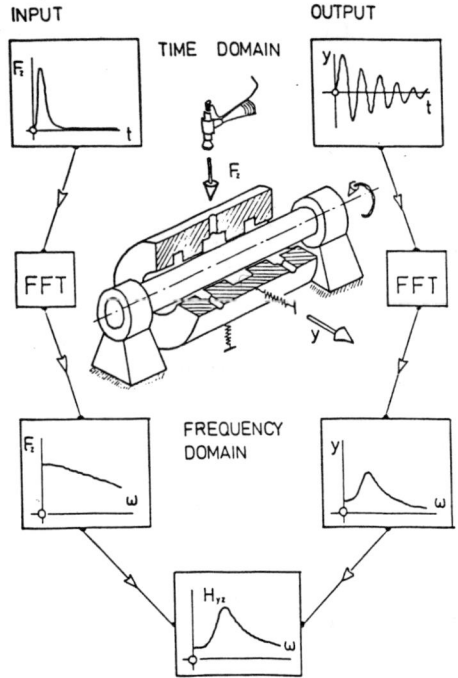

Fig. 13   Measured   Frequency
          Response Functions by
          Means of Fast Fourier
          Transformation

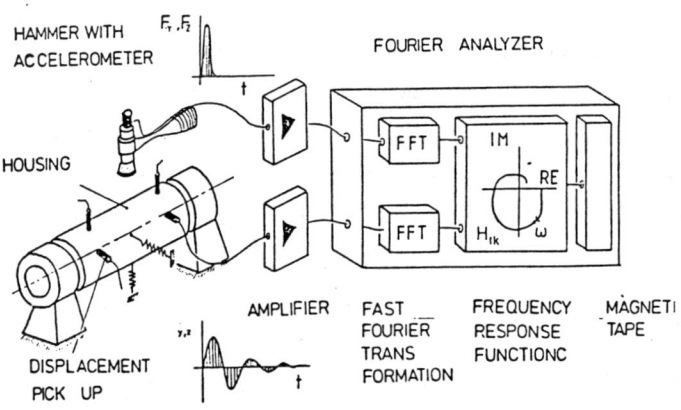

Fig. 14   Measurement Equipments

culated.

Estimation of the Dynamic Seal Coefficients. Different possibilities excist to determine the seal coefficients from measured frequency
response curves. One idea is to fit analytical flexibility functions to
the measured ones. When working with the output error, this leads to
nonlinear equations for the unknown parameters. Another method will be
presented here with the definition of an error for the input signals.

From a theoretical point of view the product of the complex mobi-
lity matrix $\underline{H}$ and the complex stiffness matrix $\underline{K}$ should be the unity
matrix $\underline{E}$. By combining the measured matrix $\underline{H}$ with the analytical matrix
$\underline{K}$ the result will be $\underline{E}$ plus an additional error matrix $\underline{S}$, caused by
measurement noise. Fig. 15 points out this fact and shows either the
complex equation or the two real equations for the unknown parameters,
concentrated in $\underline{M}$, $\underline{C}$, $\underline{K}$. The two real equations belong to one exciter
frequency. In the case of broadband excitation (impulse force) we have

PARAMETER - ESTIMATION

$$\begin{bmatrix} K_{yy} & K_{yz} \\ \hline K_{zy} & K_{zz} \end{bmatrix} \circ \begin{bmatrix} H_{yy} & H_{yz} \\ \hline H_{zy} & H_{zz} \end{bmatrix} = \begin{bmatrix} E \end{bmatrix} + \begin{bmatrix} S \end{bmatrix}$$

$$[\underline{K} - \omega^2\underline{M} + i\omega\underline{C}] \cdot \underline{H} = \underline{E} + \underline{S}$$

$$\underline{K}\,\underline{H}^r - \omega^2\underline{M}\,\underline{H}^r - \omega\underline{C}\,\underline{H}^i = \underline{E} + \underline{S}^r$$
$$\underline{K}\,\underline{H}^i - \omega^2\underline{M}\,\underline{H}^i + \omega\underline{C}\,\underline{H}^r = \underline{S}^i$$

Fig. 15   Parameter Estimation

as much equations as frequency lines, generally much more, than unknown
parameters. The overdetermined equation system is presented in Fig. 16.
The rectangular matrix $\underline{A}$ contains all information about the measured
frequency response functions ($\underline{H}^r$, $\underline{H}^i$ real and imaginary part of fre-

quency response) and the related exciter frequencies ω. $\underline{X}$ contains the
unknown matrices $\underline{M}$, $\underline{C}$, $\underline{K}$ and $\underline{E}'$ is a modified unity matrix. Applying the

> EQUATIONS FOR THE DETERMINATION
> OF UNKNOWN   SEAL PARAMETER X

$$\underline{A} \cdot \underline{X} = \underline{E}' + \underline{S}'$$

$\underline{A}$ CONTAINS THE MEASURED FREQUENCY RESPONSE DATA $H(\omega)$

$\underline{X}$ CONTAINS THE UNKNOWN PARAMETERS M, D, K

CRITERION : $\underline{S}' \longrightarrow$ MINIMAL

$$\underline{A}'\,\underline{A}\,\underline{X} = \underline{A}'\,\underline{E}' \qquad \text{NORMAL EQUATIONS}$$

Fig. 16   Normal Equations for Unknown Parameters

criterion to this equation, that $\underline{S}'$ shall become a minimum, we find the
so called normal equations. This is a determined system of equations
for the twelve unknown seal parameters in the matrices $\underline{M}$, $\underline{C}$, $\underline{K}$. Caused
by the definition of an input error the solution procedure for the li-
near system may fail. This can be avoided by introducing instrumental
variables /6/. In this case the matrix $\underline{A}^T$ is substituted by a matrix $\underline{W}^T$,
containing the instrumental variables. In our case they have the meaning
of the frequency response data, determined with the estimated dynamic
coefficients from a previous step. Then the solution is found by an iterative procedure.

Test Rig and Some Measurement Results

Seal Test Rig. The mechanical part of the test facility is shown
in Fig. 17. The main components are a very stiff rotating shaft, driven
by a speed controled motor and the stiff housing, supported in flexible
springs. The shaft is rigidly supported in roller bearings. The fluid
enters the housing in the center, flows across the two seals in axial
direction and is exciting the housing at the two ends. With removable

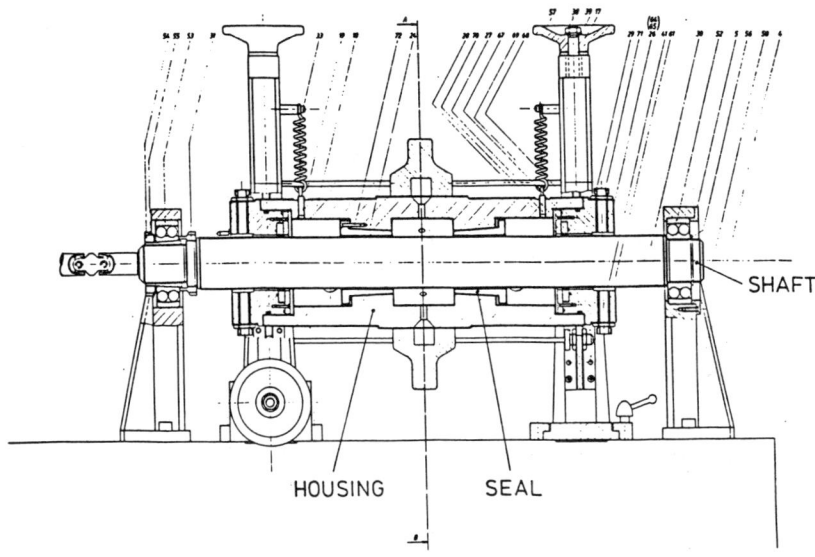

Fig. 17 Seal Test Rig

stator parts of the seal different geometrics and roughnesses can be
realized. The static position (zero eccentricity) is adjusted by a special mechanism and measured by eddy current pick ups, which are also
used to measure the shaft motions.

The range for the rotational speed is from 0 to 6000 rpm and for
the axial fluid velocity from 0 to 14 m/sec. With this we achieve Reynolds numbers up to 15000, when the fluid temperature is about 30° Celsius.

Dynamic measurements. In case of a dynamic measurement four frequency response functions have to be determined for one measurement set,
defined by a working condition with constant axial fluid velocity, rotational speed and fluid temperature. A computer takes over the measured
data and calculates the dynamic coefficients by means of the described
estimation procedure. Both sets of frequency response functions, the
measured and the fitted one, can then be displayed and plotted.

Some Measurement Results. The process of the dynamic measurement
is demonstrated for one working condition with V = 12 m/sec, n = 3450
rpm and a temperature of 30° Celsius. Fig. 18 shows the corresponding

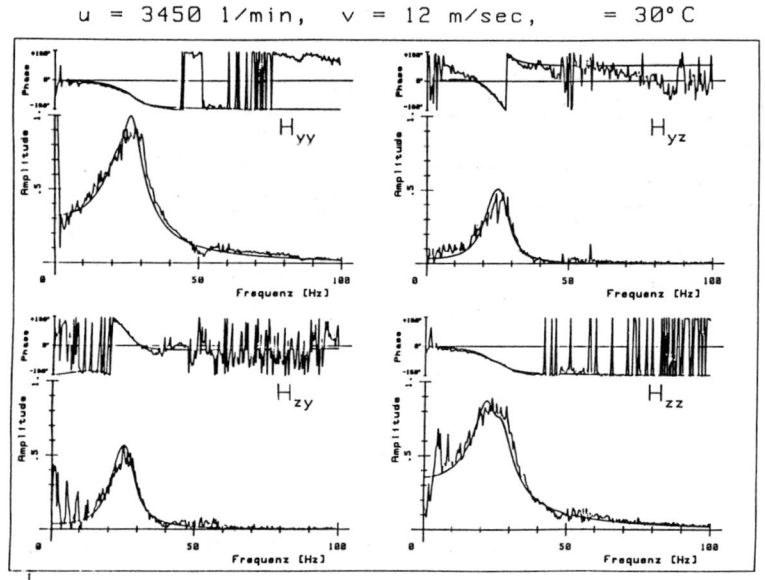

Fig. 18   Measured and Fitted Frequency Response Functions

measured and fitted response functions as magnitude and phase characteristics in the frequency range 0 to 100 Hz. The two direct functions H_{yy}
and H_{zz} should be equal and the cross coupled functions H_{yz} and H_{zy}
should be equal in magnitude but opposite in phase. The correlation
between fitted and measured functions is more or less good.

Several measurements were carried out for different rotational
speeds with constant temperature and axial velocity. For each set of
functions the inertia, damping and stiffness coefficients were calculated. They are shown in Figures 19, 20 and 21 versus the rotational
speed. The values for the complete system with two seals and all known
additional terms (mass of the housing, soft springs etc.) are presented.
It was found, that the direct coefficients are not equal, furthermore
the expected skewsymmetry could not be found exactly in the measured

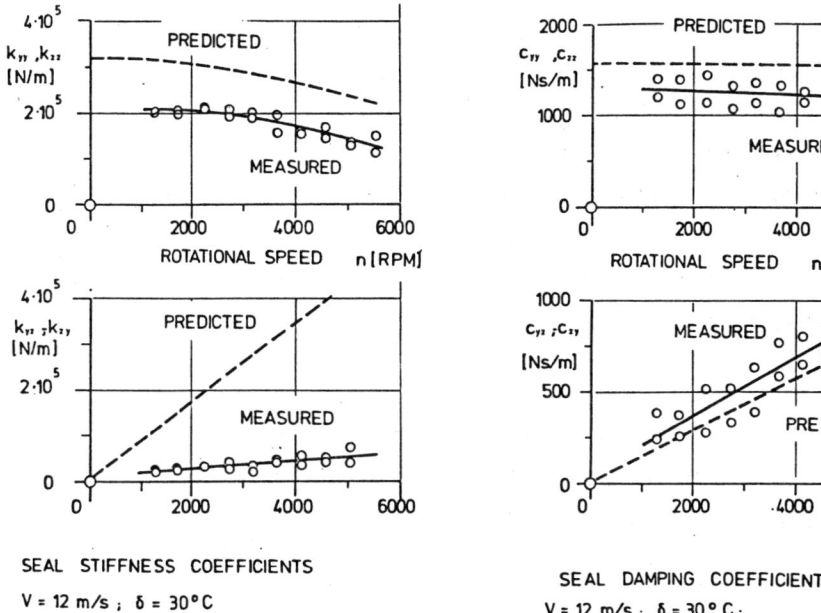

SEAL STIFFNESS COEFFICIENTS
V = 12 m/s ; δ = 30°C

Fig. 19   Stiffness Coefficients

SEAL DAMPING COEFFICIENTS
V = 12 m/s ; δ = 30°C ;

Fig. 20   Damping Coefficients

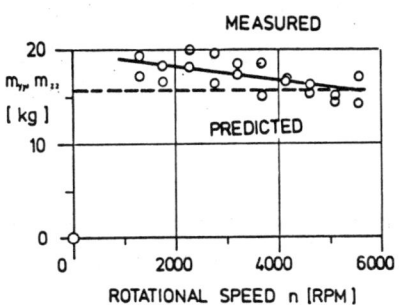

SEAL INERTIA COEFFICIENTS
V =12 m/s, =30°C

Fig. 21   Inertia Coefficients

results. The coefficients, which should be equal in magnitude are shown
in one diagram and treated as two values for the same operating condition. Besides the measured coefficients the corressonding values of the

above mentioned theoretical model are also shown.

The correlation is not that good for the stiffness terms, which
are presented in Fig. 19. The dependence on the rotational speed is
reasonable for both quantities, a parabolic decrease for the main stiffness and a linear increase for the cross coupled terms. However, the
main stiffnesses are found out 40 to 50 % to small and the cross coupled
terms much more than this. Meanwhile it is known, that the reason for
the small measured cross coupled terms is the low entry swirl in the
test rig, which has a large influence on k_{yz} and k_{zy}.

Fig. 20 shows the damping values. The theoretical model predicts
nearly constant values for the main damping and an increase with rotational speed for the cross coupled terms. The correlation between measurements and predictions are good to fair.

Finally the total inertia terms are presented in Fig. 21. The
theory predicts constant inertia terms. The correlation of measured and
calculated values looks good. However, it has to be noted, that the
mass of the housing is approximately 15 kg. This mass is included in
the results. Therefore the relative error of the seal inertia coefficients, related to the model prediction, is much higher.

References

/1/   Hirs, G.G.: "Fundamentals of a bulk-flow theory for turbulent
      lubricant films"; Diss. TH Delft, Niederlande, 1970.

/2/   Childs, D.W.: "Finite-Length solution for rotordynamic
      coefficients of turbulent annular seals"; ASME 82-Lub-42.

/3/   Nordmann, R. et. al: "Rotordynamic Coefficients and Leakage
      Flow for Smooth and Grooved Seals in Turbopumps"; IFTOMM-
      Conference Proceedings, Tokyo 1986.

/4/ Black, H.F.: "Effects of hydraulic forces in annular
     pressure seals on the vibrations of centrifugal pump rotors";
     J. Mech. Eng. Sci., Vol 11, No 2, 1969, S. 206-213.

/5/ Childs, D.W.: "Dynamic analysis of turbulent annular seals
     based on Hirs' lubrication equation"; ASME 82-Lub-41.

/6/ Maßmann, H.: "Ermittlung der dynamischen Parameter turbu-
     lent durchströmter Ringspalte bei inkompressiblen Medien";
     Diss. Universität Kaiserslautern 1986.

IDENTIFICATION OF MODAL PARAMETERS OF AN
ELASTIC ROTOR WITH OIL FILM BEARINGS

R. Nordmann
University of Kaiserslautern, Kaiserslautern, F.R.G.

Introduction

The dynamic behavior of many rotating machines e.g., turbines, compressors, pumps is influenced by stiffness and damping characteristics of nonconservative effects like oil film forces, forces in seals etc. It is important to know that besides the forced unbalance vibrations also unstable vibrations may occur, caused by the abovementioned selfexciting mechanisms /1/. To have a better understanding of the vibration behavior of rotating systems, the knowledge of the modal parameters - eigenvalues and eigenvectors - is very valuable.

Calculation procedures exist, for example, the finite element method, to determine the modal parameters in design stage /2/. But the input data for the calculation are not exactly known in any case and the predicted eigenvalues and eigenvectors may be different from those of the real machine in operation. Therefore mechanical engineers also try to find the modal parameters of built rotating machines or test rig rotors by measurements /3/.

In the past years the powerful method of experimental modal analysis has been used to measure modal parameters in many engineering
problems /4/. The method was mainly applied to nonrotating structures
without destabilizing effects. A successful application of the method
in rotating machines requires some improvements of the classical experimental modal analysis. In /5/ such improvements were described and
a very simple rigid rotor in oil film bearings was investigated. In
this paper a more complicated application is given, dealing with an
elastic test rig rotor in oil film bearings. The essential eigenvalues
and the corresponding eigenvectors (natural modes) could be found by
experimental modal analysis.

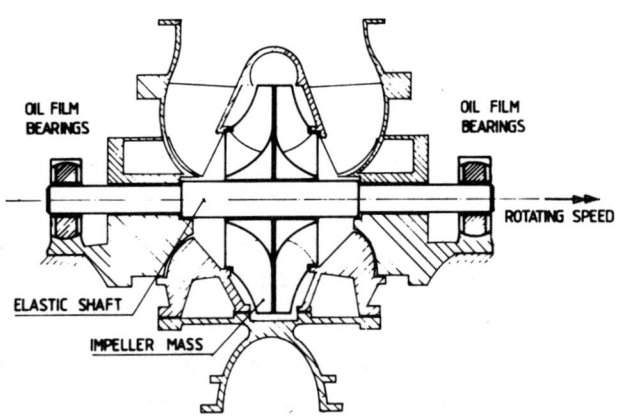

Fig. 1 Turbopump rotor

Natural Vibrations of an Elastic Rotor in Journal Bearings

Mechanical Model. Figure 1 shows a typical rotating machine, a
turbopump rotor, consisting of an elastic shaft with an impeller mass
in the middle of the shaft. The rotor is running in oil film journal
bearings. To measure modal parameters of such a rotating shaft is not
easy, because there are only a few points along the rotor to excite the

rotor and to measure the system response. A systematic investigation of
the vibration behavior of the machine during operation with all effects
(oil film forces, hydraulic forces) is difficult to realize. In a first
step we consider a simpler rotor system similar to that of the turbopump rotor. Figure 2 shows this test rig rotor with an elastic shaft,
a disk, and two oil film bearings. The modal parameters of this elastic
system can be measured in a systematic manner. All of the rotor locations can be excited and the displacements can be picked up at all locations. However, it has to be noted that not all of the effects of the
real machine can be investigated with the test rig rotor.

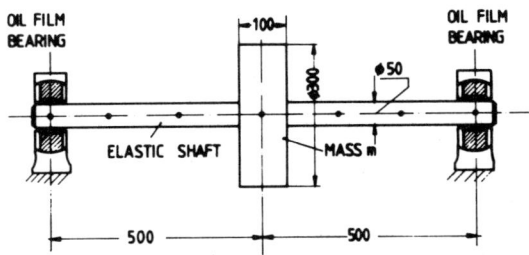

Fig. 2 Test rig rotor

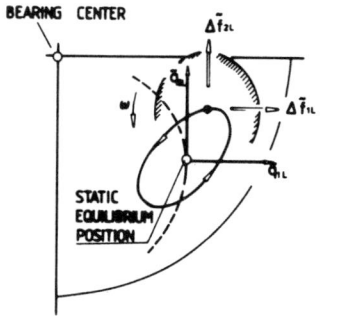

Fig. 3 Vibrations of the journal

From the theoretical considerations it is known, that for small vibrations of the journal around a static equilibrium position, the following force relationship for the oil film is true (Fig. 3).

$$
\begin{bmatrix} \Delta f_{1L} \\ \Delta f_{2L} \end{bmatrix} = - \begin{bmatrix} b_{11} & b_{12} \\ b_{21} & b_{22} \end{bmatrix} \begin{bmatrix} \dot{\tilde{q}}_{1L} \\ \dot{\tilde{q}}_{2L} \end{bmatrix} - \begin{bmatrix} k_{11} & k_{12} \\ k_{21} & k_{22} \end{bmatrix} \begin{bmatrix} \tilde{q}_{1L} \\ \tilde{q}_{2L} \end{bmatrix} \tag{1}
$$

with

k_{ik} stiffness coefficients of the bearings

b_{ik} damping coefficients of the bearings

The stiffness and damping coefficients depend on the Sommerfeld number, and on the rotational speed, respectively. Besides anisotropy, the stiffness cross coupling terms are unequal in general. This unsymmetry is the reason for selfexcited bending vibrations of the shaft.

Equation of Motion. The equations of motion for the simple shaft (Fig. 2) express the equilibrium of inertia, damping, stiffness, and external forces

$$
\underline{M} \, \ddot{\tilde{q}} + \underline{D} \, \dot{\tilde{q}} + \underline{K} \, \tilde{q} = \tilde{\underline{f}}(t) \tag{2}
$$

$\underline{M}$ mass matrix

$\underline{D}$ damping matrix

$\underline{K}$ stiffness matrix

$\tilde{\underline{q}}$ vector of displacements

$\tilde{\underline{f}}$ vector of external forces

The matrices $\underline{D}$ and $\underline{K}$ contain stiffness and damping terms from the bearings. They are nonsymmetric and depend on the running speed of the rotor.

Natural Vibrations, Eigenvalues, and Natural Modes. From the homogeneous equations of motion ($\tilde{\underline{f}} = 0$) the natural vibrations can be calculated. Assuming a solution of the form

$$\tilde{\underline{q}}(t) = \underline{\varphi}\, e^{\lambda t} \tag{3}$$

we obtain a quadratic eigenvalue problem

$$(\lambda^2 \underline{M} + \lambda\, \underline{D} + \underline{K})\, \underline{\varphi} = \underline{0} \tag{4}$$

with 2N eigenvalues λ_j and corresponding eigenvectors $\underline{\varphi}_j$. Eigenvalues as well as eigenvectors in the most cases occur in conjugate complex pairs

Eigenvalues: $\lambda_j = \alpha_j + i\omega_j$; $\bar{\lambda}_j = \alpha_j - i\omega_j$

$$\tag{5}$$

Eigenvectors: $\underline{\varphi}_j = \underline{s}_j + i\underline{t}_j$; $\underline{\bar{\varphi}}_j = \underline{s}_j - i\underline{t}_j$

We consider only the part of the solution, which belongs to a conjugate complex pair

$$\tilde{\underline{q}}_j(t) = B_j e^{\alpha_j t}\{\underline{s}_j \sin(\omega_j t + \gamma_j) + \underline{t}_j \cos(\omega_j t + \gamma_j)\} \tag{6}$$

ω_j is the circular natural frequency of this part and α_j the damping constant. If the damping constant $\alpha_j > 0$ the natural vibrations increase, for $\alpha_j < 0$ the natural vibrations decrease.

We define the expression in parantheses { } of equation (6) as natural mode. Opposite to conservative system there is no constant modal shape, proportions and relative phasing in general vary from point to point at the shaft. The natural modes of nonconservative systems represent time dependent curves in space. The plane motion of one point of the shaft is an elliptical orbit.

If we transpose the matrices $\underline{M}$, $\underline{D}$, $\underline{K}$, we obtain the so called left-hand eigenvalue problem

$$(\lambda^2 \underline{M}^T + \lambda \, \underline{D}^T + \underline{K}^T)\underline{\psi} = \underline{0} \tag{7}$$

which has the same eigenvalues λ but different eigenvectors $\underline{\psi}$. Both
eigenvector sets are needed for an expansion of the frequency response
functions in terms of the modal parameters.

Identification of Modal Parameters of Rotors

It is well known that experimental modal analysis has been often
used in many mechanical engineering problems to identify modal parameters of nonrotating elastic systems. Application of the method in rotating machines with nonconservative effects requires consideration of
some important differences, for example

- the nonsymmetry of the system matrices $\underline{K}$, $\underline{D}$
- the dependence of the modal parameters from the operating speed
- the necessity to excite the rotor and to measure the response
  during rotation

In consideration of these differences the modal analysis procedure is
also available for rotating structures (Fig. 4). The procedure consists
of the following steps:

Between a number of measurement points frequency response functions
$H_{kl}(\Omega)$ are measured at the real structure, by exciting the system at
locations l and measuring the response at locations k.

Analytical frequency response functions can be calculated in dependence of the eigenvalues, left-hand and right-hand eigenvectors.

Finally the analytical functions are fitted to the measured functions by variation of the modal parameters, resulting in a set of identified modal parameters.

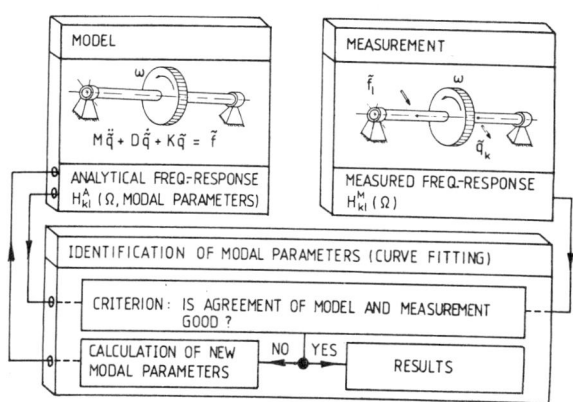

Fig. 4 Identification of modal parameters

Analytical Frequency Response Functions. If the rotor is excited in a certain point l by means of a harmonic force function $\tilde{f}_1$ with exciter frequency Ω and the displacement $\tilde{q}_k$ is measured in another point k, the response behavior of the rotor can be characterized by the complex frequency response function (Fig. 5).

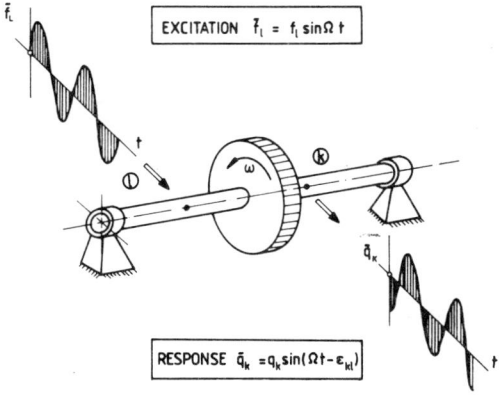

Fig. 5 Input and output signals at the rotor

$$H_{kl}(\Omega) = \frac{q_k e^{i(\Omega t - \varepsilon_{kl})}}{f_1 e^{i\Omega t}} = \frac{q_k}{f_1} e^{-i\varepsilon_{kl}} \tag{8}$$

respectively by the ratio of the amplitudes q_k/f_1 and the phase ε_{kl} between the two signals. Both are frequency dependent functions.

In /5/ is shown that the complex frequency response functions can be expressed in terms of the eigenvalues λ_j and the corresponding eigenvectors $\underline{\varphi}_j$, $\underline{\psi}_j$.

$$H_{kl}(\Omega) = \frac{\tilde{q}_k}{\tilde{f}_1} = \frac{q_k}{f_1} e^{-i\varepsilon_{kl}} = \sum_{j=1}^{2N} \frac{\varphi_{kj}\psi_{1j}}{i\Omega - \lambda_j} \tag{9}$$

For a rotor with N degrees of freedom N x N frequency response functions exist, assembled in the matrix $\underline{H}^A$ (Fig. 6). It is important to note that each row $\underline{Z}_k$ contains all of the left eigenvectors $\underline{\psi}_j$ and each column $\underline{S}_1$ contains all of the right eigenvectors $\underline{\varphi}_j$.

FREQUENCY RESPONSE MATRIX

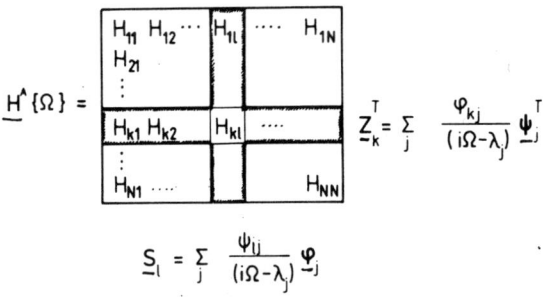

$$\underline{S}_1 = \sum_j \frac{\psi_{1j}}{(i\Omega - \lambda_j)} \underline{\varphi}_j$$

Fig. 6   Matrix of frequency response functions

$$\underline{Z}_k^T = \frac{\varphi_{k1}}{i\Omega - \lambda_1} \underline{\psi}_1^T + \frac{\varphi_{k2}}{i\Omega - \lambda_2} \underline{\psi}_2^T + \ldots \qquad (10)$$

$$\underline{S}_l = \frac{\psi_{11}}{i\Omega - \lambda_1} \underline{\varphi}_1 + \frac{\psi_{12}}{i\Omega - \lambda_2} \underline{\varphi}_2 + \ldots \qquad (11)$$

One row $\underline{Z}_k$ and one column $\underline{S}_l$ of the frequency response matrix $\underline{H}^A$ need
to be measured in order to identify all of the modal parameters λ_j, $\underline{\varphi}_j$,
$\underline{\psi}_j$ of a rotor (Fig. 7). It is sufficient to measure only one column, if
eigenvalues and right eigenvectors are required. For determination of
eigenvalues λ_j without natural modes, the whole information is contained
in one frequency response H_{kl} already. There are exceptions, if the
points of excitation or response are identical with node points of the
natural modes.

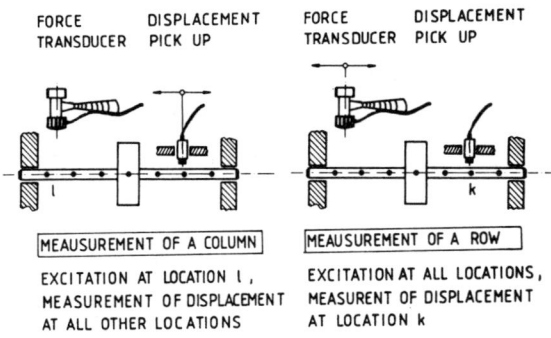

Fig. 7 Excitation and response locations

Measurements of the Frequency Response Functions. As pointed out,
the modal parameters of rotating structures can be determined from a set
of frequency response functions. The measurement of these frequency response functions is an important step of modal analysis. Several experi-

mental methods exist for acquiring this information. Usually measurements
are made by applying the force input artificially through some type of
exciter. As mentioned previously, one could excite the system with harmonic forces and measure the system response (amplitude and phase). But
sine testing is slow compared to other types of signals, since it only
excites the structure at one frequency at a time. Other types of excitation, signals like impact and random forces are considered to be the
most useful today. They have a broadband characteristic in the frequency
domain and measurements can be carried out in a relatively short time.

Basically frequency response functions may be computed directly
from the ratio of the Fourier transforms of the measured output and input signals. This Fourier Transformation theorem permits replacement of
the time consuming frequency by frequency excitation technique by the
excitation with an arbitrary signal. The only requirements are, that the
signals be Fourier transformable. Recent developments in the area of digital signal processing, especially progress of Fast Fourier Transform
algorithms enabled the successful application of this technique.

The employed measurement method consists of applying an impact
forcing function $\tilde{f}_1$ to a point 1 of the structure while picking up at
the same time the displacement $\tilde{q}_k$, of point k (Fig. 8). The time signals are transformed to the frequency domain by means of the Fast Fourier Transformation and the ratio is calculated.

The frequency content and the amplitude of the force signal can be
influenced by selection of the hammer mass, the flexibility of the impact cap and the impact velocity. With a short impulse the energy is
distributed in a wide frequency range, with a long impulse in lower frequencies.

Figure 9 shows the measurement equipment. The hammer excites the
rotating shaft and the force is measured by means of a force transducer
or an accelerometer. The displacements of the shaft can be picked up by
capacitive or inductive pickups. Force and displacement signals are amp-

lified and after analog-digital-conversion and Fast-Fourier-Transforma-
tion the frequency response functions can be calculated. They are stored
on a magnetic tape for further treatment.

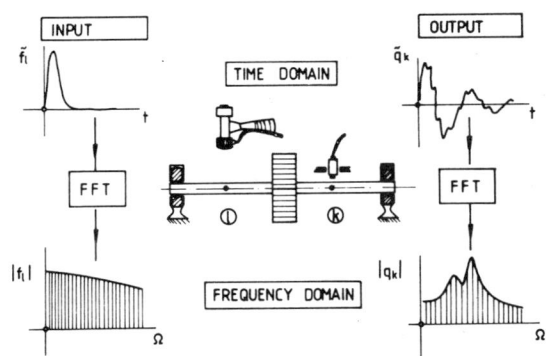

Fig. 8   Measurement of frequency response functions

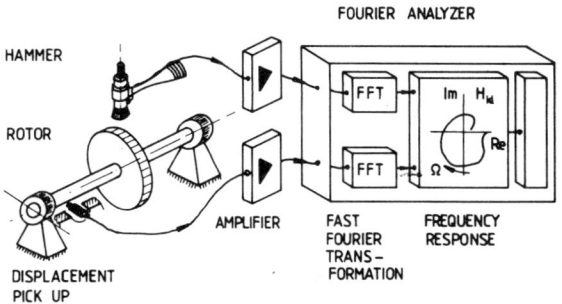

Fig. 9   Measuring device

Modal Parameter Estimation. Different shopisticated techniques are available to estimate modal parameters from frequency response data. Which one is the most suited in a special case depends on the degree of modal coupling.

At any given frequency Ω the frequency response represents the sum of all the modes of vibration which have been excited (equation (9)). Normally the contribution of a particular mode is the greatest near its natural frequency. When modal coupling is light, the frequency response can be considered in the vicinity of a resonance as if it were a single degree of freedom system. The simplest approach to determine modal parameters of such systems with well separated modes is to pick up natural frequencies and amplitudes at the resonance locations.

The investigated rotor bearing system has heavy modal coupling. In this case a more sophisticated technique to extract modal parameters is a multidegree of freedom curve fit, based on equation (9). Within a limited frequency range the goal of the procedure is to find the complex modal parameters λ_j, $\underline{\varphi}_j$. The basic idea consists of finding a best fit between the measured response plot and the generated plot from the analytical expression (9). A well known method for doing this is to use a least squares procedure. The criterion is to minimize the error function

$$
E = \sum_p^P \{RE(H_{kl}^M(\Omega_p)) - RE(H_{kl}^A(\Omega_p,\lambda_j,\underline{\varphi}_j))\}^2
$$

$$
+ \sum_p^P \{IM(H_{kl}^M(\Omega_p)) - IM(H_{kl}^A(\Omega_p,\lambda_j,\underline{\varphi}_j))\}^2
$$

(13)

H_{kl}^A analytical frequency response expression

H_{kl}^M measured frequency response

P is the number of frequency lines in the measured frequency response functions.

Differentiating E with respect to each unknown in turn and setting each result to zero, we obtain a number of equations for the unknown modal parameters. The equations are nonlinear in the eigenvalue parts, therefore they are solved by an iterative procedure (linearization). After each step the variation of the error function is controlled. At the beginning of the procedure a starting vector of the unknowns must be chosen. The process is repeated for further measured frequency curves.

Example - Modal Parameters of an Elastic Rotor in Journal Bearings

Rotor Test Rig. For testing the method, measurements were carried out at the test rig rotor shown in Fig. 2, respectively, Fig. 10. The rotor consists of a cylindrical shaft (diameter 50 mm, length 1000 mm) with a disk (mass 55 kg, diameter 300 mm, width 100 mm) at the center of the shaft. The shaft is driven by a d-c electric motor with control.

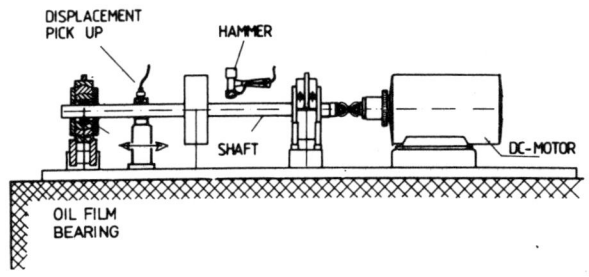

Fig. 10 Rotor test rig

The two oil film bearings are cylindrical bearings with a length-diameter ratio B/D = 0.8.

Seven measurement points were established along the rotor axis.
At each measurement point excitation by impact forces and measuring the
system response is possible in vertical as well as horizontal direction.
The hammer for pulse excitation has a mass of 1.5 kg. The displacement
pickups are movable along the rotor. Measured quantities are the exciting force, the displacements of the shaft and the rotor speed.

Measurement Results. For the nonrotating shaft (ω = 0) there is
no oilfilm effect in the bearings and the system can be considered as
a flexible shaft with rigid bearings. The results from experimental modal analysis are shown in Fig. 11. The natural frequency of the first
bending mode is 42.5 Hz. The second bending mode has a comparatively
high frequency of 221 Hz.

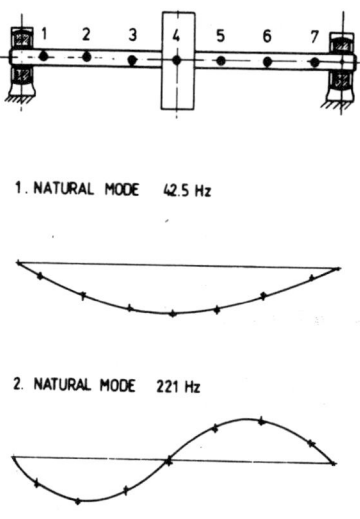

Fig. 11 Natural modes of the shaft with rigid bearings

It is important to know the first natural frequency of the rigidly supported shaft. With this information the natural frequency of the shaft in oil film bearings can be estimated. The flexibility of the oil film lowers the natural frequency. Therefore it is to be expected, that the first bending natural frequency of the rotor in journal bearings is less than 42.5 Hz. Furthermore it is known that the instability onset speed is about two times the first natural frequency.

A number of frequency spectra from 0 to 100 Hz are shown in Fig. 12 for different rotating speeds between 50 Hz and 77.5 Hz. The frequency spectrum for each rotor speed was found by impacting the rotor at the disk and measuring and analyzing the response signal near the bearings.

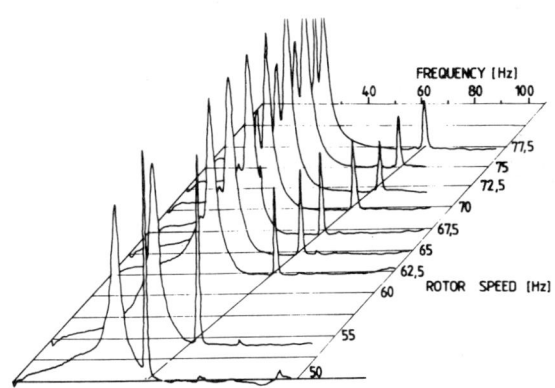

Fig. 12 Frequency spectrum of the test rig rotor

For lower rotating speeds there is one peak at 37 Hz, which is identical with a system natural frequency. The second peak is caused by unbalance excitation. The frequency at this peak is the frequency of rotation. For higher speeds the frequency spectra show two natural frequencies between 30 Hz and 40 Hz and the unbalance response peak with decreasing amplitudes at increasing rotational frequencies. The growing

peak at about 33 Hz shows that the rotor is running near the instability onset speed between 75 and 80 Hz.

Natural frequencies and damping constants can be determined approximately with the frequency spectrum. To get better results for system
eigenvalues and eigenvectors, frequency response functions are used.
For the described rotor in cylindrical bearings measurements were carried out in a speed range from 60 to 75 Hz. Because of the variation of
eigenvalues and natural modes with rotor speed, the rotational frequency
need to be constant during the measurements. For each rotor speed one
column of the frequency response matrix was calculated from the corresponding measured time signals. In Fig. 13 the amplitude frequency characteristic of one of the frequency response functions is represented
for a rotational speed of 70 Hz. The upper diagram shows the measured
function, the lower the calculated function. There are two resonance
peaks at 33 Hz and 37 Hz.

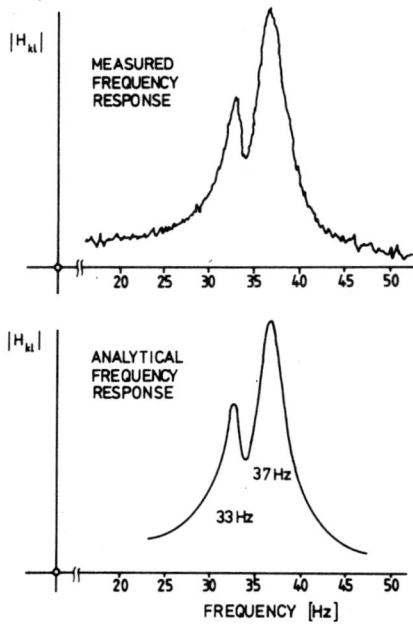

Fig. 13 Comparison of measured and calculated frequency response

From measurements for different rotating speeds the two essential
eigenvalues (natural frequencies and damping constants) were identified
by the above described curve fitting procedure. Figure 14 shows in the
complex plane the natural frequencies versus the damping constants, respectively imaginary parts of eigenvalues versus real parts. Parameter
is the rotating speed of the shaft. There is only a little influence
from the speed to the natural frequencies. The first natural frequency
is about 33 Hz, the second 37 Hz.

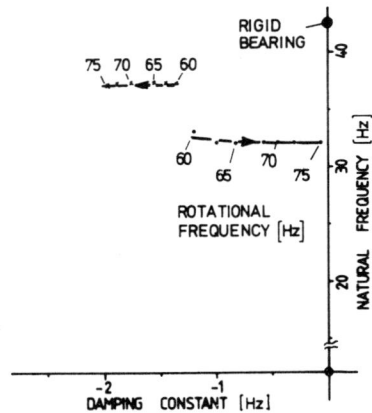

Fig. 14 Eigenvalues of the test rig rotor

It is important to note that the damping constant of the first
eigenvalue tends to zero for increasing speeds. At the stability thres-
hold speed - 78 Hz the damping constant disappears. The rotor is unstable for higher speeds.

Besides the eigenvalues also the eigenvectors were identified.
Figure 15 points out the two natural modes, as defined in equation (6),
for a frequency of rotation of 70 Hz.

The natural modes represent time dependent curves in space. The
plane motion of one point of the shaft is an elliptical orbit.

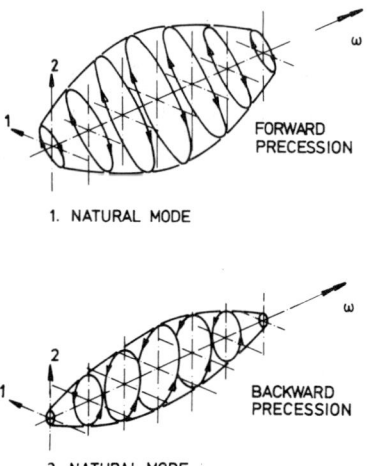

1. NATURAL MODE

2. NATURAL MODE

Fig. 15 Natural modes of the test rig rotor

Forward precession appears in the first natural mode with a natural frequency of 33 Hz. This eigenvalue leads to instability. In the second natural mode with backward precession and a natural frequency of 37 Hz the damping constant increase with the rotational frequency.

Conclusions

In this paper an application of experimental modal analysis is given for rotating machines. It is shown, that the classical method has to be improved in the case of nonconservative rotors with unsymmetric matrices. In the special case of a simple elastic shaft with oil film bearings the complex eigenvalues, the natural modes and the instability onset speed could be determined. The measured results are in good correlation with theoretical values.

The application of the suggested procedure in practical cases is

possible if suitable locations for excitation exist and the input energy is high enough to excite the essential natural vibrations.

References

1. Wachel, J.C., "Rotordynamic Instability Field Problems",
   NASA Conference Publication 2250, Rotordynamic Instability
   Problems in High Performance Turbomachinery, proceedings of
   a workshop held at Texas A&M University, May 10-12, 1982, pp. 1-19.

2. Nordmann, R., "Ein Näherungsverfahren zur Berechnung der Eigen-
   werte und Eigenformen von Turborotoren mit Gleitlagern, Spalt-
   erregung, äußerer und innerer Dämpfung", Dissertation TH Darmstadt,
   1974.

3. Morton, P.G., "The Derivation of Bearing Characteristics by Means
   of Transient Excitation Applied Directly to a Rotating Shaft",
   IUTAM-Symposium, Dynamics of Rotors, Lyngby, Denmark, Springer
   Verlag 1975, pp. 350-379.

4. Klostermann, A. L., "On the Experimental Determination and Use of
   Modal Representation of Dynamic Characteristics", Ph. D. dissertation, University of Cincinnati, 1971.

5. Nordmann, R., "Modal Parameter Identification and Sensitivity
   Analysis in Rotating Machinery", International Conference "Ro-
   tordynamic Problems in Power Plants", Proceedings of the IFTOMM
   Conference, Rome, Italy, Sept. 29-Oct. 1, 1982, pp. 95-102.